Hans-Friedrich Pfeiffer

Practical guide to coordinate transformations

Part I: Euclidian space $\mathbb{R}^n$

Practical guide to coordinate transformations

Part I: Euclidian space $\mathbb{R}^n$

Hans-Friedrich Pfeiffer

Impressum

Bibliografische Information der Deutschen Nationalbibliothek:
Die Deutsche Nationalbibliothek verzeichnet diese Publikation in der
Deutschen Nationalbibliografie; detaillierte bibliografische Daten sind im
Internet über http://dnb.dnb.de abrufbar.

Verlag: BoD · Books on Demand GmbH, Überseering 33, 22297 Hamburg,
bod@bod.de

Druck: Libri Plureos GmbH, Friedensallee 273, 22763 Hamburg

ISBN: 978-3-8192-0963-5

Contents

I.1.1 Introduction

Physical laws need to be expressed in a way that is not dependent on any particular coordinate system. Why, then, are we interested in coordinate systems and their transformations from one system to another?

Let us have a look at classical mechanics. You might encounter physical problems having constraints. An example is the movement of a particle (mass m) on the surface of a sphere with radius R. The particle's motion is constrained to the surface of the sphere. In Cartesian coordinates, this constraint is given by:

$$x^2 + y^2 + z^2 - R^2 = 0$$

Working with constraints can be extremely complicated. But if you switch your coordinate system from Cartesian to spherical coordinates, defined by

$$x = r \sin \theta \cos \phi$$

$$y = r \sin \theta \sin \phi$$

$$z = r \cos \theta,$$

then we can describe the movement easily as the constraint is implicitly contained in the equations above.

Problems in electrodynamics often require changing to spherical or cylindrical coordinates, depending on the symmetry of the problem.

In Special Relativity (SR) and General Relativity (GR) coordinate transformations are essential for simplifying problems and understanding the behavior of spacetime. AnI example is the **Schwarzschild solution** in General Relativity, which describes the gravitational field outside a spherically symmetric, non-rotating mass such as a static black hole.

Another example which we will study later in this book is the solution of integrals. We will see how a coordinate transformation can enable us to find a solution for an integral. The great question of how to find the best coordinate transformation cannot be answered definitively; it is a matter of experience. However, to gain this experience, you need to understand what coordinate transformations are about.

To understand coordinate systems, we need to know their structure, given by the metric tensor. The metric tensor enables us to calculate the distance of points in a coordinate system. We will meet Christoffel symbols which also encapsulate geometrical information. The theoretical background leads us to tensors and their correct handling.

In this book, Part I, we only examine coordinate systems which are linked to Cartesian basis. In a follow-up book, the concepts developed in this book will be extended to Riemann manifolds and the Minkowski space.

I.1.2 Required Knowledge

The reader should be familiar with basic linear algebra and analysis. In linear algebra the reader should know what an n-dimensional vector space is and what a basis is. Knowledge of vectors and matrices is also required. The reader also should know what linear mapping is and know the relationship between linear mappings and matrices.

In analysis, the reader should be familiar with derivatives and partial derivatives. The book contains a special chapter about coordinate transformation in integrals, so the reader should also know how to solve integrals.

In this book, we use the Einstein Summation Notation (ESN), therefore the reader will find an introduction chapter on how to manage ESN correctly.

I.1.4 Conventions

In the following, $(\vec{e}_1, \vec{e}_2, \ldots, \vec{e}_n)$ denotes the orthonormal standard basis for the vector space $\mathbb{R}^n$. The components of the vector $\vec{e}_j$ are all zero except for the jth position, which contains a 1.

 These vectors are perpendicular to each other (orthogonal), and all have the length 1 (orthonormal). This basis system is also called **Cartesian**.

Note, please, that we will also use basis vectors denoted by $(e_1, e_2, \ldots, e_n)$. They do not carry an arrow; they have nothing to do with the standard basis vectors. Do not mix them up!

For matrices, we will number rows and columns starting with 0 - this is done because the so-called Minkowski space is used in SR/GR. This vector space is slightly different from the $\mathbb{R}^4$, the 2nd, 3rd and 4th components behave in the same way as the vector space $\mathbb{R}^3$.

When we want to describe a vector v with its component, we also use the terminology $v = (v^1, v^2, ..., v^n)$ to describe the components of the vector. The upper index DOES NOT MEAN "to the power of" but it refers to so-called contravariant components of the vector. We will come back to contravariant and covariant vectors later.

In $\mathbb{R}^3$, the coordinate system is referred as (x, y, z). As we want to work in arbitrary n-dimensional vector spaces, we will use $(x^1, x^2, x^3, ..., x^n)$ instead.

I.1.3 Roadmap

The roadmap begins with an introduction to Einstein Summation Notation (ESN). Many equations can be easily expressed in ESN, but for practical calculations I offer – wherever it is possible – the equations also in matrix notation.

Then we have a first look at differential operators in $\mathbb{R}^n$, first only in cartesian coordinates. The concept of partial derivatives leads us to the Jacobi-matrix, which is an essential part of this book.

Then, we define what a coordinate system is, and we classify them. We introduce the metric tensor, as they play an immense role in coordinate changes. The same is true for the Christoffel-symbols, relying on a metric tensor.

We define what a tensor is and stretch out the importance of tensor calculus in physics. Having done this, we come back to the differential operators, now in curvilinear coordinates.

We have a look at integrals to find out that a change of coordinate system might help us to find a solution. The point is: if you cannot solve an integral in a certain coordinate system, change the coordinate system and try to solve the integral there.

At the end of the book, we list the most used (popular) coordinate systems, giving details for the Jacobi-matrix, the metric tensor and Christoffel symbols.

At each stage, we give examples and exercises. The solution for the exercises is at the end of the book – even if you can solve an exercise on your own: have a look at the solution because they contain additional information for practical use.

I.1.4 Sage Math

When I began writing this book, I wanted to verify my calculations, so I wrote several programs in **Sage Math**.

Sage Math is a free and powerful mathematical program system, based on Python programming language. For this book I used version 9.1, but the code I added should also work with other versions.

You can find the download of Sage Math at
https://www.sagemath.org/download.html

If you want to use Sage Math, you need to install two packages:

To do this, open a Jupiter notebook and enter the following commands:

```
pip install sympy

pip install ipywidgets
```

On my website

**http://www.hans-friedrich-pfeiffer.de/PGTCT/SageMath/
PGTCT_SageMath_Notebooks.zip**

the reader may download the indicated zip-file containing Jupiter-notebooks. The reader should also download the README-file:

**http://www.hans-friedrich-
pfeiffer.de/PGTCT/SageMath/README.txt**

The zip-file contains all calculations of the values presented in chapter $I.A$. The code contains functions for the calculation of the Jacobi-matrix, metric-tensor, squared line element, Christoffel symbols, etc. You will also find interactive Jupiter notebooks for Sage Math for the visualization of the coordinate systems.

I.1.5 About the author

The author studied mathematics and computer science at the Christian-Albrecht-University of Kiel (Germany) and at the University of Science and Techniques of Languedoc - Montpellier II (France) and finished his studies in 1987 with diploma. Since 1989 he has worked as a software developer in data warehousing.

I.1.6 Errata and contact

Should the reader find errors in my book, these can be communicated via email to **mathe@hans-friedrich-pfeiffer.de**.

You may use the email address also for questions/suggestions.

The reader may download the errata-PDF-file

https://www.hans-friedrich-pfeiffer.de/PGTCT/Errata.pdf

I.1.7 Acknowlegments

I want to express my thanks to my former collegue Dipl.-Phys. Michael Dresen (Rendsburg/Germany), astro-physicist, for reading and commenting my manuscript.

I.2 Preparations

In this chapter some preparations will be made. First, we have a closer look at the Einstein Summation Notation (ESN) – the correct dealing with the ESN is essential.

Next, we have a look at parameterization of functions. Many functions in physics are parameterized by time, so it makes sense to look at different methods for parameterizing functions. Especially the parameterization by arc length offers various advantages in contrast to other parameterizations.

We continue to look at derivatives and their generalizations in cartesian coordinates. One of the main points is the multivariate chain rule, which we cover in an own chapter.

I.2.1 Einstein summation convention

Before we begin with the Einstein Summation Notation, I would like to give a helpful tip:

Numerous equations are derived throughout the book. Read the equations in *both* directions, in particular from right to left. Especially in what will follow now, it is essential that relationships are recognized in a way that allows the reader to simplify expressions by paraphrasing.

We will give a very short introduction to Einstein Summation Convention (ESN). ESN will be used in this book very often, so you should become familiar with this notation.

The ESN is used to shorten up summations over summations over summations by leaving out the sum-sign $\sum$ and using indices instead. The main rule is:

If you see an index, let us say i, which appears in a term exactly two times, one time as a lower index, the other as an upper index, then you have to sum up over that index. It might – and will – occur that an equation contains more than only one index to sum up. The reason the index has to appear as a lower and an upper index is subtle and has to do with covariant and contravariant vectors. We will come back to these terms later.

We will see a lot of ESN equations, but we show them without proof. The proofs can be seen in [Pfeiffer].

We start with some examples. For the moment, a_i and u^i should be real numbers, i is an index running from 1 to n.

$(I.2.001)$ **Example 1**

The Einstein sum $a_i u^i$ is defined as follows:

$$a_i\, u^i \equiv \sum_{i=1}^{n} a_i\, u^i = a_1\, u^1 + \; a_2\, u^2 + \ldots + \; a_n u^n$$

The double *(and ONLY double)* index i appearing on the left side of the equation is called a **bound** or **silent** index, because it disappears after the summation is performed.

$a_i\, u^i$ is just an abbreviated notation for the summation.

$(I.2.002)$ **Example 2**

$$a_{ij} u^i \equiv \sum_{i=1}^{n} a_{ij} u^i = a_{1j} u^1 + a_{2j} u^2 + \ldots + a_{nj} u^n$$

Again, in this example, i is the bound index to sum over. The index j appears only once, it is called a **free** index. j can be freely chosen but then remains fixed after its choice. So, if j can take values from 1 to m, then the above expression results in a system of m summations:

$a_{i1} u^i$

$a_{i2} u^i$

$\ldots$

$a_{im} u^i$

Each term is also an ESN equation!

This shows that the ESN not only dispenses with the summation sign, but that a whole series of equations can be represented in just one ESN.

In the last example a_{ij} could represent rows and columns of a matrix (with i as a row and j as a column), u^i could be the components of a vector. In this case the ESN represents the multplication of the matrix $\left(a_{ij}\right)$ with the vector $u = \left(u^1, u^2, \ldots, u^n\right)$. It is important to understand that the Einstein summation **always refers only to components**.

We come to some rules for the sum convention:

$(I.2.003)$ **Rule 1**: only double-listed indices are totaled. One index must be a lower one, the other an upper one.

The last condition needs an explanation: in physics, so called contravariant and covariant vectors are distinguished. The condition says that you can only sum up if one of the indices refers to a covariant value (lower index) and the other a contravariant value (upper index). We will come back on what contravariant / covariant is in chapter $I.2.5$.

Example:

$$a_{ij}b^j = \sum_{j=1}^{n} a_{ij}b^j$$

In this example, the index j appears twice on the left. Therefore j is the summation index. The index i, on the other hand, is free.

It is also possible for an expression to contain multiple bound indices:

$$a_{ij}b^{jk}c_k$$

In this example, j and k are bound indices - both(!) must be summed:

$$a_{ij}b^{jk}c_k \equiv \sum_{j=1}^{n}\sum_{k=1}^{n} a_{ij}b^{jk}c_k$$

Here, too, there are m equations that the free index i runs through.

$(I.2.004)$ **Rule 2**: Only bound indices are allowed to be renamed:

Example:

$$a_{ij}b^j = a_{ik}b^k = a_{il}b^l$$

Here, in the second term, the double-occurring index j was renamed to k, in the third term k was then renamed to l.

On the other hand, the following ESN would be wrong:

$$a_{ij}b^j = a_{mj}b^j$$

Here the *free* index i has been renamed to m.

Free indices are not allowed to be renamed.

If a bound index is to be renamed, then two points must be observed:

- ➢ the new index name has not yet been used *in the term.*
- ➢ all occurrences of the index to be renamed must be renamed.

Important note: the renaming of indices <u>refers to a single term</u>. Suppose we have a term like

$$a_{ij}^k\, b_{ij}^k \ + \ c_{rs}^m\, d_{rs}^n$$

Then you can rename the bound indices r and s:

$$a_{ij}^k\, b_{ij}^k \ + \ c_{ij}^m\, d_{ij}^n$$

Observe, that we renamed r to i and s to j, <u>although i and j appear as bound indices in the first term</u>! This renaming of indices does not change the meaning or mathematical operations involved. It is a common practice in physics to rename indices for clarity or consistency in notation. If the indices are used consistently *within their respective terms* and with the operations applied to them, the renaming will not affect the result or validity of the expression.

$(I.2.005)$ **Rule 3**: no index may appear more than twice *per term.*

Terms are *added* to each other. This means that all constructions that are multiplicatively connected count as one term.

Examples:

$a_{ij}b^j$ is a valid expression because j occurs twice and is therefore bound.

$a_{ij}b^j + c_{ij}d^j$ is also correct: we have *two* terms here and, in each term, j is the bound index.

$a_{ii}b^{ji}$ is not correct because i is used three times in the term.

($I.2.006$) **Rule 4**: Any free index on the left-hand side of an Einstein summation equation must also appear as a free index on the right-hand side. Any free index on the right must also appear as a free index on the left. An index cannot be free on one side and a bound index on the other side.

An important note: the rule of consistency of free indices on both sides of an equation does NOT apply to bound indices. On the contrary: transformations, as we will discuss later, serve to eliminate as many bound indices as possible.

Examples:

$x_i = a_{ij}b^j$ is correct because on both sides i is a free index. The index j is bound and occurs exactly twice in the term.

$x_i = a_{ki}b^{jk}x_j$ is also correct: on the right, k and j are bound indices, and the free index i on the left also appears as a free index on the right.

$x_i = a_{ki}b^{jk}c_m x_j$ is incorrect: the free index m appears on the right but cannot be found on the left.

ESN and brackets

Consider the expression $a_{ij}(b^i + c^j + d^k)$. How can you determine the summation indices and the free indices? Remember that you can't "multiply out" the terms as you might in ordinary algebra; this does not work directly in Einstein Summation Notation (ESN).

To determine the summation and free indices, proceed as follows:

Combine a_{ij} with every term inside the parentheses and count how many times each index appears:

a_{ij} & b^i : $i \to 2, j \to 1, k \to 0$

$$a_{ij} \ \& \ c^j : \ i \to 1, j \to 2, k \to 0$$

$$a_{ij} \ \& \ d^k : \ i \to 0, j \to 0, k \to 1$$

Now, determine the highest count for each index across all terms:

$$i \to 2, j \to 2, k = 1$$

The indices i and j appear twice, so these are your summation indices. The index k appears only once: it is the free index. If any index occurs more than twice, the expression would not be a valid ESN expression.

I.2.1.E Exercises

($I.2.1.E.1$) Determine for the following equations if these are valid ESN equations:

1. $a_i + b_i = c_i$
2. $a_i \, b_i = c_i \, d_i$
3. $a_i \, b_i = c_i$
4. $t_{ij} \, v^i = w^j$
5. $a_{ij} \, b^{jk} \, c_k = d_i$
6. $a_{ijk} \, b^{ij} \, c^k = d_i$
7. $m_{ij} \, v^j = w_i$
8. $v_i \, w^i = x$
9. $\varepsilon_{ijk} \, a^i \, b^j = c_k$

I.2.1.1 Identities and non-identities

Let us look at identities and non-identities. In the case of non-identities, we will investigate why these cannot be valid.

$(\boldsymbol{I.2.007})\ x_i y^i = x_j y^j = x_k y^k$

Summation indices may be renamed arbitrarily according to $(I.2.004)$. As trivial as this may seem, it is used extensively in ESN.

For example, if two expressions need to be nested within each other - we will see an example of this shortly - then renaming the bound index is necessary.

$(\boldsymbol{I.2.008})\ a_{ij} u^j = u^j a_{ij}$

Einstein sums are commutative, the order in a term, as well as the order of additively connected terms, is arbitrary. This is because there is a summation behind it, and in this both the addition and the multiplication are commutative. However, we will discuss one restriction later: if operators (e.g., differential operators) are used in an Einstein summation, the operator cannot be interchanged with its argument.

Matrix multiplications can also be described using Einstein sums: on the one hand, it should be noted that matrix multiplication is non-commutative and, on the other hand, that the result of an Einstein sum is <u>not</u> a matrix but represents only individual components.

In the example above you can consider $A = \left[a_{ij}\right]$ as a matrix and u as a vector with the components $(u^1, u^2, \ldots, u^n)$. Then $a_{ij} u^j$ means row-by-row addition of a_{ij} with the components u^j : the result is the value of the ith component when matrix A is being multiplied with the vector u. The result is a number, but not the vector!

We will cover matrices and Einstein summation in detail in short $(I.2.1.4.4)$.

$(\boldsymbol{I.2.009})\ a_{ij}(x^i + y^j) \neq a_{ij} x^i + a_{ij} y^j$

On the left, i and j are the summation indices, so both must be summed.

On the right side of the expression i is the summation index. The index j is free. Exactly the opposite is the case in the expression $a_{ij}y^j$: here j is the summation index and i is free. Both sides are not consistent according to (*I*. 2.006).

(***I*. 2. 010**) $a_{ij}(x^j + y^j) = a_{ij}x^j + a_{ij}y^j$

Note here that x carries the same index as y. In this case you can multiply out:

$$a_{ij}(x^j + y^j) \;\equiv\; \sum_{i=1}^{n}\sum_{j=1}^{n} a_{ij}(x^j + y^j)$$

$$= \sum_{i=1}^{n}\sum_{j=1}^{n} \left(a_{ij}x^j + a_{ij}y^j \right)$$

$$= \sum_{i=1}^{n}\sum_{j=1}^{n} a_{ij}x^j + \sum_{i=1}^{n}\sum_{j=1}^{n} a_{ij}y^j$$

$$\equiv\; a_{ij}x^j + a_{ij}y^j$$

(***I*. 2. 011**) $a_{ij}x^i y^j \neq a_{ij}x^j y^i$

The inequality should be obvious: you cannot just swap the indices of objects.

(***I*. 2. 012**) $\left(a_{ij} + a_{ji}\right) x^i y^i \neq 2a_{ij}x^i y^i$

This formula is not valid: on the left-hand side j is a summation index, but on the right-hand side it is not - hence (*I*. 2.006) is violated.

(***I*. 2. 013**) On the other hand, the following applies:

$$\left(a_{ij} + a_{ji}\right) x^i x^j = 2a_{ij}x^i x^j$$

We have:

$$
\left(a_{ij} + a_{ji}\right)x^i x^j
$$

$$
\begin{aligned}
&\equiv \sum_{i=1}^{n}\sum_{j=1}^{n}(a_{ij} + a_{ji})x^i x^j \\[4pt]
&= (a_{11} + a_{11})x^1 x^1 + (\boldsymbol{a_{12} + a_{21}})\boldsymbol{x^1 x^2} + \cdots \\
&\qquad\qquad + (a_{1n} + a_{n1})x^1 x^n \\[4pt]
&+ \; (\boldsymbol{a_{21} + a_{12}})\boldsymbol{x^2 x^1} + (a_{22} + a_{22})x^2 x^2 + \cdots \\
&\qquad\qquad + (a_{2n} + a_{n2})x^2 x^n \\[4pt]
&+ \; \ldots \\[4pt]
&+ \; (a_{n1} + a_{1n})x^n x^1 + (a_{n2} + a_{2n})x^n x^2 + \cdots \\
&\qquad\qquad + (a_{nn} + a_{nn})x^n x^n \\[4pt]
&= \; 2(a_{11})x^1 x^1 + 2(a_{12} + a_{21})x^1 x^2 + \cdots \\
&\qquad\qquad + 2(a_{nn})x^n x^n \\[4pt]
&\equiv \; 2a_{ij}x^i x^j
\end{aligned}
$$

Since $(a_{il} + a_{li})x^i x^l = (a_{li} + a_{il})x^l x^i$ we can combine the expressions in bold above and get the desired result.

$(I.2.014)$ Replacements of terms

We have an equation of the form $T = b_{ij}y^i x^j$ and we want to exchange the term y^i by $y^i = a_{ij}x^j$. If we simply did this as a "copy and paste", we would get the invalid term $T = b_{ij}a_{ij}x^j x^j$, because the term indicates four occurrences of the index j. To be able to carry out a correct replacement, it is necessary to proceed as follows:

In the term $y^i = a_{ij}x^j$ we must rename the bounded index j and we get a result like $y^i = a_{ir}x^r$ (cf. $(I.2.004), (I.2.007)$).

With $T = b_{ij}y^i x^j$ and $y^i = a_{ir}x^r$ we get:

$$T = b_{ij}\, y^i\, x^j = b_{ij}\, (a_{ir}\, x_r)\, x^j = b_{ij}\, a_{ir}\, x^r\, x^j$$

$(I.2.015)$ $a_{ij}x^i x^j = a_{ji}x^i x^j$

We obtain the result by just renaming the bounded index i to j and j to i.

$$(\boldsymbol{I.2.016})\left(a_{ij} - a_{ji}\right) x^i x^j = 0$$

Use $(I.2.010)$ and $(I.2.015)$.

I.2.1.2 Kronecker symbol

The Kronecker symbol δ_{ij} occurs very frequently in tensor calculation and can be defined in different ways:

1. as a function:

$$\delta_{ij} = \left\{ \begin{array}{l} 1 \ \ if \ \ i = j \\ 0 \ \ else \end{array} \right\}$$

2. as the unit matrix:

$$\delta_{ij} \ = \ \begin{pmatrix} 1 & 0 & 0 & \ldots & 0 \\ 0 & 1 & 0 & \ldots & 0 \\ 0 & 0 & 1 & \ldots & 0 \\ \ldots & \ldots & \ldots & \ldots & \ldots \\ 0 & 0 & 0 & 1 & 0 \\ 0 & 0 & 0 & 0 & 1 \end{pmatrix}$$

3. As the scalar product of the Cartesian unit vectors:

$$\delta_{ij} = \vec{e}_i \cdot \vec{e}_j$$

If i does not equal j, the vectors are orthogonal, and the dot product is zero. However, if $i = j$, then the dot product is the square root of its length, which is 1.

What is δ_{ij} used for? Imagine, you have a *finite* list of elements $(x^1, x^2, \ldots, x^n)$. Then $\delta_{ij} x^j$ "picks up" only the element x^i, as we will see in short. Besides, there exists the Dirac-Delta-distribution $\delta(x)$, which performs a similar function for an infinite set.

We note that δ_{ij} may also be written as $\delta^{ij}, \delta^i_{\ j}$ and $\delta_j^{\ i}$. The meaning of these different spellings is subtle: in section I.6.2.3 we will show, that the Kronecker symbol is a (covariant, contravariant, mixed) tensor, so in every coordinate system it takes always the same value.

Identities:

$$(\boldsymbol{I.2.017}) \qquad \delta_{ij} = \delta_{ji}$$

$$\delta^{ij} = \delta^{ji}$$

$$\delta_{\;j}^{i} = \delta_{\;i}^{j}$$

The indices of the Kronecker symbol may be swapped. This follows directly from the definition of the Kronecker-symbol.

From now on, we write down some identities using the covariant Kronecker symbol. It should be clear now, how to establish the equations for the contravariant or mixed Kronecker symbol.

$(\boldsymbol{I.2.018})\ \delta_{ij}x^i x^j = x^i x^i = x^j x^j$

$(\boldsymbol{I.2.019})\ \delta_{ij}\,\delta^{ik} = \delta_{ji}\delta^{ik} = \delta_{\;j}^{\;k}$ **(contraction von δ)**

The summation runs over the bound index i: δ_{ij} becomes 1 only when i reaches the value of j, the same happens with δ_{ik} , so that in total there is only one non-zero term in the summation for $i\;=\;j$ and $i\;=\;k$. With this we can eliminate one δ.

Note: contraction makes a bound index "disappear".

This simple rule is very commonly used, so it deserves special attention.

$(\boldsymbol{I.2.020})\ \delta_{kk} = \delta_{11} + \delta_{22}+\ldots+\delta_{nn} = n$

Be careful with this formula: δ_{kk} does not equal to 1, as one might believe by the definition of the Kronecker symbol. The index k appears twice: the summation must therefore be observed. Which value δ_{kk} takes depends on "what" δ_{kk} is connected to.

Below, we will get to know the Levi-Civita tensor ε_{ijk} of the third order, whose indices i, j and k can assume the values 1, 2 or 3, respectively. In this case we get:

$$\varepsilon_{ijk}\,\delta_{mm} = \varepsilon_{ijk}\,3 = 3\,\varepsilon_{ijk}$$

$(I.2.021)$ $\delta_{ij}\,\delta^{ji} = \delta_{ji}\,\delta^{ij} = n$

$(I.2.022)$ $\delta_{ji}\,x^{j} = x^{i}$ **(exchange of indices)**

Again, note that the summation index vanishes.

The same applies more generally to any indexed object of the form $T_{ipq...z}$:

$$T_{\underline{i}pq...z}\delta^{ij} = T_{\underline{j}pq...z}$$

The exchange of indices is one of the most important identities that is used in many places and therefore deserves the greatest attention.

$(I.2.023)$ $\delta_{ij}\,\delta^{j}{}_{k}\,\delta^{i}{}_{n} = \delta_{kn}$

$(I.2.024)$ $x^{jmk}\delta_{nk} = x^{jmn}$

It should be noted that in the equation the summation index k on the left disappears. With summation indices, this is not only allowed, but also desired. This does not apply to free indices, they must reappear on both sides of the equation, as we have already explained in $(I.2.006)$.

I.2.1.2.E Exercises

$(I.2.1.2.E.1)$ Determine if the following equations are valid in ESN. Try to interpret the equations.

1. $\delta_{ij}\,a^{i}b^{j} = x$
2. $\delta_{ij} + \delta_{ij} = 2$
3. $a_{ijk}\,\delta^{il} = b_{jk}$

I.2.1.3 Levi-Civita Symbol

The Levi-Civita symbol ϵ_{ijk} plays an important role in the ESN. It does not show up that often in SR/GR, but more so in electrodynamics, particularly when dealing with the various differential operators, which we will discuss later.

The Levi-Civita-symbol is defined as follows with $i, j, k \in \{1,2,3\}$:

$$(\textbf{\textit{I}}.2.025)\text{:} \quad \begin{aligned} \epsilon_{ijk} &= \epsilon_{jki} = \epsilon_{kij} = 1 \\ \epsilon_{ikj} &= \epsilon_{jik} = \epsilon_{kji} = -1 \\ \epsilon_{ikk} &= \epsilon_{iki} = \epsilon_{kii} = 0 \end{aligned}$$

The indices i, j and k permute and take numbers from 1 to 3. If the permutation of the tupel $(1,2,3)$ is even, then ε takes the value 1, if the permutation is odd, ε takes the value -1. If an index appears twice, the value is zero. We have defined ε for three dimensions - it also exists for arbitrary dimensions, but this is not needed in this book.

If a permutation occurs in the string "123123" from left to right, the permutation is even and therefore taking the value 1. However, if the permutation is only from right to left, then it is an odd permutation and takes the value -1.

The Levi-Civita symbol can also be expressed as follows:

$$\epsilon_{ijk} = \det \begin{bmatrix} \delta_{i1} & \delta_{i2} & \delta_{i3} \\ \delta_{j1} & \delta_{j2} & \delta_{j3} \\ \delta_{k1} & \delta_{k2} & \delta_{k3} \end{bmatrix}$$

$$= \delta_{i1}\delta_{j2}\delta_{k3} + \delta_{i3}\delta_{j1}\delta_{k2} + \delta_{i2}\delta_{j3}\delta_{k1} - \delta_{i3}\delta_{j2}\delta_{k1} - \delta_{i1}\delta_{j3}\delta_{k2} - \delta_{i2}\delta_{j1}\delta_{k3}$$

This equation can be recalculated directly. All combinations of the indices (i, j, k) would have to be calculated (then there would be $3^3 = 27$ equations), with the following consideration we can reduce this to 6 cases:

If $i = j$ or $j = k$ or $k = i$ and thus leads to a double index, then there are (at least) two equal (row) vectors in the determinant. These are then linearly dependent, and the determinant is therefore zero. This corresponds with the definition of ϵ_{ijk}, if an index occurs (at least) twice. This leaves only six cases in which i, j and k are different from each other. In the sum above, all terms are sorted by i, j, and k. For all terms with a $(+)$ sign, it is noticeable that the order of the second index each has an even permutation, corresponding to an odd permutation for the terms with a $(-)$ sign. Depending on which permutation the first index goes through: only one term remains, all others become zero. The remaining term then also points to the correct sign of the permutation type.

This identity can be used to prove some other identities.

Below, we will note some important equations involving the Levi-Civita symbol and the Kronecker delta. When we introduced Einstein summation notation, we stated that a summation index must appear once as a lower index and once as an upper index. For the Kronecker symbol, this is not problematic, since $\delta_{ij} = \delta^{ij} = \delta^i_{\ j} = \delta^{\ j}_i$ in any coordinate system. However, the same is not true for ε_{ijk}.

In Cartesian coordinates, you can raise or lower the indices of ϵ_{ijk} without changing its value. This does not hold in curvilinear coordinates. In general, raising or lowering an index is done using the metric tensor, which we will examine in detail later. In Cartesian coordinates, the metric tensor is just the identity matrix, leaving δ_{ij} and ϵ_{ijk} invariant.

Now, consider the expression $\epsilon_{ijk}\,\epsilon_{lmk}$. Both factors have a lower index k. In Cartesian coordinates, it would be more consistent to write this as $\epsilon_{ijk}\,\epsilon^k_{lm}$ or $\epsilon^k_{ij}\,\epsilon_{lmk}$, ensuring that the repeated index is once upper and once lower.

Most of the time, in physics, you will use the Levi-Civita symbol in Cartesian coordinates. Following common practice in many textbooks, we will continue to use the (strictly speaking, incorrect) form $\epsilon_{ijk}\,\epsilon_{lmk}$.

The following equations are helpful:

$$(\boldsymbol{I.2.026}) \quad \epsilon_{ijk} = \epsilon_{jki} = \epsilon_{kij} \quad \text{(even permutation)}$$

$$\epsilon_{kji} = \epsilon_{jik} = \epsilon_{ikj} \quad \text{(odd permutation)}$$

$$\epsilon_{ijk} = -\epsilon_{kji} \quad \text{(permutation type changes)}$$

We can swap the indices, but it must be noted that changing the type of permutation (from even to odd and vice versa) entails a change in sign.

$(\boldsymbol{I.2.027})$: $\epsilon_{jik}\epsilon_{lmn} = -\epsilon_{ijk}\epsilon_{lmn}$

$(\boldsymbol{I.2.028})$: **contraction of ε with δ:**

ε and δ often come together within one term. Note that in the following equations, the ESN must be applied since each i represents a summation index:

$$\delta_{im}\epsilon_{ijk} = \epsilon_{mjk}$$

$$\delta_{im}\epsilon_{kij} = \epsilon_{kmj}$$

$$\delta_{im}\epsilon_{jki} = \epsilon_{jkm}$$

These equations become clear, considering that you sum up over i and δ_{im} takes the value 1 only if $i = m$ (see $(I.2.022)$).

$(\boldsymbol{I.2.029})$: **contraction of ε with δ:**

$$\delta_{ij}\epsilon_{ijk} = \epsilon_{iik} = 0$$

$(\boldsymbol{I.2.030})$: $\epsilon_{ijk}\epsilon_{lmn} = \delta_{il}\delta_{jm}\delta_{kn} + \delta_{im}\delta_{jn}\delta_{kl} + \delta_{in}\delta_{jl}\delta_{km} - \delta_{il}\delta_{jn}\delta_{km} - \delta_{in}\delta_{jm}\delta_{kl} - \delta_{im}\delta_{jl}\delta_{kn}$

Note that the index $(i\,j\,k)$ is passed through while the index $(l\,m\,n)$ is permuted. This is one of the most important formulas for ϵ_{ijk} and is often used to simplify expressions.

$(\boldsymbol{I.2.031})$: $\epsilon_{ijk}\epsilon_{lmk} = \delta_{il}\delta_{jm} - \delta_{im}\delta_{jl}$

Let us go into this evidence in a little more detail, as there is a stumbling block here that we'd like to examine in more detail.

First, we use $(I.2.030)$ with $n = k$. We obtain:

$$\epsilon_{ijk}\epsilon_{lmk} = \delta_{il}\delta_{jm}\delta_{kk} + \delta_{im}\delta_{jk}\delta_{kl} + \delta_{ik}\delta_{jl}\delta_{km} - \delta_{il}\delta_{jk}\delta_{km} - \delta_{ik}\delta_{jm}\delta_{kl} - \delta_{im}\delta_{jl}\delta_{kk}$$

Now comes the stopping stone: We want to replace δ_{kk}. We take up the discussion from $(I.2.020)$ again and see that δ_{kk} takes the value 3, because the indices of δ_{kk} are running from 1 to 3.

If we would take $\delta_{kk} = 1$, so we would get a wrong result by one sign! The other way around: if an index acrobatic gives a result that is wrong by sign, then you should check whether δ_{kk} appears in the expression!

So, we use $(I.2.020)$ that $\delta_{kk} = 3$ and we get:

$$\epsilon_{ijk}\epsilon_{lmk} = \delta_{il}\delta_{jm}3 + \delta_{im}\delta_{jk}\delta_{kl} + \delta_{ik}\delta_{jl}\delta_{km} - \delta_{il}\delta_{jk}\delta_{km} - \delta_{ik}\delta_{jm}\delta_{kl} - \delta_{im}\delta_{jl}3$$

We use contraction $(I.2.019)$and get:

$$
\begin{aligned}
\epsilon_{ijk}\,\epsilon_{lmk} \quad &= \quad \delta_{il}\delta_{jm}3 + \delta_{im}\delta_{jl} + \delta_{im}\delta_{jl} - \delta_{il}\delta_{jm} - \delta_{il}\delta_{jm} \\
&\qquad - \delta_{im}\delta_{jl}3 \\
&= \quad 3\delta_{il}\delta_{jm} - \delta_{il}\delta_{jm} - \delta_{il}\delta_{jm} - 3\delta_{im}\delta_{jl} - \delta_{im}\delta_{jl} \qquad \text{[reorder]} \\
&\qquad - \delta_{im}\delta_{jl} \\
&= \quad \delta_{il}\delta_{jm} - \delta_{im}\delta_{jl}
\end{aligned}
$$

$(I.2.032)$ $\epsilon_{ijk}\epsilon_{mnk} = \delta_{im}\delta_{jn} - \delta_{in}\delta_{jm}$

$(I.2.033)$ $\epsilon_{ijk}\epsilon_{imn} = \delta_{jn}\delta_{km} - \delta_{kn}\delta_{jm}$

$(I.2.034)$ The following equations apply:

$$\epsilon_{ijk}\epsilon_{imn} = \epsilon_{jki}\epsilon_{imn} = \epsilon_{kij}\epsilon_{imn} = \delta_{jn}\delta_{km} - \delta_{kn}\delta_{jm}$$

$$\epsilon_{ijk}\epsilon_{mni} = \epsilon_{jki}\epsilon_{mni} = \epsilon_{kij}\epsilon_{mni} = \delta_{jn}\delta_{km} - \delta_{kn}\delta_{jm}$$

$$\epsilon_{ijk}\epsilon_{nim} = \epsilon_{jki}\epsilon_{nim} = \epsilon_{kij}\epsilon_{nim} = \delta_{jn}\delta_{km} - \delta_{kn}\delta_{jm}$$

By exchanging the indices (observe $(I.2.019)$!), numerous other equations can be set up.

Tip: do not memorize all combinations. Reduce these to $(I.2.033)$ to $(I.2.033)$ and, for unknown index combinations, see whether you can get one of the known forms by permutation or renaming.

$(\textbf{\textit{I}.2.035})$ $\epsilon_{ijk}\epsilon_{ljk} = 2\,\delta_{li}$

$(\textbf{\textit{I}.2.036})$ $\epsilon_{ijk}\epsilon_{ijk} = 6$

$(\textbf{\textit{I}.2.037})$ $\epsilon_{imn}\epsilon_{jmn} = 2\,\delta_{ij}$ (double contraction of ε)

$(\textbf{\textit{I}.2.038})$ $\epsilon_{ijk}\epsilon_{jnl}\epsilon_{mil} = \epsilon_{nmk}$ (triple contraction of ε)

$(\textbf{\textit{I}.2.039})$ $\epsilon_{ijk}\epsilon_{klm} = \delta_{il}\delta_{jm} - \delta_{im}\delta_{jl}$

$(\textbf{\textit{I}.2.040})$ $\epsilon_{ijk}\epsilon_{lmk} = \delta_{il}\delta_{jm} - \delta_{im}\delta_{jl}$

$(\textbf{\textit{I}.2.041})$ $\epsilon_{ijk}\epsilon_{ljk} = 2\delta_{il}$

$(\textbf{\textit{I}.2.042})$ $\epsilon_{ijk}\epsilon_{jnl}\epsilon_{ilm} = \epsilon_{nmk}$

I.2.1.3.E Exercises

$(\textit{I}.2.1.3.\textit{E}.1)$ Determine if the following expressions are valid in ESN:

1. $\epsilon_{ilm}\,T_{jkl} = S_{ijm}$
2. $\epsilon_{lmn}\,a_{ijl}\,a_{kmn} = b_{ik}$
3. $\epsilon_{ijk}\,v_j\,\delta_{kl}\,w_l = x_i$
4. $\epsilon_{ijk}\,\delta_{il}\,M_{lk} = N_{ij}$

I.2.1.4 Vector calculation in index notation

In this section, we look at how vector calculus can be performed in index notation.

Let be $\vec{a}$ and $\vec{b}$ vectors with the components $(a_1, a_2, \ldots, a_n)$ respectively $(b_1, b_2, \ldots, b_3)$.

I.2.1.4.1 Representation of a vector

Let $(g_1, g_2 \ldots, g_n)$ an (arbitrary) basis, then each vector can be represented as a linear combination of the basis vectors:

$$\vec{a} = \sum_{i=1}^{n} a^i g_i$$

In index notation we write: $\vec{a} = a^i g_i$

Vectors are added component by component:

$$\vec{a} + \vec{b} = \begin{pmatrix} a^1 \\ a^2 \\ \ldots \\ a^n \end{pmatrix} + \begin{pmatrix} b^1 \\ b^2 \\ \ldots \\ b^n \end{pmatrix} = \begin{pmatrix} a^1 + b^1 \\ a^2 + b^2 \\ \ldots \\ a^n + b^n \end{pmatrix}$$

In index notation we get for the component j:

$$\left(\vec{a} + \vec{b}\right)_j = a^j + b^j$$

Please note in the index notation we can only show how the component j is calculated!

I.2.1.4.2 Scalar product in index notation

Using the previous notations, we define the scalar product as follows:

$$
\begin{aligned}
a \cdot b \quad &= \quad \sum_{i=1}^{n} a^i g_i \sum_{j=1}^{n} b^j g_j \\
&= \quad (a^1 g_1 + a^2 g_2 + \ldots + a^n g_n)(b^1 g_1 + b^2 g_2 + \ldots + b^n g_n) \\
&= \quad a^1 g_1 b^1 g_1 + a^1 g_1 b^2 g_2 + \ldots + a^1 g_1 b^n g_n
\end{aligned}
$$

$$+\ a^2 g_2 b^1 g_1 + a^2 g_2 b^2 g_2 + \ldots + a^2 g_2 b^n g_n$$

$$+\ \ldots$$

$$+\ a^n g_n b^1 g_1 + a^n g_n b^2 g_2 + \ldots + a^n g_n b^n g_n$$

$$=\ \sum_{i=1}^{n}\sum_{j=1}^{n} a^i g_i b^j g_j$$

$$=\ \sum_{i=1}^{n}\sum_{j=1}^{n} a^i b^j g_i g_j$$

$$=\ \sum_{i=1}^{n}\sum_{j=1}^{n} a^i b^j g_{ij} \qquad\qquad [g_{ij} := g_i g_j]$$

$$=\ a^i b^j g_{ij}$$

In the penultimate step we defined the scalar products of the basis vectors g_i and g_j as g_{ij} . $[g_{ij}]$ can be viewed as a matrix and is called the **metric tensor**. If it is a Cartesian basis with unit vectors, then $[g_{ij}]$ is the unit matrix. This can be described as a Kronecker delta, so the scalar product for the Cartesian basis system takes the form:

$$\vec{a} \cdot \vec{b} = a^i b^j\, \delta_{ij} \underset{\smile}{=} a^i b^i$$

$$(I.2.022)$$

I.2.1.4.3 Cross product in index notation

In the three-dimensional real vector space, the cross product of two vectors is defined as follows:

$$\vec{a} \times \vec{b} := \begin{pmatrix} a^2 b^3 - a^3 b^2 \\ a^3 b^1 - a^1 b^3 \\ a^1 b^2 - a^2 b^1 \end{pmatrix}$$

The result is a vector that is perpendicular to the plane defined by the vectors a and b . Its length corresponds to the area of a and b .

Referred to the standard orthonormal basis $(\vec{e}_1, \vec{e}_2, \vec{e}_3)$ the vector can also be represented in the following form:

$$\vec{a} \times \vec{b} = (a^2 b^3 - a^3 b^2)\vec{e}_1 + (a^3 b^1 - a^1 b^3)\vec{e}_2 + (a^1 b^2 - a^2 b^1)\vec{e}_3$$

In index notation it looks like this:

$$\left(\vec{a} \times \vec{b}\right)_k = \varepsilon_{ijk} a^i b^j$$

with i, j and k from 1 to 3, where $\left(\vec{a} \times \vec{b}\right)_k$ means the kth component. In this case, k (as the component to be considered) is to be chosen as a fixed value and only i and j are summed. When forming the sum, it should be noted that all summands where $i = j$ or $j = k$ or $k = i$ vanish after the definition of ε.

Using index notation, you can prove all major vector equations elegantly, just give it a try.

I.2.1.4.4 Matrix multiplication in index notation

Matrix multiplication can also be represented as Einstein summation. Suppose A is an $m \times n$ matrix and B is an $n \times p$ matrix. The product $C = AB$ is an $m \times p$ matrix where each element c_{ij} is computed as:

$$c_{ij} = \sum_{k=1}^{n} a_{ik}\, b_{kj}$$

In other words, the element c_{ij} is the dot product of the i-th row of matrix A and the j-th column of matrix B.

Defining the matrix A by $A^{\mu}{}_{\upsilon}$, accordingly B by $B = B^{\mu}{}_{\upsilon}$ and $C = C^{\mu}{}_{\nu}$, then we get the matrix multiplication in ESN:

$$C^{\mu}{}_{\nu} = A^{\mu}{}_{\gamma}\, B^{\gamma}{}_{\nu} = B^{\gamma}{}_{\nu}\, A^{\mu}{}_{\gamma}$$

On the one hand, it should be noted that the multiplication in ESN is commutative since it only refers to <u>one component</u>. Second, that the summation index γ one of the multipliers is the <u>inner</u> index and the other is the <u>outer</u> index! <u>The position of the indices is essential!</u>

The result here is not a matrix but calculates for each row and column the value of the component $C^{\mu}{}_{v}$. For example, the component $C^{2}{}_{1}$ results as $A^{2}{}_{\gamma}\,B^{\gamma}{}_{1} = B^{\gamma}{}_{1}\,A^{2}{}_{\gamma}$, summed up over γ. The Einstein notation always refers *only to components*, not to an entire object. If we want to denote a matrix with the Einstein notation, then we put this into brackets, i.e. $\left[C^{\mu}{}_{v}\right]$.

To denote the transpose A^{T} of the matrix $A = A^{\mu}{}_{v}$, we only need to swap the indices:

$$\left(A^{\mu}{}_{v}\right)^{T} = A^{v}{}_{\mu}$$

Now let us consider the case $C = A^{T}B$, i.e., the multiplication of the transposed matrix A by matrix B. We get:

$$C^{\mu}{}_{v} = \left(A^{\mu}{}_{\gamma}\right)^{T} B^{\gamma}{}_{v} = A^{\gamma}{}_{\mu} B^{\gamma}{}_{v}$$

Note that in this case you have to sum up over the *inner* index.

We now look at what happens at $C = A\,B^{T}$:

$$C^{\mu}{}_{v} = A^{\mu}{}_{\gamma}\left(B^{\gamma}{}_{v}\right)^{T} = A^{\mu}{}_{\gamma} B^{v}{}_{\gamma}$$

Here, it is the *outer* index (γ) that is summed up!

Summary:

It is important that, based on the location of the bound index, - in our case this is γ − it can be recognized what kind of matrix multiplication is described by the ESN:

Matrix multiplication	in component notation	position of the summation index
AB	$A^{\mu}{}_{\gamma}B^{\gamma}{}_{\nu}$ $bzw.$ $B^{\gamma}{}_{\nu}A^{\mu}{}_{\gamma}$	outer times inner index or inner times outer index
$A^{T}B$	$A^{\gamma}{}_{\mu}B^{\gamma}{}_{\nu}$ $bzw.$ $B^{\gamma}{}_{\nu}A^{\gamma}{}_{\mu}$	inner times inner index
AB^{T}	$A^{\mu}{}_{\gamma}B^{\nu}{}_{\gamma}$ $bzw.$ $B^{\nu}{}_{\gamma}A^{\mu}{}_{\gamma}$	outer times outer index

Another point: in ESN we have $A^{\mu}{}_{\gamma}B^{\gamma}{}_{\nu} = B^{\gamma}{}_{\nu}A^{\mu}{}_{\gamma}$: given such an expression: how do you know if this represents AB or BA? In *this* case look for the *outer* index: in both cases this is A which has the summation index γ as the outer index. That is the first matrix in the matrix multiplication!

In case of $A^{\gamma}{}_{\mu}B^{\gamma}{}_{\nu}$ $(A^{T}B)$ and $A^{\mu}{}_{\gamma}B^{\nu}{}_{\gamma}$ (AB^{T}) you have no chance to determine which matrix order you need to take. Let us take the first example $A^{\gamma}{}_{\mu}B^{\gamma}{}_{\nu}$: we know that one of the matrices needs to be the transposed. But which one? Let us guess that B is the candidate for the transposed matrix, so $\left(B^{\nu}{}_{\gamma}\right)^{T} = B^{\gamma}{}_{\nu}$. We get: $A^{\gamma}{}_{\mu}B^{T}{}^{\nu}{}_{\gamma}$. This translates as $B^{T}A$. Likewise, we get $A^{T}B$ when we choosed A as the candidate for the transposed matrix.

You can only hope that the writer was so kind to choose in the ESN equation the right order for converting the ESN into matrix multiplication. As far as I noticed, you will not see both equations very often.

I.2.1.4.4.1 Inverse matrix

A matrix A is invertible if there exists a matrix B with $AB = BA = I$, where I shall be the identity matrix. B is then the inverse matrix of A , A is also the inverse matrix of B . The condition in component notation is then:

$$(\textbf{\textit{I}.2.043})\ A^{\mu}{}_{\gamma}B^{\gamma}{}_{\nu} = \delta_{\mu\nu}$$

Without proof, we note that the components of the inverse of a squared matrix A can be calculated as:

$$(\textbf{\textit{I}.2.044})\ \left(A_{ij}\right)^{-1} = \frac{\epsilon_{ilm}\,\epsilon_{jnp}\,a_{nl}\,a_{mp}}{\det\left(A\right)}$$

I.2.1.4.4.2 Interpretation of ESN equations

We want to pick up another point: how do you interpret an ESN equation? Let us take the following equation in ESN as an example:

$$a_{ij}x^j = b_i$$

A system of n equations is obviously set up here, which can be represented in matrix form as follows:

$$Ax = b\ ,$$

where $A = \left[a_{ij}\right]$, $x = (x^1, x^2, \dots, x^n)$ and $b = (b_1, b_2, \dots b_n)$ shall apply. This form should be familiar from linear algebra: it describes a linear system of equations.

But what about the following expression?

$$Q = a_{ij}x^i x^j$$

Here, it is no longer clearly recognizable how this expression can be represented in matrix notation. In fact, the equation can, among other things, represent a *quadratic form with a homogeneous multi-variable polynomial of the second order*. We explain this with an example. Consider:

$$10 = 3x^2 + y^2 - z^2 - 5xy - 6yz$$

The polynomial has the variables x, y, z. *Homogeneous of the second order* now means that all terms (one also speaks of *monomials* here) must have degree 2:

For $3x^2, y^2, z^2$ this is obviously the case. For $5xy$, the degrees of the individual variables are added together, and this gives degree 2, the same for $6yz$.

Such quadratic forms can always be represented as a matrix equation:

$$Q = x^T A x$$

Detailed (for $n = 3$):

$$Q = (x_1 \quad x_2 \quad x_3) \begin{pmatrix} a_{11} & a_{12} & a_{13} \\ a_{21} & a_{22} & a_{23} \\ a_{31} & a_{32} & a_{33} \end{pmatrix} \begin{pmatrix} x_1 \\ x_2 \\ x_3 \end{pmatrix}$$

For our example we get:

$$10 = (x_1 \quad x_2 \quad x_3) \begin{pmatrix} 3 & -5 & 0 \\ 0 & 1 & 6 \\ 0 & 0 & -2 \end{pmatrix} \begin{pmatrix} x_1 \\ x_2 \\ x_3 \end{pmatrix}$$

When translating an ESN back into a matrix equation, special attention must be paid to the fact that vectors may need to be transposed (becoming a row vector) so that the multiplication steps can be performed consistently. However, this cannot be seen directly in the ESN equation!

It is therefore important that you get a feeling for whether - and above all: how - an ESN with two indices can be represented as a matrix equation. As much as an ESN can help to represent entire collections of equations elegantly: when it comes to concrete calculations, matrix multiplication is easier to carry out. As a final remark one might add that a whole range of tensors in physics (these are multi-linear maps with a very specific transformation behavior) can be and are represented in matrix form.

I.2.1.4.4.E Exercises

($I.2.1.4.4.E.1$) For a matrix A the trace of A – noted as $Tr(A)$ - is defined as the sum of its diagonal elements. Express the trace in ESN notation.

($I.2.1.4.4.E.2$) Let A, B and C be matrices, so that you can perform the product ABC. Express this multiplication in ESN.

($I.2.1.4.4.E.3$) Let A, B and C be matrices, so that the operation

$A(B + C)$ can be calculated. Express this term in ESN.

I.2.1.5 ESN and Partial Derivatives

We now look at the ESN in partial derivatives.

For the beginning, here are some equations that have to do with partial derivatives. Let $x_1, x_2, \ldots, x_n$ be linear independent functions. Then, we get the following identities:

$$(I.2.045) \qquad \frac{\partial x_p}{\partial x_k} = \delta_{pk}$$

$$(I.2.046) \qquad \frac{\partial}{\partial x_k}\left(a_{ij}x_j\right) = a_{ij}\frac{\partial}{\partial x_k}x_j = a_{ij}\delta_{jk} = a_{ik}$$

$$(I.2.047) \qquad \frac{\partial}{\partial x_k}\left(a_{11}x_1 + a_{12}x_2 + a_{13}x_3\right) = a_{1k}$$

$$(I.2.048) \qquad \frac{\partial}{\partial x_k}\left(a_{ij}x_ix_j\right) = \left(a_{ik} + a_{ki}\right)x_i$$

$$(I.2.049) \qquad \frac{\partial}{\partial x_l}\left(a_{ijk}x_ix_jx_k\right) = \left(a_{lij} + a_{ilj} + a_{ijl}\right)x_ix_j$$

Some comments on $(I.2.045)$: this equation is important and will be used very often in this book. Later, we will consider curvilinear coordinates, this means that a basis of a vector space is constructed by linear independent functions (x^i). Now consider the functions x^i and x^j:

- Case 1: $i = j$. It should be clear that $\frac{\partial x^i}{\partial x^i} = 1$.

- Case 2: $i \neq j$: then $\frac{\partial x^i}{\partial x^j}$ must be zero as x^i and x^j are linear independent.

I.2.1.5.E Exercises

$(I.2.1.5.E.1)$ Given a scalar field $\phi(x^1, x^2, x^3)$, express $\frac{\partial \phi}{\partial x^i}$ and $\frac{\partial \phi}{\partial x^i \partial x^j}$ in ESN, where $i = 1,2,3$.

$(\boldsymbol{I.2.1.5.E.2})$ Given a vector field $A^i(x^1, x^2, x^3)$, express $\frac{\partial A^i}{\partial x^j}$ in ESN, where $i = 1,2,3$.

I.2.2 Parameterization of vector functions

Functions can appear in different shapes. Consider, for example, the following equations:

(1) $y = \sqrt[2]{1 - x^2}$

(2) $x^2 + y^2 = 1$

(3) $f(t) = \begin{pmatrix} \cos{(t)} \\ \sin{(t)} \end{pmatrix}$, $t \in [0, 2\pi]$

All three functions describe the same geometrical object: the unit circle. Equation (1) describes only the (positive) semi-circle, equation (2) describes the full unit circle, the same with equation (3). Equation (1) is an **explicit** representation, equation (2) is an **implicit** function while (3) is a **parameterized** function. It should be clear that parametrized functions simplify the handling of geometric objects. For example, a 2D curve in a plane can be described with just one parameter, while a surface in 3D space requires two parameters. In physics, we often have a vector function $\vec{r}(t)$ with time as the parameter. It describes a trajectory in space. If we take the first derivate, then we get the instantaneous velocity, the second derivative is the instantaneous acceleration. *Derivates by time are notated with a dot* (for the first derivative) or two dots (for the second derivatives), so we write $\dot{r}(t)$ or shortly $\dot{r}$ for the first derivatite of $r(t)$ with respect to t and $\ddot{r}(t)$ or shortly $\ddot{r}$ for the second derivative of $r(t)$, again with respect to t.

Before we go on, just a remark on how to calculate the derivatives:
We can write the parameterization as
$r(t) = \begin{pmatrix} x(t) \\ y(t) \end{pmatrix}$, so we use component functions.

Using that notation, it is important to notice that:

$$\frac{dy}{dx} = \frac{dy}{dt}\frac{dt}{dx}$$

$$\frac{d^2y}{dx^2} = \frac{d}{dt}\left(\frac{dy}{dx}\right) \cdot \frac{dt}{dx}$$

This is the **_chain rule_**: we will come back to the chain rule a bit later (cf. $(I.2.4)$).

The parameterization of a curve is not unique:

Consider the implicit equation of a parable $4x^2 + y = 4$.
Then we have the parameterizations

$$r(\theta) = \begin{pmatrix} \cos(\theta) \\ \sin^2(\theta) \end{pmatrix} \text{ with } \theta \in [0, 2\pi]$$

$$r(t) = \begin{pmatrix} t/2 \\ 4 - t^2 \end{pmatrix} \quad \text{with } t \in \mathbb{R}$$

The parameterization with θ shows only the parabel above the x-axe, the parameterization with t shows the whole parabel.

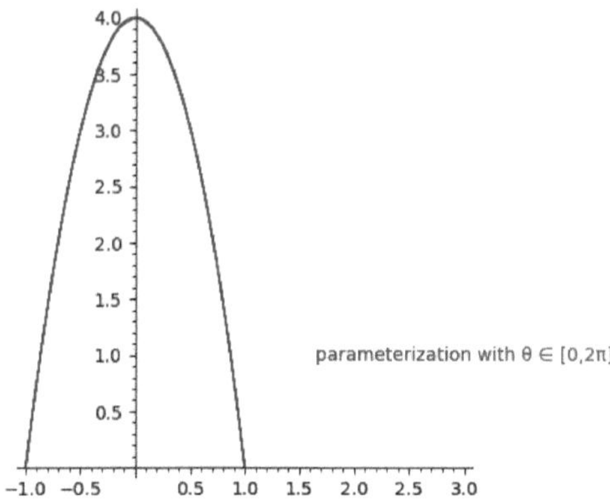

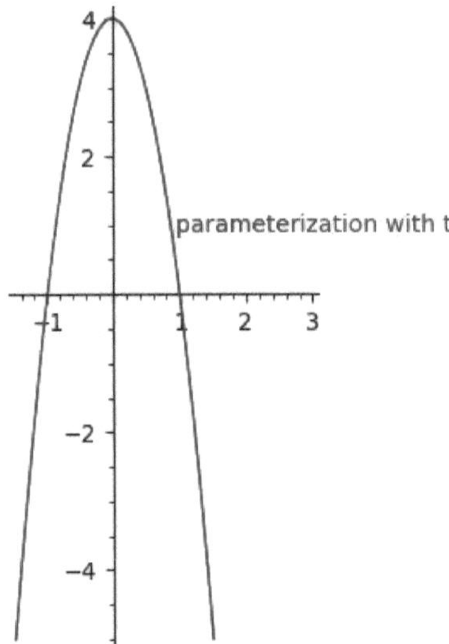

There exists a parameterization which is important if you want to calculate the curvature: the parameterization by arc length:

I.2.2.1 Parameterization by arc length

The curvature of a curve $r(t)$ at a point t measures the deviation of the curve from being a straight line at that point, where a straight line has zero curvature. To calculate the curvature of a curve accurately, it is helpful to use a parameterization where the tangent vector has a constant length of 1 at every point t on the curve. This means $| r'(t) | = 1$, and the tangent vector is called the **unit tangent vector**. Physically, parameterization by arc length implies that the curve is traversed at a constant speed.

A curve for which the tangent vector at each point is the unit tangent vector is said to be "***parameterized by arc length***", because then the integrand of arc length is always one. In standard notation we use $r(s)$ to point out that we have a parameterization by arc length. The parameterization of arc length is uniquely defined.

To parameterize a curve according to its arc length, we first determine the length of the tangent vector $| r'(t) |$. The arc length from 0 to t is then given by

$$s(t) = \int r'(u)\, du$$

If $|\, r'(t)\, |$ is a constant a, then $s(t) = at$. To parameterize by arc length, we find the inverse function $t(s)$, which is $t(s) = s/a$.

For an example, we look at the parameterization of a circle with radius r:

$$r(t) = \begin{pmatrix} r\cos(t) \\ r\sin(t) \end{pmatrix},\ t \in [0, 2\pi],\, r > 0$$

We get:

$$\dot{r}(t) = \begin{pmatrix} -r\sin(t) \\ r\cos(t) \end{pmatrix}$$

Now, the length of the unit tangent vector calculates as:

$$|\dot{r}(t)| = \sqrt[2]{r^2\sin^2(t) + r^2\cos^2(t)} = \sqrt[2]{r^2} = r$$

Since $|\dot{r}(t)| = r$, the arc length from $t_0 = 0$ to t is:

$$s(t) = \int_0^t r\, du = r\, t$$

The inverse function $t(s)$ is:

$$t(s) = \frac{s}{r}$$

Thus, we get the arc length parameterization by:

$$r(s) = \begin{pmatrix} r\cos\left(\dfrac{s}{r}\right) \\ r\sin\left(\dfrac{s}{r}\right) \end{pmatrix}$$

How can we determine if a given parameterization is a parameterization by arc length?
A parameterization of $r(t)$ is parameterized by arc length if and only if $|\dot{r}(t)| = 1$ for all t.

I.2.2.E Exercises

$(\boldsymbol{I.2.2.E.1})$ Let $r(t)$ be defined as:

$$r(t) = \begin{pmatrix} \cos 2\pi t \\ \sin 2\pi t \\ k\,t \end{pmatrix} \text{ with } k, t \in \mathbb{R}, k \text{ a fixed value, } k \neq 0.$$

Parameterize $r(t)$ by arc length.

$(\boldsymbol{I.2.2.E.2})$ Let $r(t)$ be defined as:

$$r(t) = \begin{pmatrix} \sin(2t)\cos(t) \\ \sin(2t)\sin(t) \end{pmatrix} \text{ with } t \in [0, 2\pi]$$

Determine if $r(t)$ is parameterized by arc length.

I.2.3 Derivatives and their generalizations, part 1

In this chapter, we will use the standard Euclidian basis of $\mathbb{R}^n$. The reason is that *gradient*, *divergence*, and *curl* take a simple form in this coordinate system. These operations become more complex in other coordinate systems, which we will discuss later in Chapter $I.8$.

In this section, we give an overview about the different forms of derivatives in $\mathbb{R}^n$.

I.2.3.1 Basic derivation in $\mathbb{R}$

You should be familiar with the derivative of a function $f: U \to \mathbb{R}$, where $U \subseteq \mathbb{R}, U$ open. The derivative of $f(x)$ at the point x is denotated as $f'(x)$. To be more precise: $f'(x)$ represents the slope of the tangent line to the graph of f at the point x.

It is important to remember that when we discuss derivatives, we are often referring to tangent vectors. Tangent vectors will become very important in our discussions.

I.2.3.2 Implicit derivation

In case you encounter an implicit function $F(x^1, \ldots, x^n) = 0$ and you want to find the partial derivative $\frac{\partial x^j}{\partial x^i}$, you can use the following formula, which is derived from the **implicit function theorem**:

$$(I.2.050) \quad \frac{\partial x^j}{\partial x^i} = -\frac{\dfrac{\partial F(x^1, \ldots, x^n)}{\partial x^i}}{\dfrac{\partial F(x^1, \ldots, x^n)}{\partial x^j}}$$

Observe that the indices on the left and right-side swap! This formula allows you to compute the partial derivative of one variable with respect to another, assuming the function F implicitly defines a relationship between the variables.

<u>Example:</u>

Let us take the function

$$F(x^1, x^2) := (x^1)^5 + (x^1)^2(x^2)^3 + (x^2)^6 + 7 = 0$$

Then we get:

$$\frac{\partial x^2}{\partial x^1} = -\left. 5(x^1)^4 + 2(x^1)(x^2)^3 \middle/ 3(x^2)^2(x^1)^2 + 6(x^2)^5 \right.$$

In the following example we want to give you another algorithm:

Take the equation

$$x^2 + y^2 = 25.$$

We want to calculate $\frac{dy}{dx}$.

Step 1: Differentiate both sides of the equation with respect to x, *treating y as a function of* x:

$$\frac{d}{dx}(x^2) + \frac{d}{dx}(y^2) = \frac{d}{dx}(25)$$

Step 2: Apply the chain rule for $\frac{d}{dx}(y^2)$, which is $2y\,\frac{dy}{dx}$

$$2x + 2y\,\frac{dy}{dx} = 0$$

Step 3: solve for $\frac{dy}{dx}$

$$\frac{dy}{dx} = -\frac{x}{y}$$

I.2.3.2.E Exercises

I.2.3.2.E.1 Given the equation $(x^1)^2\,(x^2)^4 - 3 = \sin(x^1\,x^2)$.

Calculate $\dfrac{dx^2}{dx^1}$.

I.2.3.3 Partial derivation

We look now at functions in $\mathbb{R}^n, n \geq 2$. So let U be an open set of $\mathbb{R}^n$ and let f be a differentiable function from U to $\mathbb{R}$, so $f : U \to \mathbb{R}$. Furthermore, let $a = (a^1, a^2, \ldots, a^n)$ be an element in U.

If

$$\frac{\partial f}{\partial x^i} := \lim_{h \to 0} \frac{f(a^1,\ a^2, \boldsymbol{a^i + h},\ \ldots, a^n) - f(a^1,\ a^2, \boldsymbol{a^i},\ \ldots, a^n)}{h}$$

exists, then we call $\frac{\partial f}{\partial x^i}$ the **partial derivative** of f with respect to x^i at point a.

In the same way, you can define partial derivatives of higher degrees, as, for example,

$$\frac{\partial^2 f}{\partial x^i \partial x^j} \quad \text{for the second degree.}$$

We get:

$$\frac{\partial^2 f}{\partial x^i \partial x^j} = \frac{\partial}{\partial x^i}\left(\frac{\partial f}{\partial x^j}\right)$$

In the special case $i = j$, we get:

$$\frac{\partial^2 f}{\partial x^i \partial x^i} = \frac{\partial^2 f}{(\partial x^i)^2} = \frac{\partial}{\partial x^i}\left(\frac{\partial f}{\partial x^i}\right)$$

We quote the **Theorem of Schwartz**: if the partial derivatives $\frac{\partial f}{\partial x^i}$ and $\frac{\partial f}{\partial x^j}$ exist and are continuous, then you can swap the indices:

$$\frac{\partial^2 f}{\partial x^i \partial x^j} = \frac{\partial^2 f}{\partial x^j \partial x^i}$$

This means, that the derivation first by x^i, then by x^j will produce the same result as if you first derivate by x^j and then by x^i. We will use this theorem without further notice.

If you set $n = 1$, then the partial derivate is just the basic derivative.

I.2.3.4 Gradient

The **gradient** of a multivariable scalar-valued function $f(x^1, x^2, \ldots, x^n): \mathbb{R}^n \to \mathbb{R}$ is defined as the vector of its partial derivatives:

$$grad\ f := \nabla f := \begin{pmatrix} \dfrac{\partial f}{\partial x^1} \\ \dfrac{\partial f}{\partial x^2} \\ \ldots \\ \dfrac{\partial f}{\partial x^n} \end{pmatrix}$$

∇ is called the **Nabla operator**, defined by

$$\nabla := \begin{pmatrix} \dfrac{\partial}{\partial x^1} \\ \dfrac{\partial}{\partial x^2} \\ \ldots \\ \dfrac{\partial}{\partial x^n} \end{pmatrix}$$

The value of $\nabla f(x)$ at a point x is the direction and rate of fastest increase. If the gradient of a function is non-zero at point x, the direction of the gradient is the direction in which the function increases most quickly from x.

If $n = 1$, then we have a function from $\mathbb{R}$ to $\mathbb{R}$. In this case the gradient is simply the derivative which we discussed earlier. In this sense, the gradient is a generalization of the (normal) derivate.

In ESN, we may write

$$(I.2.051)\ \nabla f = \vec{e}^{\,i} \frac{\partial f}{\partial x^i}$$

Note, you have to sum up over the index i.

If you want to take the **directional derivative $\nabla_v f$** of f in direction of a **unit** vector v, then you can use the formula

$$(I.2.052)\ \nabla_v f = \nabla f \cdot v$$

Notice that we mean the scalar product on the right side!

The directional derivative measures the rate of change of f in direction of v.

<u>Example</u>:

We take the function

$$f(x^1, x^2, x^3) = (x^1)^2 + (x^2)^2 + (x^3)^2$$

We want to find the directional derivative of f at the point $(1,2,3)$ in the direction of $(1, -1, 1)^T$.

We have:

$$\nabla f = (2x^1, 2x^2, 2x^3)^T \text{ and thus:}$$

$$\nabla f(1,2,3) = (2, 4, 6)^T$$

We set $v = (1, -1, 1)$, then

$$v_{unit\_vector} = \frac{v}{|v|} = \frac{v}{\sqrt{1^2 + (-1)^2 + 1^2}} = \frac{(1, -1, 1)^T}{\sqrt{3}}$$

It follows with $(I.\,2.052)$:

$$\nabla_v f(1,2,3) = \underbrace{(2,4,6)^T \cdot \frac{(1, -1, 1)^T}{\sqrt{3}}}_{scalar\ product} = \frac{2}{\sqrt{3}} - \frac{4}{\sqrt{3}} + \frac{6}{\sqrt{3}} = \frac{4}{\sqrt{3}}$$

I.2.3.4.E Exercises

I.2.3.4.E.1

Define $T(x, y, z) = T_0/(1 + x^2 + y^2 + z^2)$ with $T_0 \in \mathbb{R}$.

Calculate the gradient.

I.2.3.5 Divergence

Let U be an open subset of $\mathbb{R}^n$ and f a function from U to $\mathbb{R}^n$. We call those functions a vector field as each vector gets a vector as its image. f can then be represented by its component functions:

$$f(x) = \begin{pmatrix} f_1(x^1, x^2, \ldots, x^n) \\ f_2(x^1, x^2, \ldots, x^n) \\ \ldots \\ f_n(x^1, x^2, \ldots, x^n) \end{pmatrix}$$

Now, we can define another differential operator, the **divergence**:

$$(I.2.053) \quad div\, f := \frac{\partial f_i}{\partial x^i}$$

Sum up over i.

Using ∇, we can express the divergence as the scalar product of ∇ with f:

$$(I.2.054) \quad div\, f = \nabla \cdot f$$

As for a physical interpretation, I quote the following text from wikipedia:

In physical terms, the divergence of a vector field is the extent to which the vector field flux behaves like a source at a given point. It is a local measure of its "outgoingness" – the extent to which there are more of the field vectors exiting from an infinitesimal region of space than entering it. A point at which the flux is outgoing has positive divergence and is often called a "source" of the field. A point at which the flux is directed inward has negative divergence and is often called a "sink" of the field. The greater the flux of field through a small surface enclosing a given point, the greater the value of divergence at that point. A point at which there is zero flux through an enclosing surface has zero divergence.

A warning: do not mix up ∇f and $\nabla \cdot f$: the first is the gradient, the second is the divergence!

I.2.3.6 Curl

Especially in $\mathbb{R}^3$ we can define the differential operator *curl* as the cross product of Nabla with a vector field f:

$$curl\ \vec{V} = \nabla \times \vec{V} = \begin{pmatrix} \dfrac{\partial V_3}{\partial x_2} - \dfrac{\partial V_2}{\partial x_3} \\[2ex] \dfrac{\partial V_1}{\partial x_3} - \dfrac{\partial V_3}{\partial x_1} \\[2ex] \dfrac{\partial V_2}{\partial x_1} - \dfrac{\partial V_1}{\partial x_2} \end{pmatrix}$$

In ESN we get "the easy to check" notation:

$$(I.2.055)\quad curl\ f = \vec{e}_i\ \varepsilon_{ijk}\ \frac{\partial f_k}{\partial x^j}$$

In physics, *curl* is a mathematical concept used to describe the rotation or circulation of a vector field. Imagine you have a vector field representing the velocity of fluid flow, such as water or air. The curl of this vector field at a specific point tells you how much the fluid is rotating or circulating around that point.

I.2.3.7 Laplacian

Laplacian is another differential operator on scalar fields.

Geometrically, the Laplacian represents the divergence of the gradient of a function. It quantifies the rate at which the values of a function are changing at any given point in space. If the Laplacian of a function is zero at a particular point, it implies that the function is locally constant at that point. On the other hand, if the Laplacian is positive, it suggests that the function is increasing, while a negative Laplacian indicates a decreasing behavior.

Imagine a scalar field that represents some physical quantity, such as temperature or pressure, distributed throughout space. The Laplacian operator describes how this scalar field changes smoothly or abruptly from point to point. If the Laplacian is zero at a point, it suggests that the scalar field is uniform and smooth in that region. Conversely, if the Laplacian is nonzero, it indicates that the scalar field is changing in curvature or intensity at that point.

Think of the Laplacian as a measure of the curvature of the gradients of the scalar field. At points where the Laplacian is positive, the gradients of the scalar field are curving outward, suggesting regions of convexity. At points where the Laplacian is negative, the gradients are curving inward, indicating regions of concavity.

In its simplest form, in Cartesian coordinates, it is defined as the *sum of second partial derivatives* of a function with respect to its spatial variables:

$$(I.2.056) \quad \nabla^2 \varphi = \sum_i \frac{\partial^2 \varphi}{\partial (x^i)^2}$$

Here, $(x^i)^2$ signifies really "to the power of two."

I.2.3.8 Jacobi matrix

The Jacobi matrix is a generalization of the differential operator of a system of functions.

For this chapter, we will consider a function $f : \mathbb{R}^n \longrightarrow \mathbb{R}^m$ with the component functions $f_1, \ldots, f_n$. The component functions are presumed to be differentiable, so that all partial derivatives exist.

The Jacobi matrix is defined as:

$$(I.2.057) \quad J_f = \begin{bmatrix} \nabla f_1 \\ \ldots \\ \nabla f_m \end{bmatrix} = \begin{bmatrix} \dfrac{\partial f_1}{dx_1} & \dfrac{\partial f_1}{dx_2} & \cdots & \dfrac{\partial f_1}{dx_n} \\ \ldots & \ldots & \ddots & \ldots \\ \dfrac{\partial f_m}{dx_1} & \cdots & \cdots & \dfrac{\partial f_m}{dx_n} \end{bmatrix}$$

The rows of J_f consists of the *transposed* gradients of the component functions. J_f will play an important role for coordinate transformations – we will come back to it in subsequent chapters.

The Jacobi matrix represents the differential of f at every point where f is differentiable. In detail, if h is a displacement vector represented by a column matrix, the matrix product $J_f \cdot h$ is another displacement vector, that is the best linear approximation of the change of f in a neighborhood of x, if $f(x)$ is differentiable at x. This means that the function that maps y to $f(x) + J(x) \cdot (y - x)$ is the best linear approximation of $f(y)$ for all points y close to x. The linear map $h \to J(x) \cdot h$ is known as the *derivative* or the *differential* of f at x. If $m = 1$ the Jacobi matrix is the transposed gradient of f.

If $m = 1$, that is when $f: \mathbb{R}^n \to \mathbb{R}$ is a scalar-valued function, the Jacobian matrix reduces to the row vector ; this row vector of all first-order partial derivatives of f is the transpose of the gradient of f. Specializing further, when $m = n = 1$, that is when $f: \mathbb{R} \to \mathbb{R}$ is a scalar-valued function of a single variable, the Jacobian matrix has a single entry; this entry is the derivative of the function f.

If $n = m$, the Jacobian is a quadratic matrix, and you can calculate the determinant. The determinant of a Jacobian matrix will play an important role in the substitution for multiple variables when integrating f, as we will see later.

If $n = m$ and $f_i = x_i$, then J_f is just the unit matrix, as

$$\frac{\partial x_i}{d x_j} = \delta_{ij} \qquad (I.2.045)$$

I.2.3.8.1.1 The Jacobian determinant

When solving an integral in curvilinear coordinates, we use the determinant of the Jacobian matrix, often referred to as the **Jacobian determinant** or simply the **Jacobian**. Consider two sets of functions (x^i) and $(\bar{x}^i)$, each describing a coordinate system. We assume these function sets are invertible.

The Jacobi matrix from (x^i) to $(\bar{x}^i)$ is:

$$(I.2.058) \quad J(x^i, \bar{x}^j) = \left[\frac{\partial x^i}{d\bar{x}^j}\right] = \begin{bmatrix} \dfrac{\partial x^1}{d\bar{x}^1} & \dfrac{\partial x^1}{d\bar{x}^2} & \cdots & \dfrac{\partial x^1}{d\bar{x}^2} \\ \cdots & \cdots & \ddots & \cdots \\ \dfrac{\partial x^n}{d\bar{x}^1} & \cdots & \cdots & \dfrac{\partial x^1}{d\bar{x}^n} \end{bmatrix}$$

Similarly, we can calculate the Jacobian matrix from $(\bar{x}^i)$ to (x^i), which is the inverse:

$$J(\bar{x}^i, x^j) = \left[\frac{\partial \bar{x}^i}{dx^j}\right] = \begin{bmatrix} \dfrac{\partial \bar{x}^1}{dx^1} & \dfrac{\partial \bar{x}^1}{dx^2} & \cdots & \dfrac{\partial \bar{x}^1}{dx^2} \\ \cdots & \cdots & \ddots & \cdots \\ \dfrac{\partial \bar{x}^n}{dx^1} & \cdots & \cdots & \dfrac{\partial \bar{x}^1}{dx^n} \end{bmatrix}$$

Without proof, we state that

$$(I.2.059) \quad \det\left(J(x^i, \bar{x}^j)\right) = \frac{1}{\det\left(J(\bar{x}^i, x^j)\right)}$$

This equation is useful if you need to determine the Jacobian of the inverse Jacobian matrix. If the Jacobian matrix $J(x^i, \bar{x}^j)$ is given, you do not need to calculate its inverse. Instead, calculate the determinant of $J(x^i, \bar{x}^j)$ and take its reciprocal value.

I.2.3.8.E Exercises

I.2.3.8.E.1: Given the function

$$f(r, \theta) = \left(e^{-r} \sin(\theta), e^{-r} \cos(\theta) \right).$$

Calculate the Jacobi-matrix $J(r, \theta)$

I.2.4 The multivariate chain rule for partial derivatives

The correct handling of partial derivatives of functions depending on other functions is essential. Therefore, we cover this theme:

The following chain rule for partial derivatives is used when working with curvilinear coordinates (discussed later):

If $f(x^1, x^2, \ldots, x^n)$ is a function that depends on $x^1, x^2, \ldots, x^n$, and if these x^i are themselves dependent on some functions $u^1(u^1, u^2, \ldots, u^n)$, then the partial derivate $\frac{\partial f}{\partial u^j}$ calculates as:

$$(I.2.060) \quad \frac{\partial f}{\partial u^j} = \frac{\partial f}{\partial x^i} \frac{\partial x^i}{\partial u^j}$$

Two important notes:

1. **Einstein Summation Notation (ESN):** We use Einstein Summation Notation here, so you must sum over i.
2. **Partial derivatives are not fractions:** Never treat ∂x^i in such equations as fractions that you can cancel out. Partial derivatives are <u>limits</u> of functions and not actual fractions. Misinterpreting them can lead to significant errors in calculations.

Let us take an example:

We take a function f that depends on x^1 and x^2. Now, x^1 and x^2 are themselves functions depending on t and u. Taking the partial derivative with respect to t, we get:

$$\frac{\partial f(x^1(t,u), x^2(t,u))}{\partial t} = \frac{\partial f}{\partial x^1} \frac{\partial x^1}{\partial t} + \frac{\partial f}{\partial x^2} \frac{\partial x^2}{\partial t}$$

We will use the multivariate chain rule very often in this book.

I.2.4.E Exercises:

I.2.4.E.1 Given the function $z = \cos (x^2 (x^1)^2)$ with $x^1(t) = t^4 - 2t$, $x^2(t) = 1 - t^6$, calculate $\partial z/\partial t$.

I.2.4.E.2 Given the function

$z = 4 x^2 \sin (2 x^1)$

with $x^1(u, p) = 3u - p, \quad x^2(u, p) = p^2 u, \quad u(t) = t^2 + 1$

Calculate $\dfrac{dz}{dt}$ and $\dfrac{dz}{dv}$

I.2.4.E.3 Given the equations

$w = f(x^1, x^2, x^3)$

$x^1 = x^1(t)$

$x^2 = x^2(u, v, p)$

$x^3 = x^3(v, p)$

$v = v(r, u)$

$p = p(t, u)$

Determine $\dfrac{\partial w}{\partial t}$ and $\dfrac{\partial w}{\partial u}$.

I.2.5 Covariant and contravariant vectors

A vector is a geometric object with a magnitude and direction, both of which are invariant regardless of the coordinate system in which it is expressed. When a vector is represented in a coordinate system, it is described by components that depend on the chosen basis. If the coordinate system is changed, these components (and not the vector itself) will change accordingly.

In mathematics, the concepts of covariant and contravariant vectors are not inherently significant; they become important in the context of physics and differential geometry. To illustrate these concepts, let us consider an example:

In $\mathbb{R}^2$ we take a vector $\vec{v}$ in the Euclidian basis. Then $\vec{v}$ can be expressed as a linear combination of the basis vectors $\vec{e}_1$ and $\vec{e}_2$ with some uniquely determined coefficients v^1 and v^2:

(1) $\vec{v} = v^1\vec{e}_1 + v^2\vec{e}_2$

Now, we take another basis $\vec{e}_1^*$ and $\vec{e}_2^*$, different from the Cartesian basis. The new basis vectors can also be expressed as a linear combination of the basis vectors $\vec{e}_1$ and $\vec{e}_2$:

(2) $\vec{e}_1^* = a^1{}_1\vec{e}_1 + a^2{}_1\vec{e}_2$

(3) $\vec{e}_2^* = a^1{}_2\vec{e}_1 + a^2{}_2\vec{e}_2$

Let us come back to $\vec{v}$:

Of course, $\vec{v}$ has also a representation in the new basis with some uniquely determined coefficients v^{1*} and v^{2*}:

$$\vec{v} = v^{1*}\vec{e}_1^* + v^{2*}\vec{e}_2^*$$

By substituting the expressions for $\vec{e}_1^*$ and $\vec{e}_2^*$ using (2) and (3):

$$\begin{aligned}
\vec{v} &= v^{1*}\vec{e}_1^* + v^{2*}\vec{e}_2^* \\
&= v^{1*}(a^1{}_1\vec{e}_1 + a^2{}_1\vec{e}_2) + v^{2*}(a^1{}_2\vec{e}_1 + a^2{}_2\vec{e}_2)
\end{aligned}$$

$$= \quad v^{1*}a^1{}_1\vec{e}_1 + v^{1*}a^2{}_1\vec{e}_2 + v^{2*}a^1{}_2\vec{e}_1 + v^{2*}a^2{}_2\vec{e}_2$$

$$= \quad (v^{1*}a^1{}_1 + v^{2*}a^1{}_2)\,\vec{e}_1 + (v^{1*}a^2{}_1 + v^{2*}a^2{}_2)\,\vec{e}_2$$

With (1) we conclude for the components:

$$v^1 = a^1{}_1\, v^{1*} + a^1{}_2\, v^{2*}$$

$$v^2 = a^2{}_1\, v^{1*} + a^2{}_2\, v^{2*}$$

Notice carefully that the components v^1 and v^2 change differently from the components of the basis vectors $\vec{e}_1^{\,*}$ and $\vec{e}_2^{\,*}$ in (2) and (3). The transformation properties of these components define them as **contravariant**. Conversely, components that transform in the same way as the basis vectors are termed **covariant**.

Covariance and contravariance describe how the components of geometric or physical objects change under a change of basis. For instance, if the scaling in a Euclidean space changes from meters to centimeters (i.e., the reference axes are divided by 100), the components of a velocity vector are multiplied by 100. The vector components change inversely to the scaling of the reference axes, illustrating contravariance.

For a vector, such as a direction vector or velocity vector, to remain invariant under a change of basis, its components must transform contravariantly to compensate for the change in the basis. This means that the matrix transforming the vector components must be the inverse of the matrix transforming the basis vectors. Vectors whose components transform in this manner are called contravariant vectors.

Covariant vectors are written with lower indices, contravariant with upper indices.

How can you memorize the position of the indices? Look at the third letter: for *covariant* this is **v** and **v** points downwards, so you use lower indices. For *contravariant* this is **n** and – with a bit of imagination – **n** point upwards. You use superscripts then.

($\textbf{I.2.061}$) An important point:

In the standard cartesian coordinate system, there is no distinction between covariant and contravariant. We will come back to this remark a bit later in the chapter „*I.4.2.1 Raising and lowering indices*" ($I.4.2.1.3$).

($\textbf{I.2.062}$) Another practical point:

In this book we will write covariant vectors as a row-vector, and we use the brackets "[" and "]":

Covariant vector $v = [v_1 \ v_2 \ ... \ v_n]$ (note: there are no commas but blanks)

For contravariant vector we use column-notation:

Contravariant vector $v = \begin{pmatrix} v^1 \\ v^2 \\ ... \\ v^n \end{pmatrix}$

Why is it so important to distinguish between covariant and contravariant vectors? Distinguishing between covariant and contravariant vectors is crucial because certain physical quantities, such as velocity, need to remain invariant under coordinate transformations. Covariant components do not change with coordinate transformations, ensuring that physical quantities like velocity are consistent across different reference frames.

I.2.6 Dual basis

We consider an n-dimensional vector space V over $\mathbb{R}$. Let $g = \{\vec{g}_1, \vec{g}_2, \dots, \vec{g}_n\}$ be a basis. Note that we use lower indices for the basis vectors, so we consider them to be *covariant* vectors.

Now, based on $\{g_i\}$ we want to create another basis $g^* = \{\vec{g}^1, \vec{g}^2, \dots, \vec{g}^n\}$.

We want that $\vec{g}^i$ (fixed i) is perpendicular to all $\vec{g}_j$ with the exception to $\vec{g}_i$. This means that the scalar product of $\vec{g}_i$ and $\vec{g}^j$ is zero whenever $i \neq j$. Furthermore, the scalar product of $\vec{g}^i$ and $\vec{g}_i$ (same index!) shall have the value 1. We can put this requirement in a simplier form:

$$\vec{g}_i \cdot \vec{g}^j = \delta^j{}_i$$

(Don't bother for the moment that we write the Kronecker symbol as $\delta^j{}_i$: later, we will show that $\delta_{ij} = \delta^{ij} = \delta^j{}_i$).

The new basis g^* is called **the dual basis of** g. The dual basis exists always, and with our requirements the dual basis is unique.

We now want to show, how the new basis g^* can be calculated based on our basis vectors $\vec{g}_i$:

First, we note that $\vec{g}^i$ can be represented by $\vec{g}_j$, $j = 1, \dots, n$ with some (still unknown) coefficients A^{ij}:

(1) $\vec{g}^i = A^{ij}\,\vec{g}_j$

Now, we multiply (1) on both sides with $\vec{g}^k$ and we get:

(2) $\vec{g}^i \vec{g}^k = A^{ij}\,\vec{g}_j\,\vec{g}^k$

By construction of $\vec{g}^k$, we know that $\vec{g}_j\,\vec{g}^k = \delta_j{}^k$, so we get:

(3) $\vec{g}^i \vec{g}^k = A^{ij}\,\delta_j{}^k$

Defining $g^{ik} = \vec{g}^i \vec{g}^k$ and using the contraction rule of $\delta_j{}^k$ we get:

(4) $g^{ik} = A^{ik}$,

coming back to equation (1) , we can conclude:

(5) $\vec{g}^i \underset{(1)}{=} A^{ij}\, \vec{g}_j \underset{(4)}{=} g^{ij}\, \vec{g}_j$

We conclude: $A^{ij} = g^{ij}$

With the same argumentation we can conclude:

(6) $\vec{g}_k = g_{kl}\, \vec{g}^l$

Now, using (5) and (6) we have:

$$\underbrace{\vec{g}^i \cdot \vec{g}_k}_{=\,\delta_k{}^i} = g^{ij}\, \vec{g}_j\, g_{kl}\, \vec{g}^l = g^{ij}\, g_{kl}\, \underbrace{\vec{g}_j \cdot \vec{g}^l}_{=\,\delta_j{}^l}$$

But $\vec{g}^i \cdot \vec{g}_k = \delta_k{}^i$ and $\vec{g}_j \cdot \vec{g}^l = \delta_j{}^l$, so we get:

$$\delta_k{}^i = g^{ij}\, g_{kl}\, \delta_j{}^l$$

Using exchange of indices $(I.2.1.2.6)$ of $\delta_j{}^l$ we get:

$$\delta_k{}^i = g^{ij}\, g_{kj}$$

Now, as the scalar product is commutative, $g_{kj} = g_{jk}$, so we can write:

$$\delta_k{}^i = g^{ij}\, g_{jk}$$

In terms of matrix-notation, $\delta_k{}^i$ is the unit matrix and the equation tells us that $[g^{ij}]$ *is the inverse matrix of* g_{ij}. Looking back to equation (1), we have now the rule to calculate $\vec{g}^i$:

$$\boxed{(I.2.063) \quad \vec{g}^i = g^{ij}\, \vec{g}_j}$$

Procedure to calculate the dual basis:

Build up a matrix $[\,g_{ij}]$ by calculating the scalar products $\vec{g}_i \cdot \vec{g}_j$:

$$[\,g_{ij}] = \begin{bmatrix} \vec{g}_1 \cdot \vec{g}_1 & \vec{g}_1 \cdot \vec{g}_2 & \cdots & \vec{g}_1 \cdot \vec{g}_n \\ \vec{g}_2 \cdot \vec{g}_1 & \vec{g}_2 \cdot \vec{g}_2 & \cdots & \vec{g}_2 \cdot \vec{g}_n \\ \cdots & \cdots & \cdots & \cdots \\ \vec{g}_n \cdot \vec{g}_1 & \vec{g}_n \cdot \vec{g}_2 & \cdots & \vec{g}_n \cdot \vec{g}_n \end{bmatrix}$$

In component notation we have:

$$g_{ij} = \vec{g}_i \cdot \vec{g}_j$$

Now, build the inverse matrix $[g^{ij}]$, denoting the elements by

$$[\,g^{ij}\,] = \begin{bmatrix} g^{11} & g^{12} & \cdots & g^{1n} \\ g^{21} & g^{22} & \cdots & g^{2n} \\ \cdots & \cdots & \cdots & \cdots \\ g^{n1} & g^{n2} & \cdots & g^{nn} \end{bmatrix}$$

The vectors for the dual basis can be directly calculated:

$$\vec{g}^i = \sum_{j=1}^{n} g^{ij}\,\vec{g}_j = g^{ij}\,\vec{g}_j$$

The last expression is in ESN-notation, you need to sum up over j.

We can express the last equation also in matrix-notation:

$$\begin{bmatrix} \vec{g}^{1^T} \\ \vec{g}^{2^T} \\ \cdots \\ \vec{g}^{n^T} \end{bmatrix} = [g^{ij}]\,[\,\vec{g}_1 \;\; \vec{g}_2 \;\; \cdots \;\; \vec{g}_n\,]^T$$

Read this as follows:

Build up a matrix $[g_{ij}]$ whose component g_{ij} is the scalar product $g_i \cdot g_j$. Calculate the inverse matrix $[g^{ij}] = [g_{ij}]^{-1}$. Create another matrix whose rows(!) are the $\vec{g}_i$-vectors. Transpose this matrix, so that you get $[\,\vec{g}_1 \;\; \vec{g}_2 \;\; \cdots \;\; \vec{g}_n\,]^T$. Multiply $[g^{ij}]$ with this new matrix: the result is a matrix whose rows(!) are just the $\vec{g}^i$ – vectors.

The whole procedure can be taken in a single equation:

$(\boldsymbol{I.2.064})$ Given a basis $\{\,\vec{g}_1,\ \vec{g}_2, \ldots,\ \vec{g}_n\,\}$ of $\mathbb{R}^n$, the dual basis $\{\,\vec{g}^1, \vec{g}^2, \ldots, \vec{g}^n\,\}$ can be derived by the equation

$$
\begin{bmatrix} \vec{g}^{1^T} \\ \vec{g}^{2^T} \\ \cdots \\ \vec{g}^{n^T} \end{bmatrix} = \begin{bmatrix} \vec{g}_1 \cdot \vec{g}_1 & \vec{g}_1 \cdot \vec{g}_2 & \cdots & \vec{g}_1 \cdot \vec{g}_n \\ \vec{g}_2 \cdot \vec{g}_1 & \vec{g}_2 \cdot \vec{g}_2 & \cdots & \vec{g}_2 \cdot \vec{g}_n \\ \cdots & \cdots & \cdots & \cdots \\ \vec{g}_n \cdot \vec{g}_1 & \vec{g}_n \cdot \vec{g}_2 & \cdots & \vec{g}_n \cdot \vec{g}_n \end{bmatrix}^{-1} [\,\vec{g}_1\ \ \vec{g}_2\ \ \cdots\ \ \vec{g}_n\,]^T
$$

We give an example:

Take $\mathbb{R}^2$ with the Basis $g = \{\vec{g}_1, \vec{g}_2\} = \{\vec{e}_1, \vec{e}_1 + \vec{e}_2\} = \left\{ \begin{pmatrix} 1 \\ 0 \end{pmatrix}, \begin{pmatrix} 1 \\ 1 \end{pmatrix} \right\}$

We calculate all scalar products we need:

$g_{11} = \vec{e}_1 \cdot \vec{e}_1 = 1$

$g_{12} = \vec{e}_1 \cdot (\vec{e}_1 + \vec{e}_2) = 1$

$g_{22} = (\vec{e}_1 + \vec{e}_2) \cdot (\vec{e}_1 + \vec{e}_2) = 2$

As $g_{ij} = g_{ji}$ we do not need to calculate g_{21}.

We get:

$[g_{ij}] = \begin{bmatrix} 1 & 1 \\ 1 & 2 \end{bmatrix}$

We calculate the inverse of $[g_{ij}]$:

$[g^{ij}] = [g_{ij}]^{-1} = \begin{bmatrix} 2 & -1 \\ -1 & 1 \end{bmatrix}$

A check that $[g^{ij}][g_{ij}] = \begin{bmatrix} 1 & 0 \\ 0 & 1 \end{bmatrix}$ asserts that $[g^{ij}]$ is indeed the inverse of $[g_{ij}]$.

Now, we can calculate $\vec{g}^i$:

$\vec{g}^1 = g^{1j}\vec{g}_j = g^{11}\vec{g}_1 + g^{12}\vec{g}_2 = 2*\begin{pmatrix} 1 \\ 0 \end{pmatrix} - 1*\begin{pmatrix} 1 \\ 1 \end{pmatrix} = \begin{pmatrix} 1 \\ -1 \end{pmatrix}$

$\vec{g}^2 = g^{2j}\vec{g}_j = g^{21}\vec{g}_1 + g^{22}\vec{g}_2 = -1*\begin{pmatrix} 1 \\ 0 \end{pmatrix} + 1*\begin{pmatrix} 1 \\ 1 \end{pmatrix} = \begin{pmatrix} 0 \\ 1 \end{pmatrix}$

You get the same result in matrix-notation, as explained above:

$$\begin{bmatrix} g^{1^T} \\ g^{2^T} \end{bmatrix} = \underbrace{\begin{bmatrix} 2 & -1 \\ -1 & 1 \end{bmatrix}}_{g^{ij}} \underbrace{\begin{bmatrix} 1 & 1 \\ 0 & 1 \end{bmatrix}^T}_{[\vec{g}_1\ \vec{g}_2]^T} = \begin{bmatrix} 2 & -1 \\ -1 & 1 \end{bmatrix} \underbrace{\begin{bmatrix} 1 & 0 \\ 1 & 1 \end{bmatrix}}_{[\vec{g}_1\ \vec{g}_2]} = \underbrace{\begin{bmatrix} 1 & -1 \\ 0 & 1 \end{bmatrix}}_{\begin{bmatrix} g^{1^T} \\ g^{2^T} \end{bmatrix}}$$

Now, as we learned how to calculate the dual basis, it is time to have a different view on what a dual basis „really" is. The following explanation is a bit abstract – but important, when we want to understand later, what a tensor is.

Let V be a vector space in $\mathbb{R}^n$. Then we define V^* as the set of all linear mappings from V to $\mathbb{R}$. V^* is called the **dual space** of V:

$$V^* = \{\, f\colon V \!-\!\!> \mathbb{R}, f \ linear \,\}$$

It is easy to verify that V^* is a vector space itself, furthermore V^* and V have the same dimension and they are isomorphic. Now, the dual basis is the unique mapping $g^j\colon V \!-\!\!> \mathbb{R}$ with $g^j(\, g_i\,) = \delta^i{}_j$.

Looking back at the equation $\vec{g}^i = g^{ij}\,\vec{g}_j$: where is the mapping? Remember, that every linear mapping can be expressed as a matrix and every matrix represents a linear mapping. *It is the matrix g^{ij} which is/does the mapping!*

A final remark: the matrix $[g_{ij}]$ which we created by calculating all scalar-products is a very special matrix called „**metric tensor**". Do not worry about the term „tensor" at the moment, we will come back to tensors later.

The metric tensor turns out to be <u>very</u> important, as it describes the metric of a space, enabling us to measure distances correctly. We will explore the metric tensor later.

I.2.6.E Exercises:

($I.2.6.E.1$): Calculate the dual basis of the basis vectors

$$\left\{ \begin{pmatrix} -5 \\ 2 \\ 4 \end{pmatrix}, \begin{pmatrix} -5 \\ 1 \\ 3 \end{pmatrix}, \begin{pmatrix} 3 \\ -2 \\ -3 \end{pmatrix} \right\}$$

I.2.7 Tensor product of vectors

Let V be a n-dimensional vector space with the standard Euclidian basis.

For two vectors, there are several ways to multiply them: the one is the dot product $(\cdot)$, which takes in two vectors and returns a scalar. Then we already encountered the cross product $(\times)$, which also takes in two vectors and returns a vector.

The **dyadic product** $(\otimes)$ of two vectors, also called **tensor product**, returns a matrix (to be more precise: it returns a tensor of rank 2). Let us take two vectors $a = (a_1, a_2, \ldots, a_n)$ and $b = (b_1, b_2, \ldots, b_n)$.

Then we define the **tensor product** as follows:

$$\boxed{\begin{array}{l} (\textbf{\textit{I.2.065}}) \text{ The } \textbf{tensor product } a \otimes b \text{ is defined as:} \\[1em] a \otimes b = \begin{pmatrix} a_1 b_1 & a_1 b_2 & \cdots & a_1 b_n \\ a_2 b_1 & a_2 b_2 & \cdots & a_1 b_n \\ \cdots & \cdots & \cdots & \cdots \\ a_n b_1 & a_n b_2 & \cdots & a_n b_n \end{pmatrix} \end{array}}$$

It is necessary to make the following comments:

Sometimes, $a \otimes b$ is noted as ab, so there is no tensor cross $\otimes$ "between" them. This notation is sometimes called **"outer product of two vectors."** This can lead to misunderstandings as – sometimes – the dot in the dot product is omitted, too. For reasons of preciseness, we will use the tensor cross $\otimes$ in this book.

Let us go back to the definition with the matrix.

You might notice that the matrix above can be written as

$$\begin{pmatrix} a_1 b_1 & a_1 b_2 & \cdots & a_1 b_n \\ a_2 b_1 & a_2 b_2 & \cdots & a_1 b_n \\ \cdots & \cdots & \cdots & \cdots \\ a_n b_1 & a_n b_2 & \cdots & a_n b_n \end{pmatrix} = \underbrace{\begin{pmatrix} a_1 \\ a_2 \\ \cdots \\ a_n \end{pmatrix} (b_1, b_2, \ldots, b_n)}_{matrix\ multiplication} = ab^T$$

We can conclude:

$(\textbf{\textit{I.2.066}})\quad a \otimes b = ab^T$

You will need this formula for proving some identities involving tensor products.

As we can define the tensor product by a matrix multiplication, there is an important implication:

$(I.2.067)$ **The tensor product is NOT commutative.**

In a sense, the tensor product is a generalization of the dot product: the dot product is just the trace of the matrix (that is the summation of the diagonal elements of the matrix).

Before going on, let's take an example: we take $V = \mathbb{R}^3$ with the standard Euclidian basis $\vec{e}_1 = (1,0,0)^T$, $\vec{e}_2 = (0,1,0)^T$ and $\vec{e}_2 = (0,0,1)^T$.

Then we get:

$$\vec{e}_1 \otimes \vec{e}_1 = \vec{e}_1 \vec{e}_1^{\,T} = \begin{pmatrix} 1 \\ 0 \\ 0 \end{pmatrix}(1,0,0) = \begin{bmatrix} 1 & 0 & 0 \\ 0 & 0 & 0 \\ 0 & 0 & 0 \end{bmatrix}$$

$$\vec{e}_1 \otimes \vec{e}_2 = \vec{e}_1 \vec{e}_2^{\,T} = \begin{pmatrix} 1 \\ 0 \\ 0 \end{pmatrix}(0,1,0) = \begin{bmatrix} 0 & 1 & 0 \\ 0 & 0 & 0 \\ 0 & 0 & 0 \end{bmatrix}$$

$$\vec{e}_1 \otimes \vec{e}_2 = \vec{e}_1 \vec{e}_2^{\,T} = \begin{pmatrix} 1 \\ 0 \\ 0 \end{pmatrix}(0,0,1) = \begin{bmatrix} 0 & 0 & 1 \\ 0 & 0 & 0 \\ 0 & 0 & 0 \end{bmatrix}$$

$$\vec{e}_2 \otimes \vec{e}_1 = \vec{e}_2 \vec{e}_1^{\,T} = \begin{pmatrix} 0 \\ 1 \\ 0 \end{pmatrix}(1,0,0) = \begin{bmatrix} 0 & 0 & 0 \\ 1 & 0 & 0 \\ 0 & 0 & 0 \end{bmatrix}$$

$$\vec{e}_2 \otimes \vec{e}_2 = \vec{e}_2 \vec{e}_2^{\,T} = \begin{pmatrix} 0 \\ 1 \\ 0 \end{pmatrix}(0,1,0) = \begin{bmatrix} 0 & 0 & 0 \\ 0 & 1 & 0 \\ 0 & 0 & 0 \end{bmatrix}$$

$$\vec{e}_2 \otimes \vec{e}_2 = \vec{e}_2 \vec{e}_2^{\,T} = \begin{pmatrix} 0 \\ 1 \\ 0 \end{pmatrix}(0,0,1) = \begin{bmatrix} 0 & 0 & 0 \\ 0 & 0 & 1 \\ 0 & 0 & 0 \end{bmatrix}$$

$$\vec{e}_2 \otimes \vec{e}_1 = \vec{e}_2 \vec{e}_1^{\,T} = \begin{pmatrix} 0 \\ 0 \\ 1 \end{pmatrix}(1,0,0) = \begin{bmatrix} 0 & 0 & 0 \\ 0 & 0 & 0 \\ 1 & 0 & 0 \end{bmatrix}$$

$$\vec{e}_2 \otimes \vec{e}_2 = \vec{e}_2 \, \vec{e}_2{}^T = \begin{pmatrix} 0 \\ 0 \\ 1 \end{pmatrix} (\, 0, 1, 0 \,) = \begin{bmatrix} 0 & 0 & 0 \\ 0 & 0 & 0 \\ 0 & 1 & 0 \end{bmatrix}$$

$$\vec{e}_2 \otimes \vec{e}_2 = \vec{e}_2 \vec{e}_2{}^T = \begin{pmatrix} 0 \\ 0 \\ 1 \end{pmatrix} (\, 0, 0, 1 \,) = \begin{bmatrix} 0 & 0 & 0 \\ 0 & 0 & 0 \\ 0 & 0 & 1 \end{bmatrix}$$

The generalization to a n-dimensional vector space should be clear, so in $\mathbb{R}^n$ with the standard Euclidian basis the tensor product $e_i \otimes e_j$ is the matrix with 1 in the ith row and the jth column, all other entries are 0.

What you see above is the **standard basis of the dyad**. You have nine basis elements, which were created by 6 vectors.

So, returning to our esample:

Take the vector $a = 2 \, \vec{e}_1 \otimes \vec{e}_2 + 4 \, \vec{e}_2 \otimes \vec{e}_3 - 3 \, \vec{e}_3 \otimes \vec{e}_3$

With what we calculated above, we obtain:

$$a = 2 \begin{bmatrix} 0 & 1 & 0 \\ 0 & 0 & 0 \\ 0 & 0 & 0 \end{bmatrix} + 4 \begin{bmatrix} 0 & 0 & 0 \\ 0 & 0 & 1 \\ 0 & 0 & 0 \end{bmatrix} - 3 \begin{bmatrix} 0 & 0 & 0 \\ 0 & 0 & 0 \\ 0 & 0 & 1 \end{bmatrix} = \begin{bmatrix} 0 & 2 & 0 \\ 0 & 0 & 4 \\ 0 & 0 & -3 \end{bmatrix}$$

<u>A hint</u>: if you really need to calculate the tensor product of two vectors in an equation: do this in the *very* last step, if not: your calculations will be overboarded!

There is another representation of $a \otimes b$:

$$a \otimes b = \quad + \quad a_1 b_1 \vec{e}_1 \otimes \vec{e}_1 + a_1 b_2 \vec{e}_1 \otimes \vec{e}_2 + \cdots + a_1 b_n \vec{e}_1 \otimes \vec{e}_n$$

$$+ \quad a_2 b_1 \vec{e}_2 \otimes \vec{e}_1 + a_2 b_2 \vec{e}_2 \otimes \vec{e}_2 + \cdots + a_2 b_n \vec{e}_2 \otimes \vec{e}_n$$

$$+ \quad \ldots$$

$$+ \quad a_n b_1 \vec{e}_n \otimes \vec{e}_1 + a_n b_2 \vec{e}_n \otimes e_2 + \cdots + a_n b_n \vec{e}_n \otimes \vec{e}_n$$

or much shorter, in ESN:

$$(\boldsymbol{I.2.068}) \quad a \otimes b = a_i\, b_j\, \vec{e}_i \otimes \vec{e}_j$$

Later, we will need some identities:

$$(\boldsymbol{I.2.069}) \quad (a \otimes b)\, x = (b \cdot x)\, a \quad \text{for all } x \in V$$

This identity is most useful when you need to prove identities with respect to the tensor product.

Proof:

$$(a \otimes b)\, x \;=\; \begin{pmatrix} a_1 b_1 & a_1 b_2 & \cdots & a_1 b_n \\ a_2 b_1 & a_2 b_2 & \cdots & a_1 b_n \\ \cdots & \cdots & \cdots & \cdots \\ a_n b_1 & a_n b_2 & \cdots & a_n b_n \end{pmatrix} \begin{pmatrix} x_1 \\ x_2 \\ \cdots \\ x_n \end{pmatrix}$$

$$= \begin{pmatrix} a_1 b_1 x_1 + a_1 b_2 x_2 + \cdots + a_1 b_n x_n \\ a_2 b_1 x_1 + a_2 b_2 x_2 + \cdots + a_2 b_n x_n \\ \cdots \\ a_n b_1 x_1 + a_n b_2 x_2 + \cdots + a_n b_n x_n \end{pmatrix}$$

$$= \begin{pmatrix} a_1(b_1 x_1 + b_2 x_2 + \cdots + b_n x_n) \\ a_2(b_1 x_1 + b_2 x_2 + \cdots + b_n x_n) \\ \cdots \\ a_n(b_1 x_1 + b_2 x_2 + \cdots + b_n x_n) \end{pmatrix}$$

$$= \begin{pmatrix} a_1\,(b \cdot x) \\ a_2\,(b \cdot x) \\ \cdots \\ a_n\,(b \cdot x) \end{pmatrix}$$

$$= \quad (b \cdot x)\, a$$

∎

Let S be a $n \times n$ matrix. Then the following equation holds:

$(\boldsymbol{I.2.070})\ S(a \otimes b) = Sa \otimes b$

As a matrix, S is a linear operator. If S acts on a tensor product (left side of the equation above), then the result will be the same if S acts only on the (first) vector a, then you form the tensor product with b.

We will demonstrate how $(I.2.069)$ is used for proving identities. Therefore, we prove that $(I.2.070)$ holds for every vector x:

$$
\begin{aligned}
[\,S(a \otimes b)\,]\,x \quad &= \quad S[\,(a \otimes b)\,x\,] \qquad && S \text{ is linear} \\[2em]
&= \quad S\,[\,(b \cdot x)\,a\,] \qquad && (I.2.069) \\[2em]
&= \quad (b \cdot x)\,S\,a \qquad && b \cdot x \in \mathbb{R},\, S \text{ linear} \\[2em]
&= \quad (S\,a \otimes b)\,x \qquad && (I.2.069)
\end{aligned}
$$

Note, that we used $(I.2.069)$ in *both* directions.

∎

We already mentioned that the trace of a quadratic matrix is the sum of its diagonal elements.

As the tensor product is a $n \times n$ matrix, let's explore the trace of $\vec{e}_i \otimes \vec{e}_j$: this matrix contains 0 in all positions, except the ith row and the jth column where we find a 1. So, it should be clear that $tr(\vec{e}_i \otimes \vec{e}_j) = \delta_{ij}$.

Furthermore, we can conclude that

$$tr(a \otimes b) = tr(\,a_i\,\vec{e}_i \otimes b_j\,\vec{e}_j = a_i\,b_j\,\delta_{ij} = a_i\,b_i = a \cdot b$$

So, the trace is just the scalar product.

I.2.7.E Exercises

I.2.7.E.1

Proof the following identities:

    (a) $(a \otimes b)^T = b \otimes a$

    (b) $(a \otimes b)(c \otimes d) = (b \cdot c)\, a \otimes d$

I.3 Coordinate systems - Definitions

In this chapter, we will have a look at different kinds of coordinate systems. There is no definite definition of what a coordinate system is. According to WikiPedia:

*"A **coordinate system** is a system that uses one or more numbers, or **coordinates**, to uniquely determine the position of the points or other geometric elements."*

(https://en.wikipedia.org/wiki/Coordinate_system)

[Neutsch, p.73] states:

"A coordinate system over $K = \mathbb{R}, \mathbb{C}$ of a topological space M is an injective mapping $\varkappa: U \to V \cong K^n$, $U \subseteq M$, U open."

For our purpose, we simply understand by a coordinate system a set of basis elements for a n-dimensional vector space, so that every vector can uniquely be represented with respect to the basis elements. These basis elements can consist of fixed vectors or can also be functions.

We will define rectangular, curvilinear and − as a special case of curvilinear coordinates - affine coordinate systems.

In this chapter, we consider an n-dimensional vector space V over $\mathbb{R}$. V can be $\mathbb{R}^n$ itself or a subset of $\mathbb{R}^n$.

I.3.1 Cartesian basis

The most used basis for V is the **cartesian** basis. The basis consists of the unit vectors $\vec{e}_i = \begin{pmatrix} 0 \\ 0 \\ \dots \\ 1 \\ 0 \\ \dots \\ 0 \end{pmatrix}$ whose components are zero except for the i-th column, which takes a 1.

It is a rectangular coordinate system – the formal definition will be given below -, all the basis vectors are perpendicular to each other, and they all have the length of 1. It is an *orthonormal basis system.*

Two vectors v and w are prependicular if and only if their (Euclidian standard) scalar product equals zero: $v \cdot w = v_i\, w^i = 0$, where v_i are the coefficients of v and w^i the coefficients of w. For the moment, do not bother that we use for the coefficients superscripts and subscripts. The reason will become clear a bit later. Perpendicular vectors are called **orthogonal**.

If a basis of V consists of pairwise orthogonal vectors, the basis is called an **orthogonal basis system**. If, furthermore, all these vectors have all the length of 1, then the basis system is called an **orthonormal basis system**.

With the Kronecker symbol δ_{ij} there is a short way to describe a set of *orthonormal* vectors:

A set of vectors $\{v^1, \ldots, v^n\}$ is an **orthonormal basis** of V if and only if the following equation holds for every $i, j \in \{1, \ldots, n\}$:

$$v^i \cdot v^j = \delta_{ij}$$

If $i = j$, then $v^i \cdot v^j = v^i \cdot v^i = \delta_{(ii)} \underset{no\ summation\ here}{=} 1.$

It follows, that the length of v^i, denoted by $\left|v^i\right|$, is 1, as

$$\left|v^i\right| := \sqrt{v^i \cdot v^i} = \sqrt{1} = 1.$$

If $i \neq j$, then their scalar product is zero.

To check, if a set of vectors is a basis, you can write down the vectors in a matrix

$$A = \begin{bmatrix} v^1 & v^2 & , \ldots, & v^n \end{bmatrix}$$

The vectors are linear independent if and only if $\det(A) \neq 0$.

The concept of the standard Euclidian vector space can be generalized to rectangle coordinates, which will be explored next.

I.3.2 Rectangular coordinate system

We define what a rectangular coordinate system is:

$(I.3.001)$ **Definition (rectangular coordinates):**

Let V be a n-dimensional vector space and $x^1, x^2, \ldots, x^n$ a coordinate system. Then $(x^i), i = 1, \ldots, n$ is **rectangular** if the distance between two arbitrary points $v = (v^1, v^2, \ldots, v^n)^T$ and $w = (w^1, w^2, \ldots, w^n)^T$ is given by

$$d_{v,w} = \sqrt{(v^1 - w^1)^2 + (v^2 - w^2)^2 + \cdots + (v^n - w^n)^2}$$

What you see here is the generalization of the *Theorem of Pythagoras*, $d_{v,w}$ is called the **Euclidian distance**.

Using ESN, we can write:

$$d_{v,w} = \sqrt{\delta_{ij}\, \Delta v^i\, \Delta v^j} \ , \text{ where } \Delta v^i := v^i - w^i$$

There are infinite rectangular coordinate systems. If (x^i) is a rectangular coordinate system and $A = \left[a_{ij}\right]$ an orthogonal matrix (this means: $A^T A = I$) then $\bar{x}^i := a_{ir} x^r$ is also a rectangular coordinate system. An orthogonal matrix describes a rotation of a coordinate system.

We can conclude that we get all rectangular coordinate systems by rotating and translating the standard Euclidian basis. Rotations and translations are linear transformations – therefore they preserve the Euclidian distance.

A rectangle coordinate system has always constant basis vectors: the basis vectors consist of fixed chosen real (or complex) numbers.

I.3.3 Curvilinear coordinate system

A **coordinate line**, also known as a coordinate curve, is a curve that corresponds to one of the coordinate axes in a coordinate system. In a two-dimensional Cartesian coordinate system, for example, the coordinate lines are straight lines parallel to the x-axis (horizontal lines) and the y-axis (vertical lines). Each point on a coordinate line has coordinates where one coordinate is fixed (usually at a constant value) and the other coordinate varies.

In three-dimensional Cartesian coordinates, coordinate lines form a grid of planes parallel to the coordinate planes (xy-plane, xz-plane, and yz-plane). Each coordinate line corresponds to fixing one coordinate and allowing the other two to vary. This concept can be extended to any dimension.

A **curvilinear coordinate system** is a coordinate system in which the coordinate lines are curved rather than straight, they have no constant basis vectors. A basis system will be constructed by the tangent vectors of the coordinate lines. Thus, the basis can vary from point to point.

We need to point out that curvilinear coordinate systems are not the same as curved spaces: curvilinear coordinates are always situated in the Euclidian space $\mathbb{R}^n$ and are always tight to Cartesian coordinates. A curved space – like part of a spherical surface – is something totally different.

$(I.3.002)$ **Definition:**

Let V be a n-dimensional vector space and $x^1, x^2, \ldots, x^n$ a coordinate system. If $(\bar{x}^i)$ is another coordinate systems, so that the following conditions are fulfilled:

(1) $\bar{x}^i$ can be expressed by scalar functions $\bar{x}^i(x^1, x^2, \ldots, x^n)$, so $\mathcal{T}: \bar{x}^i = \bar{x}^i(x^1, x^2, \ldots, x^n)$

(2) For all $i \in \{1, 2, \ldots, n\}$ $\quad \bar{x}^i(x^1, x^2, \ldots, x^n)$ has continuous second-partial derivatives

Then $\mathcal{T}$ is called a **coordinate transformation**.

If $x^1, x^2, \ldots, x^n$ is a rectangular coordinate system and $\mathcal{T}$ linear, then $\mathcal{T}$ is called an **affine transformation**, the $(\bar{x}^i)$ are called **affine coordinates**.

If $x^1, x^2, \ldots, x^n$ is a rectangular coordinate system and $\mathcal{T}$ is a bijection, $\mathcal{T}$ is called a **curvilinear transformation** and $(\bar{x}^i)$ are called **curvilinear coordinates**.

If $\mathcal{T}$ is locally bijective on an open set U of $\mathbb{R}^n$ and the Jacobi-matrix of $\mathcal{T}$ is non-zero for every point of U, then $\mathcal{T}$ is called an **admissible change of coordinates**.

We need to explain the term "locally bijective": it means that for each point in the open set U, there exists a neighborhood around that point where T is bijective. This property satisfies *almost everywhere* in the domain, excluding isolated singularities.

So, an admissible coordinate transformation is a mapping represented by a sufficiently differentiable set of equations and it is invertible by having a non-vanishing Jacobian.

The set of admissible transformations form a group. If T_1 and T_2 are admissible transformations, so $T_1 T_2$ is also transmissible. $T_1 T_2$ is the concatenation of the transformation T_1 and T_2, so you first transform with T_2, then you transform the previous result with T_1. Notice: this group is NOT commutative!

If $T_3 = T_1 T_2$ and $\mathcal{J}_i$ the Jacobian of T_i, then $\mathcal{J}_3 = \mathcal{J}_1 \mathcal{J}_2$. If furthermore $T_1 T_2 = I$, where I is the identity transformation, then $\mathcal{J}_1 \mathcal{J}_2 = 1$, meaning that $\mathcal{J}_1$ and $\mathcal{J}_2$ are reciprocal. This means, that T_1 is the reciprocal transformation of T_2, and vice versa.

In physics, we are mostly interested in these *admissible changes of coordinates*, as the transformation from one coordinate system to another should follow the following rules:

- **Conservation of physical laws**

    The transformation should not alter the form of the fundamental equations that describe physical phenomena. For example, in classical mechanics, Newton's laws of motion should remain unchanged under admissible coordinate transformations.

- **Maintaining the symmetry of the physical system**

    Admissible coordinate transformations should respect the symmetries of the system. For example, if a physical system exhibits rotational symmetry, the coordinate transformation should preserve this symmetry.

- **Continuity and differentiability**

  The coordinate transformation functions should be continuous and differentiable to ensure the mathematical consistency of the transformation.

- **One-to-one mapping (bijective mapping)**

  Each point in the original coordinate system should have a unique corresponding point in the transformed coordinate system, and vice versa. This ensures that the transformation is well-defined. *As we will see a bit later, it is allowed that only some points of a coordinate transformation do not have a one-to-one mapping.*

- **Causality preservation**

  In relativistic physics, admissible coordinate transformations should not result in causality violations, such as signals traveling faster than the speed of light.

As a general remark we note that *curvilinear coordinate systems are always defined in relation to a rectangle coordinate system*. This remark is important because some equations – for example, the metric tensor as the product of the transposed Jacobi matrix with itself – require that the coordinate system be related to a rectangle coordinate system.

As a convention throughout the book: whenever we talk about coordinate transformations, we only mean *admissible coordinate transformations*.

We will have a look at some examples:

I.3.3.1 Example (polar coordinates)

Polar coordinates are one of the most used curvilinear coordinate systems in $\mathbb{R}^2$. Every point is determined by a distance from a reference point and an angle from a reference direction. Normally, $\begin{pmatrix} 0 \\ 0 \end{pmatrix}$ is taken as the reference point.

The transformation $\mathcal{T}$ for polar coordinates is:

$$\mathcal{T}:\begin{cases} \bar{x}^1(r,\theta) = r\cos(\theta) \\ \bar{x}^2(r,\theta) = r\sin(\theta) \end{cases}$$

with $r \geq 0$, $0 \leq \theta \leq 2\pi$

It should be clear that the $\bar{x}^i$- functions are differentiable to any degree. Looking at the Jacobi-matrix

$$\mathcal{J}_{\mathcal{T}} \;=\; \begin{bmatrix} \partial\bar{x}^1/\partial r & \partial\bar{x}^1/\partial\theta \\ \partial\bar{x}^2/\partial r & \partial\bar{x}^2/\partial\theta \end{bmatrix}$$

$$=\; \begin{bmatrix} \cos\theta & -r\sin\theta \\ \sin\theta & r\cos\theta \end{bmatrix},$$

we see that $\det(\mathcal{J}_{\mathcal{T}}) = r$, so we have an admissible transformation as long as $r > 0$.

For $r = 0$, we have a point which is not bijective. As long as there are only some points where the inverse function does not exist, the coordinate transformation is considered to be still admissible.

I.3.3.2 Inadmissible coordinate transformation

We want to show that not every coordinate transformation is admissible:

Let x^1, x^2 define a cartesian coordinate system in $\mathbb{R}^2$. Define the functions

$$\mathcal{T}:\begin{cases} \bar{x}^1(x^1,x^2) = (x^1)^3 \\ \bar{x}^2(x^1,x^2) = (x^2)^3 \end{cases}$$

$\bar{x}^1$ and $\bar{x}^2$ are continious and ∞-times differentiable, so we have a coordinate transformation. It is even a curvilinear coordinate transformation as it is connected to a rectangular coordinate system. But the mappings $\bar{x}^1$ and $\bar{x}^2$ are not bijective:

Consider the point $\bar{p} = \begin{pmatrix} \bar{x}^1 \\ \bar{x}^2 \end{pmatrix} = \begin{pmatrix} 1 \\ 8 \end{pmatrix}$: there are several distinct solutions:

$$\mathcal{T}(1,2) = \mathcal{T}(-1,2) = \mathcal{T}(1,-2) = \mathcal{T}(-1,-2) = \begin{pmatrix} 1 \\ 8 \end{pmatrix}$$

$\mathcal{T}$ is not bijective, so we do not have an admissible coordinate transformation.

∎

I.3.3.3 Basis vectors in curvilinear coordinate system

If $\left(x^i(u_1,\dots,u_n)\right)$ describes a curvilinear coordinate system, we want to explore on how to construct basis vectors.

The basis vectors can be calculated as the derivates of the x^i-functions. To be more precise: you use the tangents of the coordinate lines as the basis vector:

$$(I.3.003)\quad e_{u_i} = \left(\frac{\partial x^1}{\partial u_i}, \frac{\partial x^2}{\partial u_i}, \dots, \frac{\partial x^n}{\partial u_i}\right)$$

This formula represents the basis vector e_{u_i} as the partial derivatives of the coordinates x^i with respect to the curvilinear coordinates u_i.

<u>Example</u> (basis vectors in spherical coordinates):

We take the following set of functions (x^i) which depend on the variables r, θ and φ:

$$x^1 = r\sin(\theta)\cos(\varphi)$$

$$x^2 = r\sin(\theta)\sin(\varphi)$$

$$x^3 = r\cos(\theta)$$

Now, using $(I.3.003)$, we get:

$$e_r = \begin{pmatrix}\dfrac{\partial x^1}{\partial r}\\[2mm]\dfrac{\partial x^2}{\partial r}\\[2mm]\dfrac{\partial x^3}{\partial r}\end{pmatrix} = \begin{pmatrix}\sin(\theta)\cos(\varphi)\\ \sin(\theta)\sin(\varphi)\\ \cos(\theta)\end{pmatrix}$$

$$e_\theta = \begin{pmatrix} \dfrac{\partial x^1}{\partial \theta} \\[4pt] \dfrac{\partial x^2}{\partial \theta} \\[4pt] \dfrac{\partial x^3}{\partial \theta} \end{pmatrix} = \begin{pmatrix} r\cos(\theta)\cos(\varphi) \\ r\cos(\theta)\sin(\varphi) \\ -r\sin(\theta) \end{pmatrix}$$

$$e_\varphi = \begin{pmatrix} \dfrac{\partial x^1}{\partial \varphi} \\[4pt] \dfrac{\partial x^2}{\partial \varphi} \\[4pt] \dfrac{\partial x^3}{\partial \varphi} \end{pmatrix} = \begin{pmatrix} -r\sin(\theta)\sin(\varphi) \\ r\sin(\theta)\cos(\varphi) \\ 0 \end{pmatrix}$$

There is a simple way to get the basis vectors using the Jacobi-matrix:

$$J = \begin{bmatrix} \partial x^1/\partial r & \partial x^1/\partial \theta & \partial x^1/\partial \varphi \\ \partial x^2/\partial r & \partial x^2/\partial \theta & \partial x^2/\partial \varphi \\ \partial x^3/\partial r & \partial x^3/\partial \theta & \partial x^3/\partial \varphi \end{bmatrix}$$

$$= \begin{bmatrix} \cos(\varphi)\sin(\theta) & r\cos(\theta)\cos(\varphi) & -r\sin(\theta)\sin(\varphi) \\ \sin(\theta)\sin(\varphi) & r\cos(\theta)\sin(\varphi) & r\cos(\varphi)\sin(\theta) \\ \cos(\theta) & -r\sin(\theta) & 0 \end{bmatrix}$$

The first column corresponds to e_r, as all elements are derivated with respect to ∂r, the second to e_θ, finally the third column to e_φ.

Normally, you are not so much interested in e_{u_i} but in the normalized basis vectors $\hat{e}_{u_i}$ defined as:

$$\hat{e}_{u_i} := \frac{e_{u_i}}{|e_{u_i}|}$$

The only difference between $\hat{e}_{u_i}$ and e_{u_i} is their length: the length of e_{u_i} can be anything, the length of $\hat{e}_{u_i}$ is always 1.

Coming back to our example above, let us calculate the length of the basis vectors:

$$|e_r| = \sqrt{(\sin(\theta)\cos(\varphi))^2 + (\sin(\theta)\sin(\varphi))^2 + (\cos(\theta))^2} = 1$$

$$|e_\theta| = \sqrt{(r\cos(\theta)\cos(\varphi))^2 + (r\cos(\theta)\sin(\varphi))^2 + (-r\sin(\theta))^2} = r$$

$$|e_\varphi| = \sqrt{(-r\sin(\theta)\sin(\varphi))^2 + (r\sin(\theta)\cos(\varphi))^2} = r\sin(\theta)$$

Finally, we get:

$$\hat{e}_r = \frac{e_r}{|e_r|} = \frac{1}{1}\begin{pmatrix} \sin(\theta)\cos(\varphi) \\ \sin(\theta)\sin(\varphi) \\ \cos(\theta) \end{pmatrix}$$

$$\hat{e}_\theta = \frac{e_\theta}{|e_\theta|} = \frac{1}{r}\begin{pmatrix} r\cos(\theta)\cos(\varphi) \\ r\cos(\theta)\sin(\varphi) \\ -r\sin(\theta) \end{pmatrix} = \begin{pmatrix} \cos(\theta)\cos(\varphi) \\ \cos(\theta)\sin(\varphi) \\ -\sin(\theta) \end{pmatrix}$$

$$\hat{e}_\varphi = \frac{e_\varphi}{|e_\varphi|} = \frac{1}{r\sin(\theta)}\begin{pmatrix} -r\sin(\theta)\sin(\varphi) \\ -r\sin(\theta)\cos(\varphi) \\ 0 \end{pmatrix} = \begin{pmatrix} -\sin(\varphi) \\ \cos(\varphi) \\ 0 \end{pmatrix}$$

You will meet the normalized basis vectors later.

I.3.4 Curvilinear orthogonal coordinate systems

A curvilinear function systems is called curvilinear orthogonal that each family of surfaces intersects the others at right angles, we , so a system where each family of surfaces intersects the others at right angles.

How can we determine if a given set of functions $x^i = x^i(u_1, u_2, \ldots, u_m)$ is orthogonal?

In chapter *I.3.3.3* we showed how to construct the normalized basis vectors $\hat{e}_{u_i}$ for the (x^i). So $x^i = x^i(u_1, u_2, \ldots, u_m)$ is orthogonal if the equation

$$\hat{e}_{u_i} \cdot \hat{e}_{u_j} = \delta_{ij}$$

holds.

Orthogonality implies that the metric tensor is always a diagonal matrix: the metric tensor g_{ij} is constructed as the scalar product of its basis vectors:

$g_{ij} = e_{u_i} \cdot e_{u_j}$. As the basis vectors are orthogonal, all entries but the diagonal disappear.

As we will see in subsequent chapters, equations will simplify if the coordinate system is orthogonal: the metric tensor will be a diagonal matrix, thus simplifying the calculations of the scale factors (I.7). Most coordinate systems presented in appendix $I.A$ are orthogonal.

I.3.4 General law of coordinate transformation

In this chapter, we examine admissible coordinate transformations. We need to derive an important characteristic of these transformations, which will be very significant in the subsequent chapters. Therefore, we need a bit of theory:

Let V be a n-dimensional vector space over a field K ($= \mathbb{R}, \mathbb{C}$). Let $\{\, e_1, \dots, e_n \,\}$ be a basis for V and $\{\, \bar{e}_1, \dots, \bar{e}_n \,\}$ another basis of V. Then e_j can be expressed by a combination of the basis vectors $\bar{e}_1, \dots, \bar{e}_n$:

$$(1)\ e_j = a^i{}_j\, \bar{e}^i$$
with some $a^i{}_j \in K$.

In the same way, $\bar{e}_j$ can be expressed by the basis vectors $e_1, \dots, e_n$:

$$(2)\ \bar{e}^j = \bar{a}^k{}_j\, e_k$$

Note that both notations are ESN equations, so you have to sum up over i resp. k. In equation (2) we use different summation indices compared to (1) for clarity.

Notice, that $[a^i{}_j]$, as a marix, is the inverse of $[\bar{a}^i_j]$ and vice versa. To see why, we substitute $\bar{e}^i$ in equation (1) with $\bar{a}^k{}_i\, e_k$:

$$e_j = a^i{}_j\, \bar{e}^i = a^i{}_j\, \bar{a}^k{}_i\, e_k$$

On the other hand, $e_j = \delta^k{}_j\, e_k$

Therefore, $a^i_j\, \bar{a}^k_i = \delta_{jk}$, and that is the definition of the inverse matrix $(I.\,2.043)$.

Now, take an arbitrary vector $x = (x^1, \dots, x^n)^T \in V$, so $x = x^i e_i$. With $\bar{x} = (\bar{x}^1, \dots, \bar{x}^n\,)$ we denote the vector x with respect to the basis $\{\,\bar{e}_1, \dots, \bar{e}_n\,\}$: $\quad \bar{x} = \bar{x}^i \bar{e}_i$

First, we claim:

(3) $\bar{x}^j = a^j_i\, x^i$ and

(4) $x^j = \bar{a}^j_i\, \bar{x}^i$

This means that we can express the component $\bar{x}^j$ by the components x^i with coefficients from equation (1) and vice versa.

Proof:

We will only prove the equation (3), as the proof of (4) works out the same.

By definition of x we have:

$$x = x^j\, e_j$$

We replace e_j using equation (2). As previously noted, replacing ESN terms in an ESN term requires using different summation indices:

(5) $x = x^j\, a^k_j\, \bar{e}_k = \left(\, x^j\, a^k_j\,\right) \bar{e}_k$

But x can also represented by a linear combination of $\bar{x}_j\, \bar{e}^j$:

(6) $x = \bar{x}^k\, \bar{e}_k$

Thus, we have two representations of x with respect to the basis $\{\bar{e}_1, \dots, \bar{e}_n\}$. Since the representation of a vector with respect to a basis is unique, it follows that the coefficients of (5) $x^j\, a^k_j$ and (6) $\bar{x}_k$ must be the same:

$\left(\, x^j\, a^k_j\,\right) \bar{e}_k = \bar{x}_k\, \bar{e}^k$ and it follows:

$$x^j\, a^k_j = \bar{x}_k$$

In the same way we conclude

$$x^k = \bar{a}^k_j \, \bar{x}^j$$

■

Now, the interesting part:

Using $\bar{x}^k = x^j \, a^k_j$, we want to determine a^k_j.

Do not forget, this is an ESN-equation, so DO NOT DIVIDE BOTH SIDES by x^j, instead use the partial derivative:

If $\bar{x}^i = x^j \, a^j_i$ holds, then $\dfrac{\partial}{\partial x^j}(\bar{x}_i) = \dfrac{\partial}{\partial x^j}(x^j \, a^j_i)$ and vice versa.

Now, a^j_i are independant of x^j as $a^j_i \in K$, so:

$$(7)\quad \frac{\partial \bar{x}_i}{\partial x^j} = \frac{\partial}{\partial x^j}(x^j \, a^j_i) \underset{\bar{a}^j_i \, indep.from \, x^j}{=} a^j_i \frac{\partial}{\partial x^j}(x^j) = a^j_i \frac{\partial x^j}{\partial x^j} = a^j_i$$

In the same way we may conclude:

$$(8)\quad \frac{\partial x_i}{\partial \bar{x}^j} = \bar{a}^j_i$$

This procedure works as $\dfrac{\partial x^i}{\partial x^j} = \delta_{ij}$ (see $(I.2.061)$).

Now, we come back to (3) and (4):

We conclude:

$$(9)\quad \bar{x}^j = a^j_i \, x^i = \frac{\partial \bar{x}_i}{\partial x^j} \, x^i$$

$$(10)\quad x_j = \frac{\partial x_i}{\partial \bar{x}^j} \, \bar{x}^i$$

Interpretation: to express the components x^j of the vector or tensor in the new coordinate system $\bar{x}^i$, you need to multiply the components $\bar{x}^i$ by the Jacobi matrix $\dfrac{\partial x^j}{\partial \bar{x}^i}$.

What we calculated in equations (9) and (10) is the *general law of coordinate transformations*:

(I.3.004) ***General law of coordinate transformations***

Let V be a n-dimensinal vector space and $\{x^i\}, \{\bar{x}^i\}$ two different coordinate systems linked by an admissible coordinate transformation, so that

$$\bar{x}^i = \bar{x}^i(x^1, x^2, \dots, x^n).$$

Then the coordinates $\bar{x}^j$ and x^j are connected by:

$$\bar{x}^j = \frac{\partial \bar{x}_i}{\partial x_j} \, x^i$$

$$x^j = \frac{\partial x^i}{\partial \bar{x}^j} \, \bar{x}^i$$

Some remarks:

- $\bar{x}^j = \frac{\partial \bar{x}^i}{\partial x^j} \, x^i$ can be rewritten as $d\bar{x}^j = \frac{\partial \bar{x}^i}{\partial x^j} \, dx^i$

- $x^j = \frac{\partial x^i}{\partial \bar{x}^j} \, \bar{x}^i$ can be rewritten as $dx^j = \frac{\partial x^j}{\partial \bar{x}^i} \, d\bar{x}^i$

- The expression $\frac{\partial \bar{x}^i}{\partial x^j}$ is the Jacobi-matrix:

$$J = \frac{\partial \bar{x}^i}{\partial x^j} = \begin{vmatrix} \dfrac{\partial \bar{x}^1}{\partial x^1} & \dfrac{\partial \bar{x}^1}{\partial x^2} & \cdots & \dfrac{\partial \bar{x}^1}{\partial x^n} \\ \dfrac{\partial \bar{x}^2}{\partial x^1} & \cdots & \cdots & \dfrac{\partial \bar{x}^2}{\partial x^n} \\ \cdots & \cdots & \cdots & \cdots \\ \dfrac{\partial \bar{x}^n}{\partial x^1} & \dfrac{\partial \bar{x}^n}{\partial x^2} & \cdots & \dfrac{\partial \bar{x}^n}{\partial x^n} \end{vmatrix}$$

and

$$\bar{J} = \frac{\partial x^j}{\partial \bar{x}^i} = \begin{vmatrix} \dfrac{\partial x^1}{\partial \bar{x}^1} & \dfrac{\partial x^1}{\partial \bar{x}^2} & \cdots & \dfrac{\partial x^1}{\partial \bar{x}^n} \\ \dfrac{\partial x^2}{\partial \bar{x}^1} & \cdots & \cdots & \dfrac{\partial x^2}{\partial \bar{x}^n} \\ \cdots & \cdots & \cdots & \cdots \\ \dfrac{\partial x^n}{\partial \bar{x}^1} & \dfrac{\partial x^n}{\partial \bar{x}^2} & \cdots & \dfrac{\partial x^n}{\partial \bar{x}^n} \end{vmatrix}$$

Furthermore, we see:

J and $\bar{J}$ are inverse of each other:

Let $C = \left[c_{ij}\right] = J\bar{J}$, then we have in components:

$$c_{ij} = J_{rj}\,\bar{J}_{ir} = \frac{\partial \bar{x}^r}{\partial x^j}\,\frac{\partial x^i}{\partial \bar{x}^r} \underset{chain\,rule}{=} \frac{\partial x^i}{\partial x^j} = \delta_{ij}$$

The general law of transformation will be used when we define tensors.

I.4 Metric tensor

A metric has to do with measuring something, the length of a vector, the distance between two points, the length of a curve. Of course, what we measure should be independent of the coordinate system. Calculating the length of a vector or a curve may be a challenge in a certain coordinate system, while in another system it might be easier to do.

In physics, we have to deal with different spaces: flat or rectangular spaces, curved spaces, elliptical spaces, spherical spaces and so on. We need a method to define the concept of distance in all these spaces. The metric tensor will do this job for us.

In this chapter, we will look at the metric tensor $[g_{ij}]$. Although the precise definition of what a tensor is, will follow later, we will refer to $[g_{ij}]$ a **metric-tensor** for now. The metric tensor describes the structure of a vector space, as it helps us to calculate the distance between points and the angle between to vectors.

I.4.1 Definition of the metric tensor

The metric tensor is a mathematical tool that defines the geometry of a manifold/vector space, providing a measure of distances and angles between points. Its *components* vary with the choice of coordinates and encapsulate information about the intrinsic properties of the space.

Let us start with the Euclidean standard basis $(\vec{e}_1, \dots, \vec{e}_n)$. With a generalized form of the **Theorem of Pythagoras** we can compute the length s of vector x with components $(x_1, x_2, \dots, x_n)$ as

$$(I.4.001) \quad s = \sqrt{(x_1)^2 + (x_2)^2 + \dots + (x_n)^2}$$

To get rid of the root, we can square both sides, getting:

$$(I.4.002) \quad s^2 = (x_1)^2 + (x_2)^2 + \dots + (x_n)^2$$

If we want to calculate the distance between a (very, very) small displacement Δ of $\vec{x}$, we get:

$$(I.4.003) \quad \Delta s^2 = (\Delta x_1)^2 + (\Delta x_2)^2 + \dots + (\Delta x_n)^2, \text{ where}$$

$$\Delta x_1 := (\, x_1 - \Delta x_1 \,)$$

If Δ is *infinitesimal small*, the equation $(I.4.1.3)$ can be written as:

$$\boxed{(I.4.004)\quad ds^2 = (dx_1)^2 + (dx_2)^2 + \ldots + (dx_n)^2}$$

ds^2 is called the **squared line element** and will play an important role a bit later. We will come back to the squared line element.

Now, consider the case that we have a vector v in polar coordinates. So, the vector can be written as $v = v_r\,\hat{r} + v_\theta\,\hat{\theta}$ where v_r is the radial component and v_θ is the angular component, and $\hat{r}$ and $\hat{\theta}$ are unit vectors in the radial and angular directions respectively. Now, the infinitesimal displacement ds can be expressed as:

$$ds = v_r\,\hat{r}\,dr + v_\theta\,\hat{\theta}\,d\theta$$

The squared line element ds^2 is the dot-product of ds with itself:

$$ds^2 = ds \cdot ds = \left(v_r\,\hat{r}\,dr + v_\theta\,\hat{\theta}\,d\theta\right) \cdot \left(v_r\,\hat{r}\,dr + v_\theta\,\hat{\theta}\,d\theta\right)$$

Expanding and using the fact that $\hat{r}\hat{\theta} = 0$, $\hat{r}\hat{r} = 1$ and $\hat{\theta}\hat{\theta} = 1$, we get:

$$ds^2 = (v_r)^2\,(dr)^2 + (v_\theta)^2\,(r\,d\theta)^2$$

In polar coordinates, v_r represents the rate of change of the radial position with respect to time or some other parameter. It denotes the magnitude of the velocity in the radial direction. When $v_r = 1$, it implies that the vector moves with unit speed radially, which means it moves by one unit of radial distance per unit of time. This essentially sets the scale for the radial direction.

So, by convention, we often set $v_r^2 = 1$ without loss of generality, as it simply scales the radial direction. This choice simplifies calculations and aligns with the intuitive understanding of radial motion.

With the same arguments, we can conclude or define that $v_\theta^2 = 1$, so that we get finally:

$$ds^2 = (dr)^2 + r^2 \, (d\theta)^2 \text{ (polar coordinates)}$$

We will have a geometrical discussion of the squared line element in polar coordinates in Chapter $I.4.2.4$.

The calculation of a squared line element in other coordinate systems can become quite complicated. Therefore, we need a different approach to measure a distance. This is achieved by introducing the metric tensor:

$(I.4.005)$ **Definition (metric tensor)**

Let V be an n-dimensional vector space and $(g_1, g_2, ..., g_n)$ a basis of V. Then the components of the metric tensor $\left[g_{ij}\right]$ are defined as $g_{ij} = g_i \cdot g_j$ (scalar product)

[MISNER] (p.50-53, §2.4) calls the metric tensor a "machine for calculating the squared length of a single vector, or the scalar product of two different vectors".

The metric tensor can be viewed as a matrix:

$$(I.4.006) \quad \left[g_{ij}\right] = \begin{bmatrix} g_1 \cdot g_1 & g_1 \cdot g_2 & \cdots & g_1 \cdot g_n \\ g_2 \cdot g_1 & g_2 \cdot g_2 & \cdots & g_2 \cdot g_n \\ \cdots & \cdots & \cdots & \cdots \\ g_n \cdot g_1 & g_n \cdot g_2 & \cdots & g_n \cdot g_n \end{bmatrix}$$

When we talked about the dual basis $(I.2.6)$, we already encountered the construction of the metric tensor.

We can rewrite $(I.4.1.6)$ in the following form:

$$(I.4.007) \quad \left[g_{ij}\right] = [\, g_1 \ g_2 \ \cdots \ g_n \,]^T [\, g_1 \ g_2 \ \cdots \ g_n \,]$$

The last equation is what you need: you just write your g_i-vectors as a matrix $U = [\, g_1 \ g_2 \ \cdots \ g_n \,]$, transpose this matrix so that you get $U^T = [\, g_1 \ g_2 \ \cdots \ g_n \,]^T$ and calculate the matrix product of U^T with U.

Now, let us start with some examples:

We take $V = \mathbb{R}^n$ with the standard cartesian basis $\{\vec{e}_1, \vec{e}_2, \dots, \vec{e}_n\}$. Using $(I.4.1.5)$, we calculate:

$$g_{ij} := \vec{e}_i \cdot \vec{e}_j = \delta_{ij}$$

Let us put together the result in one line:

$$
[g_{ij}] =
\begin{bmatrix}
g_{11} & g_{12} & \cdots & g_{1n} \\
g_{21} & g_{22} & \cdots & g_{2n} \\
\cdots & \cdots & \cdots & \cdots \\
g_{n1} & g_{n2} & \cdots & g_{nn}
\end{bmatrix}
$$

$$
=
\begin{bmatrix}
\vec{e}_1 \cdot \vec{e}_1 & \vec{e}_1 \cdot \vec{e}_2 & \cdots & \vec{e}_1 \cdot \vec{e}_n \\
\vec{e}_2 \cdot \vec{e}_1 & \vec{e}_2 \cdot \vec{e}_2 & \cdots & \vec{e}_2 \cdot \vec{e}_n \\
\cdots & \cdots & \cdots & \cdots \\
\vec{e}_n \cdot \vec{e}_1 & \vec{e}_n \cdot \vec{e}_2 & \cdots & \vec{e}_n \cdot \vec{e}_n
\end{bmatrix}
$$

$$
=
\begin{bmatrix}
\delta_{11} & \delta_{12} & \cdots & \delta_{1n} \\
\delta_{21} & \delta_{22} & \cdots & \delta_{2n} \\
\cdots & \cdots & \cdots & \cdots \\
\delta_{n1} & \delta_{n2} & \cdots & \delta_{nn}
\end{bmatrix}
$$

$$
=
\begin{bmatrix}
1 & 0 & \cdots & 0 \\
0 & 1 & \cdots & 0 \\
\cdots & \cdots & \ddots & \cdots \\
0 & 0 & \cdots & 1
\end{bmatrix}
$$

In our example, the metric tensor of the standard cartesian basis is just the unit matrix.

∎

Let us take another example:

Let $B = \{\vec{e}_1,\ \vec{e}_1 + \vec{e}_2,\ 3\,\vec{e}_3\} = \left\{ \begin{pmatrix} 1 \\ 0 \\ 0 \end{pmatrix}, \begin{pmatrix} 1 \\ 1 \\ 0 \end{pmatrix}, \begin{pmatrix} 0 \\ 0 \\ 3 \end{pmatrix} \right\}$ be a basis for $\mathbb{R}^3$. We leave it to the reader to verify that the vectors are linear independent.

Let us calculate the scalar products:

$$g_{11} = \vec{e}_1 \cdot \vec{e}_1 = \begin{pmatrix} 1 \\ 0 \\ 0 \end{pmatrix} \cdot \begin{pmatrix} 1 \\ 0 \\ 0 \end{pmatrix} = 1 * 1 + 0 * 0 + 0 * 0 = 1$$

$$g_{12} = \vec{e}_1 \cdot (\vec{e}_1 + \vec{e}_2) = \begin{pmatrix} 1 \\ 0 \\ 0 \end{pmatrix} \cdot \begin{pmatrix} 1 \\ 1 \\ 0 \end{pmatrix} = 1*1 + 0*1 + 0*0 = 1$$

$$g_{13} = \vec{e}_1 \cdot (3\,\vec{e}_3) = \begin{pmatrix} 1 \\ 0 \\ 0 \end{pmatrix} \cdot \begin{pmatrix} 0 \\ 0 \\ 3 \end{pmatrix} = 1*0 + 0*0 + 0*3 = 0$$

$$g_{22} = \cdot (\vec{e}_1 + \vec{e}_2) \cdot (\vec{e}_1 + \vec{e}_2) = \begin{pmatrix} 1 \\ 1 \\ 0 \end{pmatrix} \cdot \begin{pmatrix} 1 \\ 1 \\ 0 \end{pmatrix} = 1*1 + 1*1 + 0*0 = 2$$

$$g_{33} = (3\,\vec{e}_3) \cdot (3\,\vec{e}_3) = \begin{pmatrix} 0 \\ 0 \\ 3 \end{pmatrix} \cdot \begin{pmatrix} 0 \\ 0 \\ 3 \end{pmatrix} = 0*0 + 0*0 + 3*3 = 9$$

We do not need to calculate g_{13} as the scalar product is symmetric, so $g_{31} = g_{13}$ and $g_{21} = g_{12}$

Building up our metric tensor, we get:

$$[g_{ij}] = \begin{bmatrix} 1 & 1 & 0 \\ 1 & 2 & 0 \\ 0 & 0 & 9 \end{bmatrix}$$

Instead of calculating the scalar products, we take ($I.4.007$):

$$\vec{g}_1 = \begin{pmatrix} 1 \\ 0 \\ 0 \end{pmatrix}, \ \vec{g}_2 = \begin{pmatrix} 1 \\ 1 \\ 0 \end{pmatrix}, \ \vec{g}_3 = \begin{pmatrix} 0 \\ 0 \\ 3 \end{pmatrix},$$

so, build up the matrix U:

$$U = [\,\vec{g}_1 \ \ \vec{g}_2 \ \ \vec{g}_3\,] = \begin{bmatrix} 1 & 1 & 0 \\ 0 & 1 & 0 \\ 0 & 0 & 3 \end{bmatrix}, \text{ thus:}$$

$$U^T = \begin{bmatrix} 1 & 0 & 0 \\ 1 & 1 & 0 \\ 0 & 0 & 3 \end{bmatrix}$$

Now, we calculate the metric tensor as matrix multiplication:

$$[g_{ij}] = U^T U = \begin{bmatrix} 1 & 0 & 0 \\ 1 & 1 & 0 \\ 0 & 0 & 3 \end{bmatrix} \begin{bmatrix} 1 & 1 & 0 \\ 0 & 1 & 0 \\ 0 & 0 & 3 \end{bmatrix} = \begin{bmatrix} 1 & 1 & 0 \\ 1 & 2 & 0 \\ 0 & 0 & 9 \end{bmatrix}$$

The concept of the metric tensor can be extended to the case that the vector space does not have a constant basis but rather a set of independent functions denoted by (x^i), which define the vector space. Basis vectors in this case can be constructed as the *tangent vectors of the coordinate curves* (see section *I.3.3.3*):

For $i = 1, \ldots, n$ let (x^i) be coordinate curves depending on some parameters $(u^1, u^2, \ldots, u^m)$. We start with the standard Euclidian basis vectors $(\vec{e}_1, \vec{e}_2, \ldots, \vec{e}_n)$. As a basis vector g_l we take the partial derivative of (x^i) with respect to u^l and we express the basis vector in terms of the standard basis:

We get:

$$(1) \; g_l = \left(\frac{\partial x^k}{\partial u^l}\right) \vec{e}_k$$

Now, we calculate the scalar product of g_i and g_j:

$$
\begin{aligned}
g_{ij} \quad &= \quad g_i \cdot g_j \\
&= \quad \left(\frac{\partial x^k}{\partial u^i} \vec{e}_k\right) \cdot \left(\frac{\partial x^m}{\partial u^j} \vec{e}_m\right) \\
&= \quad \left(\frac{\partial x^k}{\partial u^i} \frac{\partial x^m}{\partial u^j}\right) \vec{e}_k \cdot \vec{e}_m \\
&= \quad \left(\frac{\partial x^k}{\partial u^i} \frac{\partial x^m}{\partial u^j}\right) \delta_{km} \\
&= \quad \frac{\partial x^k}{\partial u^i} \frac{\partial x^k}{\partial u^j}
\end{aligned}
$$

We can express the equations in matrix form:

$$
[g_{ij}] \quad = \quad
\begin{bmatrix}
\dfrac{\partial x^k}{\partial u^1} \dfrac{\partial x^k}{\partial u^1} & \dfrac{\partial x^k}{\partial u^1} \dfrac{\partial x^k}{\partial u^2} & \cdots & \dfrac{\partial x^k}{\partial u^1} \dfrac{\partial x^k}{\partial u^n} \\
\dfrac{\partial x^k}{\partial u^2} \dfrac{\partial x^k}{\partial u^1} & \dfrac{\partial x^k}{\partial u^2} \dfrac{\partial x^k}{\partial u^2} & \cdots & \cdots \\
\cdots & \cdots & \cdots & \cdots \\
\dfrac{\partial x^k}{\partial u^m} \dfrac{\partial x^k}{\partial u^1} & \cdots & \cdots & \dfrac{\partial x^k}{\partial u^m} \dfrac{\partial x^k}{\partial u^n}
\end{bmatrix}
$$

$$
= \underbrace{\begin{bmatrix} \dfrac{\partial x^1}{\partial u^1} & \dfrac{\partial x^2}{\partial u^1} & \cdots & \dfrac{\partial x^n}{\partial u^1} \\[2mm] \dfrac{\partial x^1}{\partial u^2} & \cdots & \cdots & \cdots \\[1mm] \cdots & \cdots & \cdots & \cdots \\[1mm] \dfrac{\partial x^1}{\partial u^m} & \cdots & \cdots & \dfrac{\partial x^n}{\partial u^m} \end{bmatrix}}_{= J^T} \underbrace{\begin{bmatrix} \dfrac{\partial x^1}{\partial u^1} & \dfrac{\partial x^1}{\partial u^2} & \cdots & \dfrac{\partial x^1}{\partial u^m} \\[2mm] \dfrac{\partial x^2}{\partial u^1} & \cdots & \cdots & \cdots \\[1mm] \cdots & \cdots & \cdots & \cdots \\[1mm] \dfrac{\partial x^n}{\partial u^1} & \cdots & \cdots & \dfrac{\partial x^n}{\partial u^m} \end{bmatrix}}_{= J}
$$

$$= \; J^T J$$

The ESNs in the $\left[g_{ij}\right]$-matrix can be expressed by the multiplication of the transposed Jacobi-matrix with itself!

This is the main formula for calculating the metric tensor in curvilinear coordinates:

$$\boxed{(I.4.008) \quad \left[g_{ij}\right] = J^T J}$$

There is no faster way to calculate the metric tensor in curvilinear coordinates! Every mathematical program as Sage Math, Maple etc. can calculate the Jacobi-matrix and the matrix multiplication should be no problem.

Example:

We take $\mathbb{R}^3$ as our vector space and we define the following basis:

$$x^1(x,y,z) = {}^1\!/_2 \, (x^2 - y^2)$$

$$x^2(x,y,z) = x\,y + z^3$$

$$x^3(x,y,z) = z$$

We first calculate the derivatives of the (x^i) with respect to x, y and z:

$$\frac{\partial x^1}{\partial x} = x, \frac{\partial x^1}{\partial y} = -y, \frac{\partial x^1}{\partial z} = 0$$

$$\frac{\partial x^2}{\partial x} = y, \frac{\partial x^2}{\partial y} = x, \frac{\partial x^2}{\partial z} = 3\,z^2$$

$$\frac{\partial x^3}{\partial x} = 9, \frac{\partial x^3}{\partial y} = 0, \frac{\partial x^3}{\partial z} = 1$$

The Jacobi matrix calculates as:

$$J = \begin{vmatrix} \dfrac{\partial x^1}{\partial x} & \dfrac{\partial x^1}{\partial y} & \dfrac{\partial x^1}{\partial z} \\[2mm] \dfrac{\partial x^2}{\partial x} & \dfrac{\partial x^2}{\partial y} & \dfrac{\partial x^2}{\partial z} \\[2mm] \dfrac{\partial x^3}{\partial x} & \dfrac{\partial x^3}{\partial y} & \dfrac{\partial x^3}{\partial z} \end{vmatrix} = \begin{bmatrix} x & -y & 0 \\ y & x & 3\,z^2 \\ 0 & 0 & 1 \end{bmatrix}$$

Now, we use $(I.4.008)$:

$$[g_{ij}] = J^T J$$

$$= \begin{bmatrix} x & y & 0 \\ -y & x & 0 \\ 0 & 3\,z^2 & 1 \end{bmatrix} \begin{bmatrix} x & -y & 0 \\ y & x & 3\,z^2 \\ 0 & 0 & 1 \end{bmatrix}$$

$$= \begin{bmatrix} x^2 + y^2 & 0 & 3\,z^2 y \\ 0 & x^2 + y^2 & 3\,z^2 x \\ 3\,z^2 y & 3\,z^2 x & 9\,z^4 + 1 \end{bmatrix}$$

Let us verify the result by calculating the scalar product of the partial derivatives:

$$(1) \quad \frac{\partial}{\partial x} = \begin{pmatrix} x \\ y \\ 0 \end{pmatrix}$$

$$(2) \quad \frac{\partial}{\partial y} = \begin{pmatrix} -y \\ x \\ 0 \end{pmatrix}$$

$$(3) \quad \frac{\partial}{\partial z} = \begin{pmatrix} 0 \\ 3\,z^2 \\ 1 \end{pmatrix}$$

So, we get now:

$$g_{11} = \frac{\partial}{\partial x} \cdot \frac{\partial}{\partial x} = \begin{pmatrix} x \\ y \\ 0 \end{pmatrix} \cdot \begin{pmatrix} x \\ y \\ 0 \end{pmatrix} = x^2 + y^2$$

$$g_{12} = \frac{\partial}{\partial x} \cdot \frac{\partial}{\partial y} = \begin{pmatrix} x \\ y \\ 0 \end{pmatrix} \cdot \begin{pmatrix} -y \\ x \\ 0 \end{pmatrix} = -yx + yx = 0$$

$$g_{13} = \frac{\partial}{\partial x} \cdot \frac{\partial}{\partial z} = \begin{pmatrix} x \\ y \\ 0 \end{pmatrix} \cdot \begin{pmatrix} 0 \\ 3\,z^2 \\ 1 \end{pmatrix} = 3\,z^2 y$$

$$g_{22} = \frac{\partial}{\partial y} \cdot \frac{\partial}{\partial y} = \begin{pmatrix} -y \\ x \\ 0 \end{pmatrix} \cdot \begin{pmatrix} -y \\ x \\ 0 \end{pmatrix} = x^2 + y^2$$

$$g_{23} = \frac{\partial}{\partial y} \cdot \frac{\partial}{\partial z} = \begin{pmatrix} -y \\ x \\ 0 \end{pmatrix} \cdot \begin{pmatrix} 0 \\ 3\,z^2 \\ 1 \end{pmatrix} = 3\,z^2 x$$

$$g_{33} = \frac{\partial}{\partial z} \cdot \frac{\partial}{\partial z} = \begin{pmatrix} 0 \\ 3\,z^2 \\ 1 \end{pmatrix} \cdot \begin{pmatrix} 0 \\ 3\,z^2 \\ 1 \end{pmatrix} = 9\,z^2 + 1$$

Because of symmetry, we do not need to calculate the other g_{ij}.

Finally, we get the same entries for the metric tensor.

∎

We take a second example and want to calculate the metric tensor for polar coordinates:

A vector $\vec{r}$ with the cartesian components $x = x^1, y = x^2$ can be expressed in polar coordinates by:

$$\vec{r} = \begin{pmatrix} x \\ y \end{pmatrix} = \begin{pmatrix} x^1 \\ x^2 \end{pmatrix} = \begin{pmatrix} r\cos\varphi \\ r\sin\varphi \end{pmatrix} \quad \text{with } r \in \mathbb{R} \text{ and } \varphi \in [0, 2\pi]$$

We calculate first the differentials:

$$\frac{\partial \vec{r}}{\partial r} = \begin{pmatrix} \cos\varphi \\ \sin\varphi \end{pmatrix}, \qquad \frac{\partial \vec{r}}{\partial \varphi} = \begin{pmatrix} -r\sin\varphi \\ r\cos\varphi \end{pmatrix}$$

Now, build up the metric tensor:

$$[g_{r\varphi}] \;=\; \begin{bmatrix} \dfrac{\delta \vec{r}}{\delta r}\cdot\dfrac{\delta \vec{r}}{\delta r} & \dfrac{\delta \vec{r}}{\delta r}\cdot\dfrac{\delta \vec{r}}{\delta \varphi} \\[2ex] \dfrac{\delta \vec{r}}{\delta \varphi}\cdot\dfrac{\delta \vec{r}}{\delta r} & \dfrac{\delta \vec{r}}{\delta \varphi}\cdot\dfrac{\delta \vec{r}}{\delta \varphi} \end{bmatrix}$$

$$= \begin{bmatrix} \begin{pmatrix} \cos\varphi \\ \sin\varphi \end{pmatrix}\cdot\begin{pmatrix} \cos\varphi \\ \sin\varphi \end{pmatrix} & \begin{pmatrix} \cos\varphi \\ \sin\varphi \end{pmatrix}\cdot\begin{pmatrix} -r\sin\varphi \\ r\cos\varphi \end{pmatrix} \\[2ex] \begin{pmatrix} -r\sin\varphi \\ r\cos\varphi \end{pmatrix}\cdot\begin{pmatrix} \cos\varphi \\ \sin\varphi \end{pmatrix} & \begin{pmatrix} -r\sin\varphi \\ r\cos\varphi \end{pmatrix}\cdot\begin{pmatrix} -r\sin\varphi \\ r\cos\varphi \end{pmatrix} \end{bmatrix}$$

$$= \begin{bmatrix} \cos^2\varphi + \sin^2\varphi & -r\cos\varphi\sin\varphi + r\cos\varphi\sin\varphi \\ -r\cos\varphi\sin\varphi + r\cos\varphi\sin\varphi & r^2\sin\varphi + r^2\cos\varphi \end{bmatrix}$$

$$= \begin{bmatrix} 1 & 0 \\ 0 & r^2 \end{bmatrix}$$

We would get the same result as the Jacobi matrix for polar coordinates calculates as:

$$\mathcal{J}_{(r,\varphi)} \;=\; \begin{bmatrix} \cos\varphi & -r\sin\varphi \\ \sin\varphi & r\cos\varphi \end{bmatrix}$$

∎

Let us now explore some properties of the metric tensor:

1. the metric tensor is symmetric: $g_{ij} = g_{ji}$, because the standard Euclidean scalar product is symmetric.

2. the inverse matrix $[g_{ij}]^{-1}$ always exists and is denoted by $[g^{ij}]$. Later, we will see why it is convenient to use upper indices for the inverse matrix and why/where we need the inverse metric tensor. A hint: if the metric tensor has only non-zero values in the diagonal, it is easy to compute the inverse:

$$g^{ij} = \begin{cases} 0 & if\ i \neq j \\ 1/g_{ij} & if\ i = j \end{cases}$$

3. in standard cartesian basis the length of a vector $\vec{v}$ is the square root of scalar product from $\vec{v}$ with itself:
$$|\vec{v}| = \sqrt[2]{\vec{v}\cdot\vec{v}} \;\Longrightarrow\; |\vec{v}|^2 = \vec{v}\cdot\vec{v}$$
As in the diagonal of the metric tensor $[g_{ij}]$ we have $\vec{g}_i \cdot \vec{g}_i$, the diagonal represents *the square* of the length of $\vec{g}_i$.

There is more:

We know that $\vec{v}\cdot\vec{w} = |\vec{v}||\vec{w}|\cos\sphericalangle(\vec{v},\vec{w})$

It follows that $g_{ij} = \vec{g}_i\cdot\vec{g}_j = \sqrt[2]{g_{ii}}\,\sqrt[2]{g_{jj}}\,\cos\sphericalangle(\vec{g}_i,\vec{g}_j)$

4. If all components g_{ij} are constant, the space is flat. If one or more of the components are not constant, the space is curved.

I.4.2 Applications for the metric tensor

So far, we have defined the metric tensor, and we know how to build up the metric matrix. Now, we want to explore where we can use the metric tensor.

First, we need to understand that the metric tensor can be used to convert a covariant vector into a contravariant vector and vice versa:

I.4.2.1 Raising and lowering indices

Given a contravariant vector with components x^j we can lower the index to a covariant vector with components x_i by multiplying the metric tensor with x^j:

$$(I.4.009)\quad x_i = g_{ij}\,x^j$$

in matrix-notation: $\begin{pmatrix} x_1 \\ x_n \\ ... \\ x_n \end{pmatrix} = [g_{ij}] \begin{pmatrix} x^1 \\ x^2 \\ ... \\ x^n \end{pmatrix}$

With the inverse metric tensor $[g^{ij}] = [g_{ij}]^{-1}$ we get:

$$(I.4.010)\quad x^i = g^{ij}\,x_j\,,$$

in matrix-notation: $\begin{pmatrix} x^1 \\ x^2 \\ ... \\ x^n \end{pmatrix} = [g^{ij}] \begin{pmatrix} x_1 \\ x_n \\ ... \\ x_n \end{pmatrix},$

so here we convert from covariant to contravariant.

Earlier (see $(I.2.061)$) we already stated:

In the standard Euclidean space covariant and contravariant is the same:

With $(I.4.009)$ we have:

$$(I.4.011) \quad x_i = g_{ij}\, x^j \underset{\text{euclidian space: } g_{ij}=\delta_{ij}}{=} \delta_{ij}\, x^j \underset{(I.2.022)}{=} x^i$$

,

So $x_i = x^i$ ∎

<u>Example:</u>

We take $x = \begin{bmatrix} x^1 \\ x^2 \end{bmatrix} = \begin{bmatrix} 3/5 \\ 4/(5x^1) \end{bmatrix}$, $[g_{ij}] = \begin{bmatrix} 1 & 0 \\ 0 & (x^1)^2 \end{bmatrix}$, so we defined x with contravariant components, having upper indices. To convert x to a covariant vector (lower indices), we need to multiply $[g_{ij}]$ with $\begin{bmatrix} x^1 \\ x^2 \end{bmatrix}$:

$$\begin{bmatrix} x_1 \\ x_2 \end{bmatrix} = [g_{ij}] \begin{pmatrix} x^1 \\ x^2 \end{pmatrix} = \begin{bmatrix} 1 & 0 \\ 0 & (x^1)^2 \end{bmatrix} \begin{pmatrix} 3/5 \\ 4/(5x^1) \end{pmatrix} = \begin{bmatrix} 3/5 \\ 4x^1/5 \end{bmatrix}$$

I.4.2.2 Generalized inner product (scalar product)

We define now the ***generalized inner product (scalar product)***:

$(I.4.012)$ **Definition (generalized scalar product)**

To each pair of contravariant vectors $\vec{v} = (v^i)$ and $\vec{w} = (w^i)$ is associated the real number

$$\vec{v} \cdot \vec{w} = g_{ij}\, v^i\, w^j = v_i\, w^i$$

In matrix notation we get:

$$\vec{v} \cdot \vec{w} = (v^1 \; v^2 \; \dots \; v^n)\, [g_{ij}] \begin{bmatrix} w^1 \\ \dots \\ w^n \end{bmatrix}$$

In the same way, we define the inner product of two covariant vectors as

$$\vec{v} \cdot \vec{w} = g^{ij}\, v_i\, w_j = v^i\, w_j$$

In matrix notation we get:

$$\vec{v} \cdot \vec{w} = (v_1 \ v_2 \ \ldots \ v_n) \, [g^{ij}] \begin{bmatrix} w_1 \\ \ldots \\ w_n \end{bmatrix}$$

$\vec{v} \cdot \vec{w}$ is called the **(generalized) inner product (scalar product)**.

Some notes on the definition:

We used $(I.4.009)$ and $(I.4.010)$ to convert the contravariant vector to covariant and vice versa. Pay attention to the indices: the index j of g_{ij}, respectively g^{ij}, disappears!

The inner product is invariant to coordinate changes, meaning: calculating the product in different coordinate systems produces always the same result.

Of course, $\vec{v} \cdot \vec{w}$ should produce the same result if we take the contravariant and covariant components:

$(I.4.012)$ The inner product of $\vec{v} \cdot \vec{w}$ with contravariant components equals the inner product of $\vec{v} \cdot \vec{w}$ with covariant components.

<u>Proof</u>:

We show that $g_{ij} \, v^i \, w^j = g^{rs} \, v_r \, w_s$:

$$
\begin{aligned}
g_{ij} \, v^i \, w^j \quad &= \quad g_{ij} \, (g^{ir} \, v_r) \, (g^{js} \, w_s) && (I.4.010) \\[2mm]
&= \quad g_{ij} \, g^{ir} g^{js} \, v_r \, w_s && \text{reorder} \\[2mm]
&= \quad g_{ji} \, g^{ir} g^{js} \, v_r \, w_s && \text{symmetry of } g_{ji} \\[2mm]
&= \quad \delta^r_{\ j} \, g^{js} \, v_r \, w_s && \text{as } [g^{ij}]^{-1} = [g_{ij}] \\[2mm]
&= \quad g^{rs} \, v_r \, w_s && (I.2.022)
\end{aligned}
$$

($I.4.013$) Two vectors are **orthogonal** if and only if their inner product $g_{ij}\, v^i w^j$ is 0.

In the following discussion we take $\mathbb{R}^n$ as a vector space and g_{ij} as a metric tensor. Explicitly, we allow *any* metric tensors, in particular curvilinear metrics.

I.4.2.3 Length of a vector
Using the metric tensor enables us to define the length of a vector as:

$$(I.4.014)\quad |v| \underset{ESN\ notation}{=} \sqrt[2]{g_{ij}\, v^i v^j} \underset{matrix\ notation}{=} \sqrt{v^T\, g\, v}$$

So, the length of the vector is the square root of the generalized inner product of the vector with itself. For direct calculation it might be easier to apply the *matrix notation.*

With ($I.4.012$) it should be clear that the following equation holds:

$$|v| = \sqrt{g_{ij}\, v^i v^j} = \sqrt{g^{ij}\, v_i\, v_j}\ ,$$

so, you may compute the length with covariant or contravariant components, just make sure if you need to take g_{ij} or its inverse g^{ij}.

Taking the Euclidian standard basis, so $g_{ij} = \delta_{ij}$, we get:

$$|v| = \sqrt{\delta_{ij}\, v^i v^j} = \sqrt{v_i\, v^i} \underset{(I.4.011)}{=} \sqrt{v_i v_i}$$

Remember: in the Euclidean standard basis $v_i = v^i$, there is no distinction between covariant and contravariant.

<u>Example:</u>

Let us take the metric tensor of polar coordinates:

$$[g_{ij}] = \begin{bmatrix} 1 & 0 \\ 0 & (x^1)^2 \end{bmatrix}$$

Let us calculate the length of the vector $\vec{v} = \begin{bmatrix} v^1 \\ v^2 \end{bmatrix} = \begin{bmatrix} 3/5 \\ 4/5x^1 \end{bmatrix}$

We get:

$$
\begin{aligned}
|\vec{v}|^2 &= g_{ij}\, v^i v^j \\
&= g_{11}\, v^1\, v^1 + g_{12}\, v^1\, v^2 + g_{21}\, v^2\, v^1\, g_{22}\, v^2\, v^2 \\
&= 1 * {}^3/_5 * {}^3/_5 + 0 * {}^3/_5 * {}^4/_{5x^1} + 0 * {}^4/_{5x^1} * {}^3/_5 + (x^1)^2 * {}^4/_{5x^1} * {}^4/_{5x^1} \\
&= {}^9/_{25} + {}^{16}/_{25} \\
&= 1
\end{aligned}
$$

Notice: we calculated *the square* of the length, so do not forget to take the square root of the calculation:

$$|\vec{v}|^2 = 1 \quad \Rightarrow \quad |\vec{v}| = +\sqrt[2]{1} = 1$$

We repeat the computation by using matrix calculation:

$$
\begin{aligned}
|\vec{v}|^2 &= \vec{v}^T \left[g_{ij} \right] \vec{v} \\
&= \begin{bmatrix} {}^3/_5 & {}^4/_{5x^1} \end{bmatrix} \begin{bmatrix} 1 & 0 \\ 0 & (x^1)^2 \end{bmatrix} \begin{pmatrix} {}^3/_5 \\ {}^4/_{5x^1} \end{pmatrix} \\
&= \begin{bmatrix} {}^3/_5 & {}^4/_{5x^1} \end{bmatrix} \begin{pmatrix} {}^3/_5 \\ 4(x^1)^2/_{5x^1} \end{pmatrix} \\
&= \frac{9}{25} + \frac{16}{25} = 1
\end{aligned}
$$

We got $|\vec{v}|^2 = 1$, so $|\vec{v}| = \sqrt{1} = 1$

Summing up so far, we get:

$(\boldsymbol{I.4.015})$ The length of a vector $\vec{v}$ with the metric tensor $[\,g_{ij}\,]$ is calculated as:

$$|\vec{v}| \;=\; \sqrt{g_{ij}\,v^i v^j} \qquad\qquad \{ESN$$

$$=\; \sqrt{g^{ij}\,v_i w_i}$$

$$=\; \sqrt{(v^1\; v^2\; \dots\; v^n)\,[\,g_{ij}\,]\begin{pmatrix} v^1 \\ v^2 \\ \dots \\ v^n \end{pmatrix}}$$

$$\{matrix\ notation$$

$$=\; \sqrt{(v_1\; v_2\; \dots\; v_n)\,[\,g^{ij}\,]\begin{pmatrix} v_1 \\ v_2 \\ \dots \\ v_n \end{pmatrix}}$$

A hint: calculate first $[\,g_{ij}\,]\begin{pmatrix} v^1 \\ v^2 \\ \dots \\ v^n \end{pmatrix}$. The result is a vector, then you can calculate the (cartesian) inner product with this vector and $(v_1\; v_2\; \dots\; v_n)$.

The squared line element provides a way to measure distances and intervals in a given space, which is essential for formulating physical laws. In the formulas given below, you will notice that the measures are only valid for infinitesimal changes along the basis and that you need to use the metric tensor:

In ESN, the **squared line element** s^2 is defined as

$(\boldsymbol{I.4.016})$ $ds^2 = g_{ij}\,dx^i\,dx^j$

In matrix notation, we get:

$$(\boldsymbol{I.4.017})\ ds^2 = (dx^1\ dx^2\ ...\ dx^n)\left[g_{ij}\right]\begin{pmatrix} dx^1 \\ dx^2 \\ ... \\ dx^n \end{pmatrix}$$

In a vector space, $\boldsymbol{ds^2}$ represents the squared length (or squared distance) between two vectors. This geometric interpretation remains consistent regardless of the coordinate system used to express the vectors. Thus, ds^2 is invariant under coordinate transformations because it represents an intrinsic geometric quantity in the vector space.

Let us take an example:

Define the metric tensor $\left[g_{ij}\right]$ for polar coordinates (see: *I*.A.1.1.1) as:

$$\left[g_{ij}\right] = \begin{bmatrix} 1 & 0 \\ 0 & r^2 \end{bmatrix}$$

The correspondent functions are:

$$x^1 = r\ cos\ (\varphi)$$

$$x^2 = r\ sin\ (\varphi)$$

$$\begin{aligned} ds^2 \ &=\ (dx^1\ dx^2\ ...\ dx^n)\left[g_{ij}\right]\begin{pmatrix} dx^1 \\ dx^2 \\ ... \\ dx^n \end{pmatrix} \\[2em] &=\ (dr\ d\varphi)\begin{bmatrix} 1 & 0 \\ 0 & r^2 \end{bmatrix}\begin{pmatrix} dr \\ d\varphi \end{pmatrix} \\[2em] &=\ (dr\ d\varphi)\begin{pmatrix} dr \\ r^2 d\varphi \end{pmatrix} \\[2em] &=\ dr^2 + r^2 d\varphi^2 \end{aligned}$$

Let us have a look at the picture for a geometrical interpretation of the squared line element:

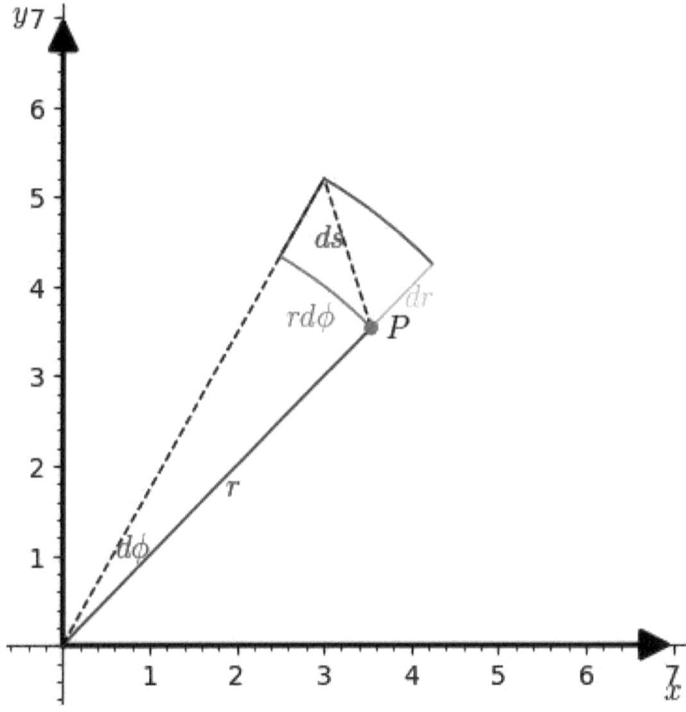

We see a point P at the position (r, ϕ) where ϕ is the angle between r and the x-axis. Consider dr, which is an infinitesimal displacement of r. In the same way we have an infinitesimal displacement of the angle ϕ, denoted by $d\phi$. Now, we connect point P with the displaced vector and we call this displacement ds. Then you can see that (roughly)

$$ds^2 \approx dr^2 + r^2 d\varphi^2$$

For polar coordinates this might seem simple to derive, for other coordinate systems the geometrical derivation can be very complex. The metric tensor encapsulates the geometry of a coordinate system in a way that the squared line element can be calculated very easily.

I.4.2.5 Angle between vectors

We know that in the standard cartesian coordinate system the angle between two non-null vectors U and V is defined by

$$\cos\theta = \frac{vw}{|v||w|}$$

with $0 \leq \theta \leq \pi$.

More generally, the angle between two non-null contravariant vectors v and w can be defined as:

$$(\textit{I}.4.018)\ \cos\theta = \frac{g_{ij}v^{i}w^{j}}{\sqrt[2]{g_{pq}v^{p}w^{q}}\ \sqrt[2]{g_{rs}v^{r}w^{s}}}$$

with $0 \leq \theta \leq \pi$. Note that all indices are summation indices.

I.4.2.6 Arc length and line element of a curve

We consider now parameterized curves. As many equations in physics are time-dependent, we choose t as the parameter.

Let us look at a vector function $r(t)$ and a small displacement Δr which can be represented as

$$\Delta r = r(\,t + \Delta t\,) - r(t)$$

Then

$$|\Delta r| = \frac{\Delta r}{\Delta t}\,\Delta t \approx |r'(t)|\,\Delta t$$

Summing up all Δr, we conclude that the length s of $r(t)$ between $r(a)$ and $r(b)$ equals:

$$(\textit{I}.4.019)\ s \quad = \quad \lim_{n\to\infty}\sum_{i=0}^{n-1}|\Delta r| =$$

$$= \lim_{n \to \infty} \sum_{i=0}^{n-1} \frac{|\Delta r|}{\Delta t} \, \Delta t$$

$$= \lim_{n \to \infty} \sum_{i=0}^{n-1} |r'(t)| \, \Delta t$$

$$= \int_a^b |r'(t)| \, dt$$

Example:

Define

$$r(t) = (\cos(t), \sin(t), t\,).$$

This function describes a helix:

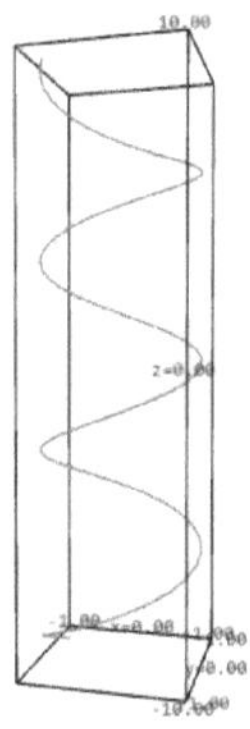

We calculate $r'(t) = \dfrac{dr}{dt}$ as

$$\frac{dr}{dt} = (-\sin(t), \cos(t), 1\,),$$

so

$$\left| \frac{dr}{dt} \right| = \sqrt{sin^2(t) + cos^2(t) + 1} = \sqrt{2}$$

Now we calculate the length of the helix between 0 and $b \geq 0$:

$$s = \int_0^b \sqrt{2}\, dt = b\sqrt{2}$$

The **generalized definition of the arc length of a curve** $x^i = x^i(t)$ between the points a and b is defined as

$$(I.4.020)$$

$$s \quad = \quad \int_a^b \sqrt{\left| g_{ij} \frac{dx^i}{dt} \frac{dx^j}{dt} \right|}\, dt \qquad \textit{ESN-notation}$$

$$= \quad \int_a^b \sqrt{\left| \left(\frac{dx^1}{dt} \quad \frac{dx^2}{dt} \quad \cdots \quad \frac{dx^n}{dt} \right)^T [g_{ij}] \begin{pmatrix} \dfrac{dx^1}{dt} \\ \dfrac{dx^2}{dt} \\ \cdots \\ \dfrac{dx^n}{dt} \end{pmatrix} \right|}\, dt \qquad \textit{Matrix notation}$$

$$= \quad \int_a^b \sqrt{|ds^2|}\, dt \qquad (I.4.017)$$

where $[g_{ij}]$ is the metric tensor. The first equation is expressed in ESN, so do not forget to sum up over i and j, the second in matrix notation. In general, g_{ij} will consist of functions of the coordinates. The third equation is formulated by the squared line element, so here we have the connection between the arc length and the squared line element.

Let us come back to our helix and see how this works. In the example we are in the standard Euclidian space, so $g_{ij} = \delta_{ij}$.

Then $(I.4.2.6.2)$ can be written as:

$$(I.4.021) \quad s \underset{g_{ij}=\delta_{ij}}{=} \int_a^b \sqrt{\left| g_{ij} \frac{dx^i}{dt} \frac{dx^j}{dt} \right|} \, dt$$

$$= \int_a^b \sqrt{\left| \delta_{ij} \frac{dx^i}{dt} \frac{dx^j}{dt} \right|} \, dt$$

$$= \int_a^b \sqrt{\left| \frac{dx^i}{dt} \frac{dx^i}{dt} \right|} \, dt$$

For the helix, we have the following set of functions:

$$x^1(t) = \cos(t), \quad x^2(t) = \sin(t), \quad x^3(t) = t$$

Calculating the derivates:

$$\frac{dx^1}{dt} = -\sin(t)$$

$$\frac{dx^2}{dt} = \cos(t)$$

$$\frac{dx^3}{dt} = 1$$

We insert $\dfrac{dx^i}{dt}$ into the last integral $(I.4.021)$ and get:

$$s = \int_a^b \sqrt{\frac{dx^i}{dt} \frac{dx^i}{dt}} \, dt = \int_a^b \sqrt{\underbrace{\sin^2(t) + \cos^2(t)}_{=1} + 1} \, dt = \sqrt{2}(b-a)$$

Here we used to sum up directly, using ESN. An example where we will be calculating the arc length by matrix notation will be given in exercise

$(I.4.E.10)$. Have a closer look at the solution, as more details are provided on how to proceed.

A hint and a warning: you cannot use the squared line element to measure distances of points. You can only measure the distance of points which are *very* close: dx^i denotes an *infinitesimal* displacement! If you want to measure the distance of points, then you MUST use the integral.

A remark:

If you have a function $y = f(x)$, then you can parmeterize the function by $f(t) = (t, f(t))$. It follows easily now that the length of f between the points a and b is given by:

$$s = \int_a^b \sqrt{1 + (f'(x))^2}\, dx$$

Let us take some examples:

Define the metric tensor $[g_{ij}]$ as

$$[g_{ij}] = \begin{bmatrix} (x^1)^2 - 1 & 1 & 0 \\ 1 & (x^2)^2 & 0 \\ 0 & 0 & 64/9 \end{bmatrix}$$

where $[(x^1)^2 - 1]\,(x^2)^2 \neq 1$

Now, define a curve $\mathfrak{C}$ as:

$$\mathfrak{C}: \begin{cases} x^1 = 2t - 1 \\ x^2 = 2t^2 \\ x^3 = t^3 \end{cases}$$

with $0 \leq t \leq 1$.

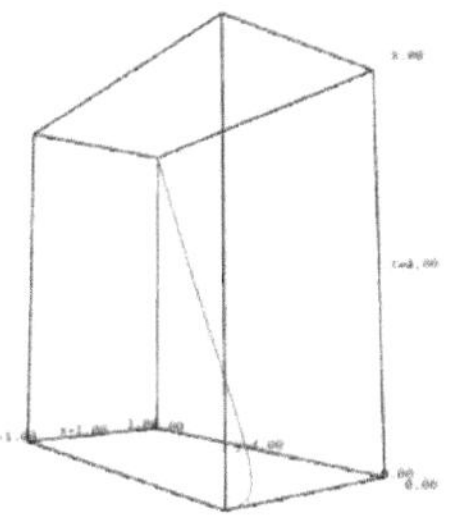

Calculating the differentials of x^i gives us:

$$\frac{dx^1}{dt} = 2$$

$$\frac{dx^2}{dt} = 4t$$

$$\frac{dx^3}{dt} = 3t^2$$

$$
\left(\frac{ds}{dt}\right)^2 = \left(\frac{dx^i}{dt}\right)^T [g_{ij}] \left(\frac{dx^i}{dt}\right)
$$

$$
= \begin{bmatrix} 2 & 4t & 3t^2 \end{bmatrix}
\begin{bmatrix} (2t-1)^2 - 1 & 1 & 0 \\ 1 & (2t^2)^2 & 0 \\ 0 & 0 & \tfrac{64}{9} \end{bmatrix}
\begin{bmatrix} 2 \\ 4t \\ 3t^2 \end{bmatrix}
$$

$$
= \begin{bmatrix} 2 & 4t & 3t^2 \end{bmatrix}
\begin{bmatrix} 2[(2t-1)^2 - 1] + 4t \\ 2 + 4t(2t^2)^2 \\ \dfrac{64}{9} 3t^2 \end{bmatrix}
$$

$$
= \begin{bmatrix} 2 & 4t & 3t^2 \end{bmatrix}
\begin{bmatrix} 8t^2 - 4t \\ 16t^5 + 2 \\ \dfrac{64}{9} 3t^2 \end{bmatrix}
$$

$$
= 64t^6 + 64t^4 + 16t^2
$$

$$
= (8t^3 + 4t)^2
$$

We note that $(8t^3 + 4t)^2 \geq 0$, so $\left(\frac{ds}{dt}\right)^2 = \left|\left(\frac{ds}{dt}\right)^2\right|$.

With this result, we can finally calculate the arc length parameter:

$$s(t) = \int_0^t (8u^3 + 4u)^2 \, du = [2u^4 + 2u^2]_0^t = 2t^4 + 2t^2$$

Notice that in the metric matrix we replaced the x^i by the definition of the curve.

I.4.3 Other ways of expressing the metric tensor

I.4.3.1 Metric tensor and squared line element

So far, we represented the metric tensor as a matrix, and we found a simple way to calculate the metric tensor by using the Jacobi matrix.

We already saw that the squared line element can be expressed by the metric tensor:

$$(\boldsymbol{I.4.022}) \quad ds^2 = g_{ij} \, dx^i dx^j$$

where (x^i) is a coordinate system and $\left[g_{ij}\right]$ the associated metric tensor.

But we can use $(I.4.022)$ to describe the metric tensor by knowing the squared line element.

<u>Example</u>:

For polar coordinates, we already calculated the metric tensor as

$$\left[g_{r\varphi}\right] = \begin{bmatrix} 1 & 0 \\ 0 & r^2 \end{bmatrix}$$

So, we get:

$$ds^2 \;=\; \underbrace{(1)}_{=\,g_{rr}} \underbrace{dr dr}_{position\,[r,r]\,in\,[g_{ij}]} \;+\; \underbrace{(r^2)}_{=\,g_{\varphi\varphi}} \underbrace{d\varphi d\varphi}_{position\,[\varphi,\varphi]\,in\,[g_{ij}]}$$

$$= \quad \mathbf{1}\, dr^2 + r^2 d\varphi^2$$

The coefficients of the metric tensor are contained in the squared line element. It should be obvious how we can reconstruct the metric as a matrix by a given squared line element: the expressions $dxdy$ show you where to put the coefficients in the tensor matrix. **Caution:** if ds^2 contains mixed terms as $c\, dx^1\, dx^2$, $c \in \mathbb{R}$, you need to pay attention to put $\frac{c}{2}$ in the position of dx^1, dx^2 and $\frac{c}{2}$ in position dx^2, dx^1, because the metric tensor is symmetric.

<u>Example:</u>

Take $ds^2 = 5\,(dx^1)^2 + 8\,(dx^2)^2 + 12\,(dx^3)^2 - 6\,dx^1\,dx^2 + 4\,dx^2\,dx^3$

The terms $-\,6\,dx^1\,dx^2$ and $4\,dx^2\,dx^1$ are mixed.

We rewrite the equation in a way that every entry point in the matrix is given:

$$ds^2 = 5\,dx^1\,dx^1 + 8\,dx^2dx^2 + 12\,dx^3dx^{3^2} - \frac{6}{2}\,dx^1\,dx^2 - \frac{6}{2}\,dx^2\,dx^1$$
$$+\,\frac{4}{2}\,dx^2\,dx^3 + \frac{4}{2}\,dx^3\,dx^2$$

Thus, we get:

$$g_{ij} = \begin{bmatrix} 5 & -3 & 0 \\ -3 & 8 & 2 \\ 0 & 2 & 12 \end{bmatrix}$$

I.4.3.2 Metric tensor and tensor products

We only want to state, that the equation $(I.\,4.022)$ can also be written as

$$(\mathbf{I.4.023})\ \ ds^2 = g_{ij}\,dx^i \otimes dx^j$$

In fact, you will find in literature that $dx^i\,dx^j$ is defined as follows:

$$dx^i \, dx^j \; = \; \frac{1}{2} \, (dx^i \otimes dx^j + dx^j \otimes dx^i)$$

Therefore, we get:

$(I.4.024)$:

$$
\begin{aligned}
g_{ij} \, dx^i \otimes dx^j \;\; &= \;\; g_{ij} \, \frac{1}{2} \, (\, dx^i \otimes dx^j + dx^j \otimes dx^i) \\[2ex]
&= \;\; g_{ij} \, dx^i dx^j
\end{aligned}
$$

I.4.4 Transformation law of the metric tensor

In this chapter, we explore the transformation behavior of the metric tensor when we change from a coordinate system (x^i) to a coordinate system $(\bar{x}^i)$.

$(I.4.025)$ _Transformation law for metric tensor:_

Let g_{ij} be the metric tensor of some coordinates (x^i). Let $(\bar{x}^i)$ be another coordinate system related by

$$\bar{x}^i = \bar{x}^i(\, x^1, x^2, \dots, x^n \,)$$

And let $\bar{g}_{ij}$ be the metric related to $(\bar{x}^i)$.

Then, the following transformation holds:

$$\bar{g}_{ij} = g_{kl} \, \frac{\partial x^k}{\partial \bar{x}^i} \, \frac{\partial x^l}{\partial \bar{x}^j}$$

In matrix notation:

$$[\bar{g}_{ij}] = J \, [g_{kl}] \, J^T$$

where J is the Jacobi-matrix of (x^i) with respect to $(\bar{x}^i)$

What we state here is that g_{ij} is a covariant tensor of order 2. As we still need to define what a tensor is, we deliver the proof of the theorem without using tensor properties.

Proof:

We noted in section $(I.4.2.3)$ that the squared line element is an invariant in coordinate transformations and is defined as:

$(1)\ ds^2 = g_{ij}\, dx^i\, dx^j$

In the $(\bar{x}^i)$-coordinate system we have:

$(2)\ d\bar{s}^2 = \bar{g}_{ij}\, d\bar{x}^i\, d\bar{x}^j$

As ds^2 is an invariant, we have $ds^2 = d\bar{s}^2$, so we get:

$(3)\ g_{ij}\, dx^i\, dx^j = \bar{g}_{kl}\, d\bar{x}^k\, d\bar{x}^l$

Since dx^i is a contravariant vector, we get the transformation law (see $(I.3.004)$):

$(4)\ d\bar{x}^k = \dfrac{\partial \bar{x}^k}{\partial x^i}\, dx^i$

Replacing (4) in (3) results in:

$$(5)\ g_{ij}\, dx^i\, dx^j = \bar{g}_{kl}\, \frac{\partial \bar{x}^k}{\partial x^i}\, dx^i\, \frac{\partial \bar{x}^l}{\partial x^j}\, dx^j$$

This means:

$$(6)\ g_{ij}\, dx^i\, dx^j - \bar{g}_{kl}\, \frac{\partial \bar{x}^k}{\partial x^i}\, dx^i\, \frac{\partial \bar{x}^l}{\partial x^j}\, dx^j = 0$$

We can rewrite (6) as

$$(7)\ dx^i\, dx^j \left(g_{ij} - \bar{g}_{kl}\, \frac{\partial \bar{x}^k}{\partial x^i}\, \frac{\partial \bar{x}^l}{\partial x^j} \right) = 0$$

As $dx^i\, dx^j \neq 0$, we can conclude that

$$\boxed{\;(I.4.026)\ \ g_{ij} = \bar{g}_{kl}\, \frac{\partial \bar{x}^k}{\partial x^i}\, \frac{\partial \bar{x}^l}{\partial x^j} = \bar{g}_{kl}(J^T)_{ki}\, J_{lj} = J^T\, [\bar{g}_{kl}]\, J\;}$$

where

$$J = \left[\frac{\partial \bar{x}^i}{\partial x^j}\right]$$

■

Of course, in the same way it can be shown that

$$(I.4.027)\quad g^{ij} = \bar{g}^{kl}\,\frac{\partial x^i}{\partial \bar{x}^k}\,\frac{\partial x^j}{\partial \bar{x}^l}$$

Both equations are not so much for practical use. In general, being in the Euclidian space, we mostly take as the departing basis the standard basis and want to transfer to another coordinate system, spherical or polar. But it is also possible to make a transformation from, let us say, spherical to helix coordinates. We need the equations later when we are talking about derivatives of tensors.

I.4.5 Derivation of metric tensor

As an abbreviation we define

$$(I.4.028)\quad g_{ijk} := \frac{\partial g_{ij}}{\partial x^k}$$

We add a final subscript k for the derivation of g_{ij} with respect to x^k.

Referring to the transformation law of g_{ij} $(I.4.025)$, we get

$$\bar{g}_{ij} = g_{rs}\,\frac{\partial x^r}{\partial \bar{x}^i}\,\frac{\partial x^s}{\partial \bar{x}^j}$$

$$\begin{aligned}
\bar{g}_{ijk} &= \frac{\partial}{\partial \bar{x}^k}\left(g_{rs}\,\frac{\partial x^r}{\partial \bar{x}^i}\,\frac{\partial x^s}{\partial \bar{x}^j}\right) \\
&= \frac{\partial g_{rs}}{\partial \bar{x}^k}\,\frac{\partial x^r}{\partial \bar{x}^i}\,\frac{\partial x^s}{\partial \bar{x}^j} + g_{rs}\,\frac{\partial^2 x^r}{\partial \bar{x}^k \partial \bar{x}^i}\,\frac{\partial x^s}{\partial \bar{x}^j} + g_{rs}\,\frac{\partial x^r}{\partial \bar{x}^i}\,\frac{\partial^2 x^s}{\partial \bar{x}^k \partial \bar{x}^j}
\end{aligned}$$

We used the product rule for derivation here. Now, we use the chain rule for the expression $\dfrac{\partial g_{rs}}{\partial \bar{x}^k}$, resulting in:

$$\frac{\partial g_{rs}}{\partial \bar{x}^k} = \frac{\partial g_{rs}}{\partial x^t}\,\frac{\partial x^t}{\partial \bar{x}^k} \underset{(I.4.025)}{=} g_{rst}\,\frac{\partial x^t}{\partial \bar{x}^k}$$

and we replace it in the last equation:

$(\mathbf{\textit{I}.4.029})$:

$$\bar{g}_{ijk} = g_{rst}\frac{\partial x^t}{\partial \bar{x}^k}\frac{\partial x^r}{\partial \bar{x}^i}\frac{\partial x^s}{\partial \bar{x}^j} + g_{rs}\frac{\partial^2 x^r}{\partial \bar{x}^k \partial \bar{x}^i}\frac{\partial x^s}{\partial \bar{x}^j} + g_{rs}\frac{\partial x^r}{\partial \bar{x}^i}\frac{\partial^2 x^s}{\partial \bar{x}^k \partial \bar{x}^j}$$

We will need this formula later.

I.4.6 The way back

Let $[g_{ij}]$ be a given metric tensor, viewed as a matrix. Is there a way to reconstruct the basis vectors $\vec{g}_i$?

First, remember that $[g_{ij}] = U^T U$, where $U = [\ \vec{g}_1\ \ \vec{g}_2\ \ \dots\ \ \vec{g}_3\ ]$. So, if we achieve to decompose $[g_{ij}]$ as the product of a matrix product of the form $U^T U$, then we get a set of $\vec{g}_i$-vectors.

Interestingly, there exists an algorithm, named **Cholesky decomposition**, which decomposes a *Hermitian, symmetric and positive definite matrix A* in the form $A = L\,L^*$, where L^* denotes the conjugate transpose of L (we will not explain here what hermitian and positive definite mean, we only state that the metric tensor satisfies these properties). L is then a lower triangular matrix with positive elements in the diagonal, uniquely defined.

Unfortunately, there may be more than one set of $\vec{g}_i$-vectors producing the same metric tensor:

Take the following metric tensor:

$$[g_{ij}] = \begin{bmatrix} 45 & 39 & -31 \\ 39 & 35 & -26 \\ -31 & -26 & 22 \end{bmatrix}$$

We can compute this metric tensor by using the vectors

$$\vec{g}_1 = \begin{pmatrix} -5 \\ 2 \\ 4 \end{pmatrix},\ \vec{g}_2 = \begin{pmatrix} -5 \\ 1 \\ 3 \end{pmatrix},\ \vec{g}_3 = \begin{pmatrix} 3 \\ -2 \\ -3 \end{pmatrix},\ \text{see exercise } (I.4.E.3).$$

Now consider the vectors which have been calculated by **Cholesky decomposition**:

$$\breve{\vec{g}}_1 = \begin{pmatrix} 6.70820393249937 \\ 0 \\ 0 \end{pmatrix}$$

$$\breve{\vec{g}}_2 = \begin{pmatrix} 5.81377674149945 \\ 1.09544511501033 \\ 0 \end{pmatrix}$$

$$\vec{\breve{g}}_3 = \begin{pmatrix} -4.62120715349957 \\ 0.791154805285242 \\ 0.136082763487943 \end{pmatrix}$$

Building up the metric tensor using $\vec{\breve{g}}_i$, we obtain the SAME metric tensor built up by g_i! And, appearantly, there is no visible relationship between $\vec{g}_i$ and $\vec{\breve{g}}_i$.

So, there is no way to reconstruct the basis vectors for a given metric tensor matrix.

I.4.E Exercises

$(\textbf{\textit{I.4.E.1}})$ Given the following ESN expressions: simplify them in a way that the metric coefficients g_{ij} vanish.

    (a) $a^m g_{lm}$

    (b) $g^{pk} u_p$

    (c) $g^{ij} b^k g_{kj}$

    (d) $g_{ij} g^{jk}$

$(\textbf{\textit{I.4.E.2}})$ Let V be a 2-dimensional vector space and e^1, e^2 a basis. Let g be the metric tensor associated to e^1 and e^2. Suppose we know that

$$g(e^1, e^1) = 3, \quad g(e^1, e^2) = 1 \text{ and } g(e^2, e^2) = 2,$$

where $g(v, w) := v^T g\, v$

Calculate

    (a) $|e^1|$ and $|e^2|$
    (b) $|e^1 + e^2|$
    (c) $|e^1 - 2e^2|$

$(I.4.E.3)$ Given the basis vectors

$$\vec{g}_1 = \begin{pmatrix} -5 \\ 2 \\ 4 \end{pmatrix}, \qquad \vec{g}_2 = \begin{pmatrix} -5 \\ 1 \\ 3 \end{pmatrix}, \qquad \vec{g}_3 = \begin{pmatrix} 3 \\ -2 \\ -3 \end{pmatrix},$$

calculate the metric tensor.

$(I.4.E.4)$ Show that the vectors $\vec{v} = \begin{pmatrix} 0 \\ 1 \\ 2\,b\,r\,sin(\theta) \end{pmatrix}$ and $\vec{w} = \begin{pmatrix} 0 \\ -2\,b\,rsin(\theta) \\ r^2\,cos^2(\theta) \end{pmatrix}$ are orthogonal under cylindrical coordinates (see

also I.A.1.1.2) defined as

$$x^1(r,\theta,h\,) = r\cos\theta$$

$$x^2(r,\theta,h\,) = r\sin\theta$$

$$x^3(r,\theta,h) = h$$

With $r \geq 0,\ 0 \leq \theta \leq 2\pi,\ -\infty < h < \infty.$

$(I.4.E.5)$ Show that the contravariant vectors $\vec{v} = \begin{pmatrix} -x^1/\,x^2 \\ 1 \\ 0 \end{pmatrix}$ and $\vec{w} = \begin{pmatrix} 1/x^2 \\ 0 \\ 0 \end{pmatrix}$ in curvilinear coordinates (x^i) are orthogonal if is related to rectangular coordinates $(\bar{x}^i)$ by

$$\bar{x}^1 = x^2, \quad \bar{x}^2 = x^3, \quad \bar{x}^3 = x^1x^2$$

$(I.4.E.6)$ (Building the metric tensor for spherical coordinates):

Let the cartesian coordinates x^1, x^2 and x^3 be expressed by:

$$x^1 = x^1(r,\theta,\phi\,) = r\sin\theta\cos\phi$$

$$x^2 = x^2(r,\theta,\phi\,) = r\sin\theta\sin\phi$$

$$x^3 = x^3(r, \theta, \phi) = r \, \cos \theta$$

with $r \geq 0$, $0 \leq \theta \leq \pi$ and $0 \leq \phi \leq 2\pi$.

    (a) Calculate the metric tensor.

    (b) Represent the squared line element s^2 as a linear combination of tensor products.

$(I.4.E.7)$ (arc length)

Calculate the arc length of the curve

$$\mathfrak{C}: \begin{cases} x^1 = \ln\left(\sqrt[2]{1+t^2}\right) \\ x^2 = \arctan t \end{cases}$$

with $0 \leq t \leq 2$.

$(I.4.E.8)$ (arc length of implicit function)

Determine the length of $\ \left(2x^1 + (x^1)^2 + (x^2)^2\right)^2 = 4\left((x^1)^2 + (x^2)^2\right)$

$(I.4.E.9)$:

Verify that the curve

$$\Phi(t) = \begin{pmatrix} \sin(t) \\ \dfrac{\sin^2(t)}{2} \\ \dfrac{1}{2}\left(t - \sin(t)\cos(t)\right) \end{pmatrix}$$

is already parameterized by its arc length.

$(I.4.E.10)$:

Let $x^1 = t$, $x^2 = arc\,sin\left(\frac{1}{t}\right)$, $x^3 = \sqrt{t^2 - 1}$ be a curve in spherical coordinates. Determine the length of the arc with

$1 \leq t \leq 2$

I.5 Christoffel symbols and metric tensor

Christoffel symbols, especially those of the second kind, are mathematical quantities used in the field of differential geometry, particularly in the study of curved spaces and the theory of general relativity in physics. They were named after the German mathematician Elwin Bruno Christoffel (1829-1900).

Christoffel symbols are a set of coefficients that describe how basis vectors change as one moves along a curved manifold, such as a curved space-time in general relativity. They are essential for understanding the connection between differential geometry and physics, specifically in the context of Einstein's theory of general relativity.

I.5.1 Definition of the Christoffel symbols

We start with a coordinate system (x^i) on an n-dimensional vector space. We can calculate the differentials of x^i and we define

$$g_i := \frac{\partial}{\partial x^i} \quad , i = 1, 2, \ldots, n$$

Doing this, we get with $\{\, g_1, g_2, \ldots, g_n \,\}$ a basis for the vector space (see I.3.3.3). In general, these basis vectors are functions, so that the basis vectors change from point to point.

In the previous chapter about the metric tensor, we defined $g_{ij} = g_i \cdot g_j$ and denoted g_{ij} as the metric tensor. As we alreday showed, the inverse $g^{ij} := (g_{ij})^{-1}$ exists, so that we were able to define the dual basis as $g^i := g_j\, g^{ji}$.

Now, we define the Christoffel symbols of the second kind as follows:

$$(I.5.001) \quad \Gamma^k{}_{ij} := \frac{\partial g_i}{\partial x^j}\, g^k$$

Notice very carefully the index positions and that we use g_i **and** the dual space vector g^k.

To understand, what $(I.5.1.1)$ means, we multiply both sides with g_k:

$$\Gamma^k{}_{ij}\, g_k := \frac{\partial g_i}{\partial x^j}\, g^k g_k$$

but $g^k g_k = 1$, so we get:

$$(\boldsymbol{I.5.002})\quad \Gamma^k{}_{ij}\, g_k := \frac{\partial g_i}{\partial x^j}$$

Do not forget you have to sum up over k.

We can interpret this equation as follows: *the Christoffel symbols $\Gamma^k{}_{ij}$ are the <u>coefficients</u> of the partial derivative of g_i with respect to x^j represented in the basis vectors g_k. With these coefficients, the changes in the basis vectors are expressed in each point in space as they progress in the direction of the coordinate lines.*

Let us view an example:

As (x^i) we take polar corrdinates:

$$x = x^1(r,\theta) = r\cos(\theta)$$

$$y = x^2(r,\theta) = r\sin(\theta)$$

The Jacobi-matrix calculates as:

$$J = \begin{bmatrix} \partial x^1/\partial r & \partial x^1/\partial \theta \\ \partial x^2/\partial r & \partial x^2/\partial \theta \end{bmatrix} = \begin{bmatrix} \cos\theta & -r\sin\theta \\ \sin\theta & r\cos\theta \end{bmatrix}$$

So, we get the basis vectors

$$e_r = \begin{bmatrix} \cos\theta \\ \sin\theta \end{bmatrix} = \cos(\theta)\, e_x + \sin(\theta)\, e_y$$

$$e_\theta = \begin{bmatrix} -r\sin\theta \\ r\cos\theta \end{bmatrix} = -r\sin(\theta)\, e_x + r\cos(\theta)\, e_y,$$

where $e_x = \begin{pmatrix} 1 \\ 0 \end{pmatrix}$ and $e_y = \begin{pmatrix} 0 \\ 1 \end{pmatrix}$

We note that $\dfrac{\partial e_x}{\partial r} = 0,\ \dfrac{\partial e_y}{\partial r} = 0,\ \dfrac{\partial e_x}{\partial \theta} = 0,\ \dfrac{\partial e_y}{\partial \theta} = 0$

Now, we calculate:

$$\frac{\partial e_r}{\partial r} = \frac{\partial}{\partial r}\left(\cos(\theta)\,e_x + \sin(\theta)\,e_y\right)$$

$$= \frac{\partial}{\partial r}\left(\cos(\theta)\,e_x\right) + \frac{\partial}{\partial r}\left(\sin(\theta)\,e_y\right)$$

$$= \underbrace{\frac{\partial}{\partial r}(\cos(\theta))}_{=0}\,e_x + \cos(\theta)\underbrace{\frac{\partial}{\partial r}(e_x)}_{=0} + \underbrace{\frac{\partial}{\partial r}(\sin(\theta))}_{=0}\,e_y$$

$$+ \sin(\theta)\underbrace{\frac{\partial}{\partial r}(e_y)}_{=0}$$

$$= 0\cdot e_x + \cos(\theta)\cdot 0 + 0\cdot e_y + \sin(\theta)\cdot 0$$

$$= 0$$

$$\frac{\partial e_r}{\partial \theta} = \frac{\partial}{\partial \theta}\left(\cos(\theta)\,e_x + \sin(\theta)\,e_y\right)$$

$$= \frac{\partial}{\partial \theta}\left(\cos(\theta)\,e_x\right) + \frac{\partial}{\partial \theta}\left(\sin(\theta)\,e_y\right)$$

$$= \underbrace{\frac{\partial}{\partial \theta}(\cos(\theta))}_{=-\sin(\theta)}\,e_x + \cos(\theta)\underbrace{\frac{\partial}{\partial \theta}(e_x)}_{=0} + \underbrace{\frac{\partial}{\partial \theta}(\sin(\theta))}_{=\cos(\theta)}\,e_y$$

$$+ \sin(\theta)\underbrace{\frac{\partial}{\partial \theta}(e_y)}_{=0}$$

$$= -\sin(\theta)\,e_x + \cos(\theta)\,e_y$$

$$= \frac{1}{r}\,e_\theta$$

Likewise, you get:

$$\frac{\partial e_\theta}{\partial r} = -\sin(\theta)\,e_x + \cos(\theta)\,e_y = \frac{1}{r}\,e_\theta$$

$$\frac{\partial e_\theta}{\partial \theta} = -r\cos(\theta)\,e_x + r\sin(\theta)\,e_y = -r\,e_r$$

Now, we can rewrite the equations in the following way:

$$\frac{\partial e_r}{\partial r} = 0 = \Gamma^r_{rr}\,e_r + \Gamma^\theta_{rr}\,e_\theta$$

$$\frac{\partial e_r}{\partial \theta} = \frac{1}{r}\,e_\theta = \Gamma^r_{r\theta}\,e_r + \Gamma^\theta_{r\theta}\,e_\theta$$

$$\frac{\partial e_\theta}{\partial r} = \frac{1}{r}\,e_\theta = \Gamma^r_{\theta r}\,e_r + \Gamma^\theta_{\theta r}\,e_\theta$$

$$\frac{\partial e_\theta}{\partial r} \quad = \quad -r\, e_r \quad\quad = \quad\quad \Gamma^r{}_{\theta\theta}\, e_r + \quad \Gamma^\theta{}_{\theta\theta}\, e_\theta$$

It follows that

$$\Gamma^r{}_{rr} = \quad \Gamma^\theta{}_{rr} = \quad \Gamma^r{}_{r\theta} = \quad \Gamma^r{}_{\theta r} = \quad \Gamma^\theta{}_{\theta\theta} = 0$$

$$\Gamma^\theta{}_{\theta r} = \quad \Gamma^\theta{}_{r\theta} = \frac{1}{r}$$

$$\Gamma^r{}_{\theta\theta} = -r \qquad\qquad \blacksquare$$

Nothing magic so far; they are just coefficients. We will see how powerful the Christoffel symbols will turn out to be.

Remembering that g_i is a differential of x^i, along with the **Theorem of Schwartz**, we can conclude immediately:

$$(\boldsymbol{I.5.003}) \quad \Gamma^k{}_{ij} = \Gamma^k{}_{ji}$$

Therefore, you can swap the subscripts as you want. For the calculations of the symbols, it is therefore not necessary to compute all subscript indices.

There are also the Christoffel symbols of the *first kind* defined as:

$$(\boldsymbol{I.5.004}) \quad \Gamma_{kij} := \frac{\partial g_i}{\partial x^j}\, g_k$$

Again, notice the position of the indices: for the first kind of Christoffel symbols, they are all subscripts, and the definition does not use the dual vector.

The Christoffel symbol of the first kind can be calculated easily by the second kind: just lower the superscript index with the metric tensor:

$$(\boldsymbol{I.5.005}) \quad \Gamma_{kij} = \Gamma^m{}_{ij}\, g_{mk}$$

It follows:

$$(\boldsymbol{I.5.006}) \quad \Gamma_{kij}\, g^{mk} = \Gamma^m{}_{ij}$$

I.5.2 Calculating the Christoffel symbols

The definition of $\Gamma^k{}_{ij}$ is not very suitable to calculate the values. Indeed, $\Gamma^k{}_{ij}$ can be expressed by the inverse of the metric tensor and the derivatives of the metric tensor itself. Before we derive this result, we need to prove the following equation:

$$(\textbf{\textit{I}.5.007}) \quad \Gamma^k{}_{ij}\, g_{kl} + \Gamma^k{}_{lj}\, g_{ki} = \frac{\partial g_{il}}{\partial x^j}$$

<u>Proof:</u>

We start with $(\textit{I}.5.002)$**:**

$$(1) \quad \Gamma^k{}_{ij}\, g_k := \frac{\partial g_i}{\partial x^j}$$

As we did before, we multiply each side with another basis vector g_l:

$$(2) \quad \Gamma^k{}_{ij}\, g_k\, g_l = \frac{\partial g_i}{\partial x^j}\, g_l$$

We may write

$$(3) \quad \frac{\partial g_i}{\partial x^j}\, g_l = \frac{\partial(\, g_i\, g_l)}{\partial x^j} - \frac{\partial(\, g_l\,)}{\partial x^j}\, g_i :$$

applying the product rule on $\dfrac{\partial(\, g_i\, g_l)}{\partial x^j}$ we get $\dfrac{\partial(\, g_i\, g_l)}{\partial x^j} = \dfrac{\partial(\, g_l)}{\partial x^j}\, g_i + \dfrac{\partial(\, g_i)}{\partial x^j}\, g_l$

Replacing (3) in (2) gives us:

$$(4) \quad \Gamma^k{}_{ij}\, g_k\, g_l = \frac{\partial(\, g_i\, g_l)}{\partial x^j} - \frac{\partial(\, g_l\,)}{\partial x^j}\, g_i$$

Now, $g_k\, g_l = g_{kl}$ (left side), the same with $g_i\, g_l = g_{il}$ on the right side. Furthermore, using $(\textit{I}.5.002)$ we get $\dfrac{\partial(\, g_l)}{\partial x^j}\, g_i = \Gamma^k{}_{lj}\, g_k\, g_i$, so we may write

$$(5) \quad \Gamma^k{}_{ij}\, g_{kl} = \frac{\partial(\, g_{il})}{\partial x^j} - \Gamma^k{}_{lj}\, g_{ki} \qquad \blacksquare$$

In the same way, we get the following equations:

$$(6)\quad \Gamma^{k}_{jl}\, g_{ki} + \Gamma^{k}_{il}\, g_{kj} = \frac{\partial g_{ij}}{\partial x^{l}}$$

$$(7)\quad \Gamma^{k}_{li}\, g_{kj} + \Gamma^{k}_{ji}\, g_{kl} = \frac{\partial g_{lj}}{\partial x^{i}}$$

Swapping the subscript indices ($I.5.003$) gives us:

$$(8)\quad \Gamma^{k}_{lj}\, g_{ki} + \Gamma^{k}_{il}\, g_{kj} = \frac{\partial g_{ij}}{\partial x^{l}}$$

$$(9)\quad \Gamma^{k}_{il}\, g_{kj} + \Gamma^{k}_{ij}\, g_{kl} = \frac{\partial g_{lj}}{\partial x^{i}}$$

We add ($I.5.002$) to (9) and substract (8). Thus, we get:

$$(10)\quad 2\,\Gamma^{k}_{ij}\, g_{kl} = \frac{\partial g_{il}}{\partial x^{j}} + \frac{\partial g_{lj}}{\partial x^{i}} - \frac{\partial g_{ji}}{\partial x^{l}}$$

Now, we use the fact that $g^{ij}\, g_{jk} = \delta^{i}{}_{k}$ and multiply the last equation by g^{ml}. Doing this, we get:

$$(11)\quad 2\,\Gamma^{k}_{ij}\, \underbrace{g_{kl}\, g^{ml}}_{=\delta^{m}{}_{k}} = g^{ml}\left(\frac{\partial g_{il}}{\partial x^{j}} + \frac{\partial g_{lj}}{\partial x^{i}} - \frac{\partial g_{ji}}{\partial x^{l}} \right),$$

leading to

$$(12)\quad 2\,\Gamma^{k}_{ij}\, \delta^{m}{}_{k} = g^{ml}\left(\frac{\partial g_{il}}{\partial x^{j}} + \frac{\partial g_{lj}}{\partial x^{i}} - \frac{\partial g_{ji}}{\partial x^{l}} \right)$$

Now we contract the left side and we get finally:

$$\boxed{(I.5.008)\quad \Gamma^{m}_{ij} = \frac{1}{2}\, g^{ml}\left(\frac{\partial g_{il}}{\partial x^{j}} + \frac{\partial g_{lj}}{\partial x^{i}} - \frac{\partial g_{ji}}{\partial x^{l}} \right)}$$

That is the formular we wanted to have: you can calculate Γ^{m}_{ij} by knowing the metric tensor and its inverse.

For the Christoffel symbols of the first kind, there exists also a formula which can be used for direct calculation:

$$(I.5.009) \quad \Gamma_{ijk} = \frac{1}{2}\left(\frac{\partial g_{jk}}{\partial x^i} + \frac{\partial g_{ki}}{\partial x^j} - \frac{\partial g_{ij}}{\partial x^k}\right)$$

As the metric tensor is symmetric, we conclude:

$$(I.5.010) \quad \Gamma_{ijk} = \Gamma_{jik}$$

Before we go on, let us take an example:

Again, we calculate the Christoffel symbol for polar coordinates:

We already know that the metric tensor and its inverse are

$$[g_{ij}] = \begin{bmatrix} 1 & 0 \\ 0 & (r)^2 \end{bmatrix}, \quad [g^{ij}] = [g_{ij}]^{-1} = \begin{bmatrix} 1 & 0 \\ 0 & \frac{1}{(r)^2} \end{bmatrix}$$

We calculate the derivates as:

$$\left[\frac{\partial g_{ij}}{\partial r}\right] = \begin{bmatrix} \dfrac{\partial g_{rr}}{\partial r} & \dfrac{\partial g_{r\varphi}}{\partial r} \\[2ex] \dfrac{\partial g_{\varphi r}}{\partial r} & \dfrac{\partial g_{\varphi\varphi}}{\partial r} \end{bmatrix} = \begin{bmatrix} 0 & 0 \\ 0 & 2r \end{bmatrix}$$

$$\left[\frac{\partial g_{ij}}{\partial \varphi}\right] = \begin{bmatrix} \dfrac{\partial g_{rr}}{\partial \varphi} & \dfrac{\partial g_{r\varphi}}{\partial \varphi} \\[2ex] \dfrac{\partial g_{\varphi r}}{\partial \varphi} & \dfrac{\partial g_{\varphi\varphi}}{\partial \varphi} \end{bmatrix} = \begin{bmatrix} 0 & 0 \\ 0 & 0 \end{bmatrix}$$

Now, we can calculate the Christoffel symbols:

$$\Gamma^{\varphi}_{\ r\varphi} \;=\; \frac{1}{2}g^{\varphi l}\left(\frac{\partial g_{rl}}{\partial \varphi} + \frac{\partial g_{l\varphi}}{\partial r} - \frac{\partial g_{\varphi r}}{\partial l}\right) \tag{1*}$$

$$= \frac{1}{2}\underbrace{g^{\varphi r}}_{=0}\Bigg(\underbrace{\frac{\partial g_{rr}}{\partial \varphi}}_{=0} + \underbrace{\frac{\partial g_{r\varphi}}{\partial r}}_{=0} - \underbrace{\frac{\partial g_{\varphi r}}{\partial r}}_{=0}\Bigg) \tag{2*}$$

$$+ \frac{1}{2}\underbrace{g^{\varphi\varphi}}_{=\frac{1}{r^2}}\Bigg(\underbrace{\frac{\partial g_{r\varphi}}{\partial \varphi}}_{=0} + \underbrace{\frac{\partial g_{\varphi\varphi}}{\partial r}}_{=2r} - \underbrace{\frac{\partial g_{\varphi r}}{\partial \varphi}}_{=0}\Bigg)$$

$$= \quad 0 + \frac{1}{2r^2}(2r)$$

$$= \quad \frac{1}{r}$$

(1^*): we use $(I.5.008)$

(2^*): the summation index l runs through r and θ. The only non-zero derivative is $\frac{\partial g_{\theta\theta}}{\partial r} = 2r$

With the same procedure we find:

$$\left[\Gamma^r_{ij}\right] = \begin{bmatrix} 0 & 0 \\ 0 & -r \end{bmatrix}$$

$$\left[\Gamma^\varphi_{ij}\right] = \begin{bmatrix} 0 & \dfrac{1}{r} \\ \dfrac{1}{r} & 0 \end{bmatrix}$$

∎

There are some special notes on the Christoffel symbols:

$(I.5.011)$ In rectangular coordinate systems all Christoffel symbols are zero. The reason is that the basis vectors are constant, the same with the metric tensor. For constants, the derivate is always zero.

Even if we found a practical method to calculate the Christoffel symbols, we should take a closer look at the case that the metric tensor is a diagonal matrix:

($I.5.012$) If $\left[g_{ij}\right]$ is a diagonal matrix, then the following equations hold for every subscripts α and β $\boxed{\text{with } \alpha \neq \beta}$ (no summation here!):

   a) $\Gamma_{\alpha\alpha\alpha} = \dfrac{1}{2}\dfrac{\partial g_{\alpha\alpha}}{\partial x^{\alpha}}$

   b) $-\Gamma_{\alpha\alpha\beta} = \Gamma_{\alpha\beta\alpha} = \Gamma_{\beta\alpha\alpha} = \dfrac{1}{2}\dfrac{\partial g_{\alpha\alpha}}{\partial x^{\beta}}$

   c)   All other symbols are zero. In particular: Γ_{ijk} vanishes if all indices are pairwise distinct.

<u>Proof</u>:

Case a)

Using ($I.5.009$), we have:

$$\Gamma_{ijk} = \frac{1}{2}\left(\frac{\partial g_{jk}}{\partial x^{i}} + \frac{\partial g_{ki}}{\partial x^{j}} - \frac{\partial g_{ij}}{\partial x^{k}}\right)$$

Now, consider $\Gamma_{\alpha\alpha\alpha}$:

$$\Gamma_{\alpha\alpha\alpha} = \frac{1}{2}\left(\frac{\partial g_{\alpha\alpha}}{\partial x^{\alpha}} + \underbrace{\frac{\partial g_{\alpha\alpha}}{\partial x^{\alpha}} - \frac{\partial g_{\alpha\alpha}}{\partial x^{\alpha}}}_{=0}\right) = \frac{1}{2}\frac{\partial g_{\alpha\alpha}}{\partial x^{\alpha}}$$

Notice, that we did not use the fact that g_{ij} is diagonal.

Case b)

Using ($I.5.009$), we have:

$$\Gamma_{\alpha\alpha\beta} = \frac{1}{2}\left(\frac{\partial g_{\alpha\beta}}{\partial x^{\alpha}} + \frac{\partial g_{\beta\alpha}}{\partial x^{\alpha}} - \frac{\partial g_{\alpha\alpha}}{\partial x^{\beta}}\right) = -\frac{1}{2}\frac{\partial g_{\alpha\alpha}}{\partial x^{\beta}}$$

Here we used the fact that the metric tensor is diagonal as $g_{\beta\alpha} = g_{\alpha\beta} = 0$, provided that $\alpha \neq \beta$. Their derivatives, then is also zero.

Now, consider $\Gamma_{\alpha\beta\alpha}$:

$$\Gamma_{\alpha\beta\alpha} = \frac{1}{2}\left(\frac{\partial g_{\beta\alpha}}{\partial x^{\alpha}} + \frac{\partial g_{\alpha\alpha}}{\partial x^{\beta}} - \frac{\partial g_{\alpha\beta}}{\partial x^{\alpha}}\right) = \frac{1}{2}\frac{\partial g_{\alpha\alpha}}{\partial x^{\beta}}$$

Considering now $\Gamma_{\beta\alpha\alpha}$, we get:

$$\Gamma_{\alpha\beta\alpha} \underset{(I.5.2.3a)}{=} \Gamma_{\beta\alpha\alpha} = \frac{1}{2}\frac{\partial g_{\alpha\alpha}}{\partial x^\beta}$$

Case c)

If $i \neq j \neq k$, then $g_{ij} = 0$ and therefore $\dfrac{\partial g_{ij}}{\partial x^k} = 0$, resulting in $\Gamma_{ijk} = 0$.

∎

$(I.5.013)$ If $\left[g_{ij}\right]$ is a diagonal matrix, then the following equations hold for every subscript α and β (again: no summation here!):

(1) $\quad \Gamma^\alpha{}_{\alpha\beta} = \Gamma^\alpha{}_{\beta\alpha} = \frac{\partial}{\partial x^\beta}\left(\frac{1}{2}\ln |g_{\alpha\alpha}|\right) = \frac{1}{2}\,g^{\alpha\alpha}\,\frac{\partial}{\partial x^\beta}\,g_{\alpha\alpha}$

(2) $\quad \Gamma^\alpha{}_{\beta\beta} = -\frac{1}{2\,g_{\alpha\alpha}}\frac{\partial}{\partial x^\alpha}\,g_{\beta\beta} = -\frac{1}{2}\,g^{\alpha\alpha}\,\frac{\partial}{\partial x^\alpha}\,g_{\beta\beta} \quad (\alpha \neq \beta)$

(3) $\quad$ All other symbols vanish. Again: $\Gamma^i{}_{jk}$ vanishes if the indices are pairwise distinct.

One important note: In $(I.5.012)$ $\alpha = \beta$ was excluded. Here only in (2)!!

Proof of (1):

First, we claim that

$$(I.5.014)\quad \frac{\partial \ln(g_{\alpha\alpha})}{\partial x^\beta} = \frac{1}{g_{\alpha\alpha}}\frac{\partial g_{\alpha\alpha}}{\partial x^\beta}$$

This is simply the fact that we use the chain rule on ln and its argument $g_{\alpha\alpha}$ together with $\dfrac{\partial \ln(x)}{\partial x} = \dfrac{1}{x}$.

Now, we get:

$$\begin{aligned}
\Gamma^\alpha{}_{\alpha\beta} &= g^{\alpha j}\,\Gamma_{\alpha\beta j} &\quad (I.5.006)\\
&= g^{\alpha\alpha}\,\Gamma_{\alpha\beta\alpha} &\quad (1^*)
\end{aligned}$$

$$= \frac{1}{g^{\alpha\alpha}} \Gamma_{\alpha\beta\alpha} \qquad\qquad (2^*)$$

$$= \frac{1}{g^{\alpha\alpha}} \frac{1}{2} \frac{\partial g_{\alpha\alpha}}{\partial x^b} \qquad\qquad (I.5.012,a)$$

$$= \frac{\partial}{\partial x^\beta} \left(\frac{1}{2} \ln|g_{\alpha\alpha}| \right) \qquad\qquad (I.5.014)$$

(1^*) As g_{ij} is diagonal, so also g^{ij}. Now, in summing up over j only the diagonal elements survive.

(2^*) As g^{ij} is diagonal, its inverse is just $\dfrac{1}{g_{ij}}$ (we mean the components here)

Proof of (2):

$$\Gamma^{\alpha}{}_{\beta\beta} = g^{\alpha\alpha} \Gamma_{\beta\beta\alpha} \qquad\qquad (I.5.006)$$

$$= \frac{1}{g_{\alpha\alpha}} \left(-\frac{1}{2} \frac{\partial g_{\beta\beta}}{\partial x^\alpha} \right) \qquad\qquad (I.5.012,b), \text{exchange } \alpha \text{ and } \beta$$

Proof of (3):

We recall from $(I.5.012,c)$ that Γ_{jki} vanishes if i, j, k are distinct. So, we have:

$$\Gamma^{i}{}_{jk} = g^{ir} \Gamma_{jkr} \underset{g \ diagonal}{=} g^{ii} \underbrace{\Gamma_{jki}}_{=0} = 0$$

∎

With the following example we want to show the steps you have to take if you are given a parameterized surface:

Example: Calculating the Christoffel symbols of the second kind for the unit sphere (radius of 1 centered at $(0,0)$).

The sphere can be represented by the following parameterized function:

$$x^i(u,v) = \begin{pmatrix} x^1(u,v) \\ x^2(u,v) \\ x^3(u,v) \end{pmatrix} = \begin{pmatrix} sin(u)\,cos(v) \\ sin(u)\,sin(v) \\ cos(u) \end{pmatrix}$$

for $(u,v) \in [0,\pi] \times [0,2\pi]$.

Note that those are the same functions as the ones defining spherical coordinates, except we have replaced φ by u and θ by v.

To calculate the Christoffel symbols, we take two ways: on the one hand the calculation using ($I.5.008$) and the shorter way using ($I.5.013$).

Solution 1 (using ($I.5.008$)):

Step 1:

Build the metric tensor. As we have a given function, we should build up the metric tensor by the Jacobi matrix of (x^i):

$$J = \begin{bmatrix} \dfrac{\partial x^1}{du} & \dfrac{\partial x^1}{dv} \\ \dfrac{\partial x^2}{du} & \dfrac{\partial x^2}{dv} \\ \dfrac{\partial x^3}{du} & \dfrac{\partial x^3}{dv} \end{bmatrix}$$

$$= \begin{bmatrix} cos(u)\,cos\,(v) & -sin(u)\,sin\,(v) \\ cos(u)\,sin\,(v) & cos(v)\,sin\,(u) \\ -sin\,(u) & 0 \end{bmatrix}$$

Now, we build up the metric tensor (we use c for cos and s for sin):

$$[g_{ij}] = J^T J$$

$$= \begin{bmatrix} c(u)\,c\,(v) & c(u)\,s(v) & -s(u) \\ -s(u)\,s(v) & c(v)\,(u) & 0 \end{bmatrix} \begin{bmatrix} c(u)\,c(v) & -s(u)\,s(v) \\ c(u)\,s(v) & c(v)\,s(u) \\ -s(u) & 0 \end{bmatrix}$$

$$= \begin{bmatrix} 1 & 0 \\ 0 & sin^2(u) \end{bmatrix}$$

To be more precise:

$$[g_{ij}] = \begin{bmatrix} g_{11} & g_{12} \\ g_{21} & g_{22} \end{bmatrix} = \begin{bmatrix} 1 & 0 \\ 0 & sin^2(u) \end{bmatrix}$$

As $[g_{ij}]$ is a diagonal matrix, the inverse $[g^{ij}]$ calculates as:

$$[g^{ij}] = \begin{bmatrix} g^{11} & g^{12} \\ g^{21} & g^{22} \end{bmatrix} = \begin{bmatrix} 1 & 0 \\ 0 & \dfrac{1}{sin^2(u)} \end{bmatrix}$$

<u>Step 2</u>: calculates the derivatives of $[g_{uv}]$:

$$\left[\frac{\partial g_{ij}}{\partial u}\right] = \begin{bmatrix} 0 & 0 \\ 0 & 2\cos(u)\sin(u) \end{bmatrix}$$

$$\left[\frac{\partial g_{ij}}{\partial v}\right] = \begin{bmatrix} 0 & 0 \\ 0 & 0 \end{bmatrix}$$

Step 3: calculate the Christoffel symbols using ($I.\,5.008$):

We will use the formula

$$\Gamma^m_{ij} = \frac{1}{2} g^{ml} \left(\frac{\partial g_{il}}{\partial x^j} + \frac{\partial g_{lj}}{\partial x^i} - \frac{\partial g_{ji}}{\partial x^l} \right)$$

Using this formula, do not forget to sum up over l, which in our case takes the values 1 and 2!

We see that all differentials of $\frac{\partial g_{ij}}{\partial u}$ are zero with the exception $\frac{\partial g_{22}}{\partial u} = 2\cos(u)\sin(u)$. Furthermore, all differentials of $\frac{\partial g_{ij}}{\partial v}$ are zero.

We will only calculate some of the Christoffel symbols, just to give you an idea on how to proceed. Notice in the following calculations that $x^1 = u$ and $x^2 = v$.

$$\Gamma^v_{uv} = \frac{1}{2} g^{2l} \left(\frac{\partial g_{1l}}{\partial u} + \frac{\partial g_{l2}}{\partial u} - \frac{\partial g_{11}}{\partial x^l} \right)$$

$$= \frac{1}{2} \underbrace{g^{22}}_{=1/sin^2(u)} \left(\underbrace{\frac{\partial g_{11}}{\partial v}}_{=0} + \underbrace{\frac{\partial g_{12}}{\partial u}}_{=0} - \underbrace{\frac{\partial g_{21}}{\partial u}}_{=0} \right)$$

$$+ \frac{1}{2} \underbrace{g^{22}}_{=1/sin^2(u)} \left(\underbrace{\frac{\partial g_{12}}{\partial u}}_{=0} + \underbrace{\frac{\partial g_{22}}{\partial u}}_{=2\cos(u)\sin(u)} - \underbrace{\frac{\partial g_{21}}{\partial v}}_{=0} \right)$$

$$= \frac{cos(u)}{sin(u)}$$

$$\Gamma^{u}_{\ vv} = \frac{1}{2} g^{1l} \left(\frac{\partial g_{1l}}{\partial u} + \frac{\partial g_{l2}}{\partial u} - \frac{\partial g_{11}}{\partial x^{l}} \right)$$

$$= \frac{1}{2} \underbrace{g^{11}}_{=1} \left(\underbrace{\frac{\partial g_{11}}{\partial v}}_{=0} + \underbrace{\frac{\partial g_{12}}{\partial v}}_{=0} - \underbrace{\frac{\partial g_{22}}{\partial u}}_{=2\cos(u)\sin(u)} \right)$$

$$+ \frac{1}{2} \underbrace{g^{12}}_{=0} \left(\underbrace{\frac{\partial g_{22}}{\partial v}}_{=0} + \underbrace{\frac{\partial g_{22}}{\partial v}}_{=2\cos(u)\sin(u)} - \underbrace{\frac{\partial g_{22}}{\partial v}}_{=0} \right)$$

$$= -cos(u)\,sin(u)$$

Because of the symmetry of the Christoffel symbol in the subscripts, we get immediately:

$$\Gamma^{v}_{\ uv} = \Gamma^{v}_{\ vu} = \frac{cos\,(u)}{sin\,(u)}$$

All other symbols calculate to zero.

<u>Solution 2 (using ($I.5.013$)):</u>

As we have already seen in the first solution, the metric tensor is given by

$$[g_{ij}] = \begin{bmatrix} g_{11} & g_{12} \\ g_{21} & g_{22} \end{bmatrix} = \begin{bmatrix} 1 & 0 \\ 0 & sin^2(u) \end{bmatrix}$$

Obviously, the matrix is diagonal, so we can apply ($I.5.013$).

We have then:

$$\Gamma^{u}_{\ uv} = \frac{\partial}{\partial v}\left(\frac{1}{2}\ln |g_{11}|\right)$$

$$= \frac{\partial}{\partial v}\left(\frac{1}{2}\ln|1|\right)$$

$$= 0$$

$$\Gamma^{v}_{vu} = \frac{\partial}{\partial u}\left(\frac{1}{2}\ln|g_{22}|\right)$$

$$= \frac{\partial}{\partial u}\left(\frac{1}{2}\ln|\sin^2(u)|\right)$$

$$= \frac{\partial}{\partial u}\left(\frac{1}{2}\ln(\sin^2(u))\right)$$

$$= \frac{1}{2}\frac{1}{\sin^2(u)}\frac{\partial}{\partial u}(\sin^2(u))$$

$$= \frac{1}{2}\frac{1}{\sin^2(u)}\,2\cos(u)\sin(u)$$

$$= \frac{\cos(u)}{\sin(u)}$$

$$\Gamma^{u}_{vv} = \;\; = -\frac{1}{2g_{11}}\frac{\partial}{\partial u}g_{22}$$

$$= -\frac{1}{2*1}\frac{\partial}{\partial u}\sin^2(u)$$

$$= -\sin(u)\cos(u)$$

$$\Gamma^{v}_{uu} = -\frac{1}{2g_{22}}\frac{\partial}{\partial u}g_{11}$$

$$= -\frac{1}{2\sin^2(u)}\underbrace{\frac{\partial}{\partial u}1}_{=0}$$

$$= 0$$

All other symbols vanish, following $(I.5.013, c)$.

As you might see, the calculations are not difficult, you should only think about using the chain rule to get the correct derivatives. Do not forget to check that the metric matrix is diagonal.

■

There is another way to calculate the Christoffel symbols based on a given set of basis functions. We have already explained the procedure to obtain the metric tensor using the Jacobian matrix of the function, followed by direct calculations of the Christoffel symbols (of the second kind) using the metric tensor. Now, we will skip the summation of all the metric tensor coefficients and use only the Jacobian matrix. You will see that we will only use matrix multiplication to calculate the Christoffel symbols of the second kind.

The procedure is just the following equations, where i and j represent the column / row <u>of a matrix</u> and l runs through 1 to n (n = dimension of the vector space):

$$(\textbf{\textit{I}}.5.\textbf{015}) \quad \left[\Gamma^i{}_{jl}\right] = \left(\frac{\partial}{\partial x^l} J^T\right) J^{T-1}$$

Notice that the subscript l on the left side appears also on the right side, where the transpose Jacobi matrix is derived with respect to x^l.

As a result, we get back a matrix of the following shape for each l:

$$\left[\Gamma^i_{jl}\right] = \begin{bmatrix} \Gamma^1_{1l} & \Gamma^2_{1l} & \cdots & \Gamma^n_{11} \\ \Gamma^1_{2l} & \Gamma^2_{2l} & \cdots & \Gamma^2_{nl} \\ \cdots & \cdots & \cdots & \cdots \\ \Gamma^1_{nl} & \Gamma^2_{nl} & \cdots & \Gamma^n_{nl} \end{bmatrix}$$

It is important to notice that the upper index i defines the column (and NOT the row!), whereas j defines the row.

We make a demonstration and come back to the Christoffel symbols of the second kind for polar coordinates:

Example (Christoffel symbols of the 2nd kind for polar coordinates):

The function f is given by:

$$f(r,\varphi) = (r\cos(\varphi), r\sin(\varphi))$$

We begin to calculate the Jacobi matrix of f:

$$J_{f(r,\varphi)} = \begin{bmatrix} \cos(\varphi) & -r\sin(\varphi) \\ \sin(\varphi) & r\cos(\varphi) \end{bmatrix}$$

So, we get the transpose by:

$$J^T{}_{f(r,\varphi)} = \begin{bmatrix} \cos(\varphi) & \sin(\varphi) \\ -r\sin(\varphi) & r\cos(\varphi) \end{bmatrix}$$

The inverse of the transposed Jacobi matrix calculates as:

$$J^{T^{-1}}{}_{f(r,\varphi)} = \begin{bmatrix} \cos(\varphi) & -\dfrac{\sin(\varphi)}{r} \\[2mm] \sin(\varphi) & \dfrac{\cos(\varphi)}{r} \end{bmatrix}$$

Now we need to calculate the differentials of the transposed Jacobi matrix:

$$\frac{\partial}{\partial r}J^T{}_{f(r,\varphi)} = \begin{bmatrix} 0 & 0 \\ -\sin(\varphi) & \cos(\varphi) \end{bmatrix}$$

$$\frac{\partial}{\partial \varphi}J^T{}_{f(r,\varphi)} = \begin{bmatrix} -\sin(\varphi) & \cos(\varphi) \\ -r\cos(\varphi) & -r\sin(\varphi) \end{bmatrix}$$

Now we calculate:

$$\left[\Gamma^i{}_{jr}\right] = \left[\Gamma^i{}_{j1}\right] = \left(\frac{\partial}{\partial r}J^T\right)J^{T^{-1}}$$

$$= \underbrace{\begin{bmatrix} 0 & 0 \\ -\sin(\varphi) & \cos(\varphi) \end{bmatrix}}_{=\frac{\partial}{\partial r}J^T} \begin{bmatrix} \cos(\varphi) & -\dfrac{\sin(\varphi)}{r} \\[2mm] \sin(\varphi) & \dfrac{\cos(\varphi)}{r} \end{bmatrix}$$

$$= \begin{bmatrix} 0 & 0 \\ -\sin(\varphi)\cos(\varphi) + \sin(\varphi)\cos(\varphi) & \dfrac{\sin^2(\varphi) + c\dots}{r} \end{bmatrix}$$

$$= \begin{bmatrix} 0 & 0 \\ 0 & \dfrac{1}{r} \end{bmatrix} = \begin{bmatrix} \Gamma^1_{11} & \Gamma^2_{11} \\ \Gamma^1_{21} & \Gamma^2_{21} \end{bmatrix} = \begin{bmatrix} \Gamma^r_{rr} & \Gamma^\varphi_{rr} \\ \Gamma^r_{\varphi r} & \Gamma^\varphi_{2r} \end{bmatrix}$$

$$\left[\Gamma^i{}_{j\varphi}\right] = \left[\Gamma^i{}_{j2}\right] = \left(\frac{\partial}{\partial \varphi}J^T\right)J^{T^{-1}}$$

$$= \underbrace{\begin{bmatrix} -\sin(\varphi) & \cos(\varphi) \\ -r\cos(\varphi) & -r\sin(\varphi) \end{bmatrix}}_{=\frac{\partial}{\partial \varphi}J^T} \begin{bmatrix} \cos(\varphi) & -\dfrac{\sin(\varphi)}{r} \\[2mm] \sin(\varphi) & \dfrac{\cos(\varphi)}{r} \end{bmatrix}$$

$$= \begin{bmatrix} 0 & 1/r \\ -r & 0 \end{bmatrix} = \begin{bmatrix} \Gamma^1_{12} & \Gamma^2_{12} \\ \Gamma^1_{22} & \Gamma^2_{22} \end{bmatrix} = \begin{bmatrix} \Gamma^r_{r\varphi} & \Gamma^\varphi_{r\varphi} \\ \Gamma^r_{\varphi\varphi} & \Gamma^\varphi_{\varphi\varphi} \end{bmatrix}$$

So, the only non-zero Christoffel symbols are:

$$\Gamma^2_{12} = 1/r$$

$$\Gamma^2_{21} = 1/r$$

$$\Gamma^1_{22} = -r$$

That is what we already calculated earlier.

∎

Using $(I.5.015)$, you should be aware of two weak points:

- The Jacobi-matrix needs to be quadratic, otherwise you cannot calculate the inverse. We will see such an example in the exercises.
- Calculating the inverse of a matrix can be a real challenge, as the Jacobi-matrix, in general, is not a diagonal matrix. <u>However</u>: if the Jacobi-matrix of the curvilinear basis functions is orthogonal, then we have:

$$JJ^T = J^TJ = I, \text{ so } J^{T^{-1}} = J$$

In that case, you do not need to calculate the inverse matrix and $(I.5.2.18)$ can be written as

$$(I.5.016)\quad \left[\Gamma^i{}_{jl}\right] = \left(\frac{\partial}{\partial x^l}J^T\right)J \quad \textbf{[only if } J \textbf{ is orthogonal !!]}$$

Let us stop here for a moment: under which conditions is the Jacobi-matrix orthogonal? To ensure the orthogonality of the Jacobi-matrix, the derivatives of the functions (x^i) according to the coordinates would have to be chosen so that the derivatives are **pairwise orthogonal** to each other, and the metric of the coordinate system is a **diagonal** matrix. In other words, the functions would have to be chosen such that the basis vectors of the coordinate system are orthogonal, and the length of the basis vectors is the same at every point.

Last not least, there exists still another way to calculate the Christoffel symbols of the 2nd kind, which we will state without proof:

$$(\boldsymbol{I.5.017}) \quad \Gamma^m_{ij} = \frac{\partial^2 x^l}{\partial \bar{x}^i \partial \bar{x}^j} \frac{\partial \bar{x}^m}{\partial x^l}$$

Once more, we take polar coordinates as an example on how to apply the equation above:

$$x^1 = x^1(r, \varphi) = r\cos(\varphi)$$

$$x^2 = x^2(r, \varphi) = r\sin(\varphi)$$

First, let us expand $(I.5.017)$ to our example:

$$\Gamma^m_{ij} = \frac{\partial^2(r\cos(\varphi))}{\partial \bar{x}^i \partial \bar{x}^j} \frac{\partial \bar{x}^m}{\partial r} + \frac{\partial^2(r\sin(\varphi))}{\partial \bar{x}^i \partial \bar{x}^j} \frac{\partial \bar{x}^m}{\partial \varphi}, \quad m, i, j \in \{r, \varphi\}$$

In the first step, we calculate $\dfrac{\partial^2 x^1}{\partial \bar{x}^i \partial \bar{x}^j}$ and $\dfrac{\partial^2 x^2}{\partial \bar{x}^i \partial \bar{x}^j}$ for $i, j \in \{r, \varphi\}$:

$$\frac{\partial^2(r\cos(\varphi))}{\partial r \partial r} = 0$$

$$\frac{\partial^2(r\cos(\varphi))}{\partial r \partial \varphi} = \frac{\partial^2(r\cos(\varphi))}{\partial \varphi \partial r} = -\sin(\varphi)$$

$$\frac{\partial^2(r\cos(\varphi))}{\partial \varphi \partial \varphi} = -r\cos(\varphi)$$

$$\frac{\partial^2(r\sin(\varphi))}{\partial r \partial r} = 0$$

$$\frac{\partial^2(r\sin(\varphi))}{\partial r \partial \varphi} = \frac{\partial^2(r\sin(\varphi))}{\partial \varphi \partial r} = \cos(\varphi)$$

$$\frac{\partial^2(r\sin(\varphi))}{\partial \varphi \partial \varphi} = -r\sin(\varphi)$$

Now, what about $\dfrac{\partial \bar{x}^m}{\partial x^1}$ and $\dfrac{\partial \bar{x}^m}{\partial x^2}$? These are the inverse functions of x^1 and x^2. If you know these functions: fine for you, but quite often you do not know them. You can easily deal with the situation that you do not know the $\bar{x}^m$-functions: you only need to know their derivatives, **so take the inverse of the Jacobi-matrix!** In this case, however, the Jacobi-matrix must be a square matrix, otherwise the inverse does not exist.

We have:

$$J_{f(r,\varphi)} = \begin{bmatrix} \cos(\varphi) & -r\sin(\varphi) \\ \sin(\varphi) & r\cos(\varphi) \end{bmatrix}$$

with its inverse

$$J^{-1}{}_{f(r,\varphi)} = \begin{bmatrix} \cos(\varphi) & \sin(\varphi) \\ -\dfrac{\sin(\varphi)}{r} & \dfrac{\cos(\varphi)}{r} \end{bmatrix}$$

Now, we can calculate:

$$\Gamma^r_{rr} = \underbrace{\frac{\partial^2(r\cos(\varphi))}{\partial r \partial r}}_{=0}\frac{\partial \bar{x}^m}{\partial r} + \underbrace{\frac{\partial^2(r\sin(\varphi))}{\partial r \partial r}}_{=0}\frac{\partial \bar{x}^m}{\partial \varphi} = 0$$

$$\Gamma^r_{r\varphi} = \underbrace{\frac{\partial^2(r\cos(\varphi))}{\partial r \partial \varphi}}_{=-\sin(\varphi)}\underbrace{\frac{\partial \bar{x}^r}{\partial r}}_{=\cos(\varphi)} + \underbrace{\frac{\partial^2(r\sin(\varphi))}{\partial r \partial \varphi}}_{=\cos(\varphi)}\underbrace{\frac{\partial \bar{x}^r}{\partial \varphi}}_{=\sin(\varphi)} = 0$$

$$\Gamma^\varphi_{rr} = \underbrace{\frac{\partial^2(r\cos(\varphi))}{\partial r \partial r}}_{=0}\frac{\partial \bar{x}^\varphi}{\partial r} + \underbrace{\frac{\partial^2(r\sin(\varphi))}{\partial r \partial r}}_{=0}\frac{\partial \bar{x}^\varphi}{\partial \varphi} = 0$$

$$\Gamma^r_{\varphi\varphi} = \underbrace{\underbrace{\frac{\partial^2(r\cos(\varphi))}{\partial \varphi \partial \varphi}}_{=-r\cos(\varphi)}\underbrace{\frac{\partial \bar{x}^r}{\partial r}}_{=\cos(\varphi)}}_{=-r\cos^2(\varphi)} + \underbrace{\underbrace{\frac{\partial^2(r\sin(\varphi))}{\partial \varphi \partial \varphi}}_{-r\sin(\varphi)}\underbrace{\frac{\partial \bar{x}^r}{\partial \varphi}}_{=\sin(\varphi)}}_{=-r\sin^2(\varphi)} = -r$$

$$= -r\left(\underbrace{\cos^2(\varphi)+\sin^2(\varphi)}_{=1}\right)$$

$$\Gamma^\varphi_{r\varphi} = \underbrace{\underbrace{\frac{\partial^2(r\cos(\varphi))}{\partial r \partial \varphi}}_{=-\sin(\varphi)}\underbrace{\frac{\partial \bar{x}^\varphi}{\partial r}}_{=-\frac{\sin(\varphi)}{r}}}_{=\frac{\sin^2(\varphi)}{r}} + \underbrace{\underbrace{\frac{\partial^2(r\sin(\varphi))}{\partial r \partial \varphi}}_{=\cos(\varphi)}\underbrace{\frac{\partial \bar{x}^\varphi}{\partial \varphi}}_{=\frac{\cos(\varphi)}{r}}}_{=\frac{\cos^2(\varphi)}{r}} = \frac{1}{r}$$

$$\Gamma^\varphi_{\varphi\varphi} = \underbrace{\underbrace{\frac{\partial^2(r\cos(\varphi))}{\partial \varphi \partial \varphi}}_{=-r\cos(\varphi)}\underbrace{\frac{\partial \bar{x}^\varphi}{\partial r}}_{=-\frac{\sin(\varphi)}{r}}}_{=\cos(\varphi)\sin(\varphi)} + \underbrace{\underbrace{\frac{\partial^2(r\sin(\varphi))}{\partial \varphi \partial \varphi}}_{=-r\sin(\varphi)}\underbrace{\frac{\partial \bar{x}^\varphi}{\partial \varphi}}_{=\frac{\cos(\varphi)}{r}}}_{=-\cos(\varphi)\sin(\varphi)} = 0$$

Anyway, if you have to build up the inverse of the Jacobi-matrix, you should better use $(I.5.015)$ for the calculation.

I.5.3 Transformation law

The Christoffel symbols of the 2nd kind transform as follows:

$(I.5.018)$ Transformation law for $\bar{\Gamma}^i_{jk}$

$$\bar{\Gamma}^s_{lm} = \Gamma^p_{ij}\,\frac{\partial \bar{x}^s}{\partial x^p}\,\frac{\partial x^i}{\partial \bar{x}^l}\,\frac{\partial x^j}{\partial \bar{x}^m} + \frac{\partial^2 x^j}{\partial \bar{x}^l \partial \bar{x}^m}\,\frac{\partial \bar{x}^s}{\partial x^j}$$

We want to draw attention to the second term on the right side: it contains partial derivatives of the second degree. These derivatives only disappear if the transformation from (x^i) to $(\bar{x}^i)$ is linear.

The verification of the transformation law is long and tedious, but not particularly complicated. Therefore, the equation is not derived here.

As a forecast to the next chapter, the transformation law can be interpreted in the way that, in general, **Christoffel symbols are NOT tensors**.

When do you use this transformation law? You use it when you make a change of coordinate system – the transformation law enables you to calculate the new Christoffel symbols. In addition, by transforming the Christoffel symbols, various geometric properties of a curved space can be analyzed, including its curvature and topological structure.

To get an important equation, we need to rewrite $(I.5.018)$

$$\frac{\partial^2 x^j}{\partial \bar{x}^l \partial \bar{x}^m}\,\frac{\partial \bar{x}^s}{\partial x^j} = \bar{\Gamma}^s_{lm} - \Gamma^p_{ij}\,\frac{\partial \bar{x}^s}{\partial x^p}\,\frac{\partial x^i}{\partial \bar{x}^l}\,\frac{\partial x^j}{\partial \bar{x}^m}$$

Now, we perform an inner multiplication with $\dfrac{\partial x^r}{\partial \bar{x}^s}$:

$$\frac{\partial^2 x^j}{\partial \bar{x}^l \partial \bar{x}^m}\,\frac{\partial \bar{x}^s}{\partial x^j}\frac{\partial x^r}{\partial \bar{x}^s} = \bar{\Gamma}^s_{lm}\,\frac{\partial x^r}{\partial \bar{x}^s} - \Gamma^p_{ij}\,\frac{\partial \bar{x}^s}{\partial x^p}\,\frac{\partial x^i}{\partial \bar{x}^l}\,\frac{\partial x^j}{\partial \bar{x}^m}\frac{\partial x^r}{\partial \bar{x}^s}$$

We note that $\dfrac{\partial \bar{x}^s}{\partial x^j}\dfrac{\partial x^r}{\partial \bar{x}^s} = \dfrac{\partial x^r}{\partial x^j}$ by chain rule. Now $\dfrac{\partial x^r}{\partial x^j} = \delta_{rj}$, so in the sum only those elements survive where $r = j$. We get:

$$\frac{\partial^2 x^j}{\partial \bar{x}^l \partial \bar{x}^m} \frac{\partial \bar{x}^s}{\partial x^j} \frac{\partial x^r}{\partial \bar{x}^s} = \frac{\partial^2 x^j}{\partial \bar{x}^l \partial \bar{x}^m} \delta_{rj} = \frac{\partial^2 x^j}{\partial \bar{x}^l \partial \bar{x}^m}$$

With the same argumentation, we get on the right side:

$$\Gamma_{ij}^p \frac{\partial \bar{x}^s}{\partial x^p} \frac{\partial x^i}{\partial \bar{x}^l} \frac{\partial x^j}{\partial \bar{x}^m} \frac{\partial x^r}{\partial \bar{x}^s} = \Gamma_{ij}^p \frac{\partial x^i}{\partial \bar{x}^l} \frac{\partial x^j}{\partial \bar{x}^m} \delta_{rp} = \Gamma_{ij}^r \frac{\partial x^i}{\partial \bar{x}^l} \frac{\partial x^j}{\partial \bar{x}^m}$$

Finally, putting all together, we have the following identity:

$$(I.5.019) \qquad \frac{\partial^2 x^j}{\partial \bar{x}^l \partial \bar{x}^m} = \bar{\Gamma}_{lm}^s \frac{\partial x^r}{\partial \bar{x}^s} - \Gamma_{ij}^r \frac{\partial x^i}{\partial \bar{x}^l} \frac{\partial x^j}{\partial \bar{x}^m}$$

I.5.E Exercises

(**I.5.E.1**) Given the set of functions for *helix coordinate system* (see $(I.A.2.7)$)

$$x^1(\rho, \psi, \varsigma) = \rho \cos (\psi + a\varsigma)$$

$$x^2(\rho, \psi, \varsigma) = \rho \sin (\psi + a\varsigma)$$

$$x^3(\rho, \psi, \varsigma) = \varsigma \quad \text{with } \rho \geq 0,\ \psi \in [0, 2\pi],\ -\infty < \varsigma < \infty$$

Calculate the Christoffel symbols of the 2nd kind by using

1. ($I.5.015$)

2. ($I.5.008$)

3. ($I.5.013$)

(**I.5.E.2**) Calculate the Christoffel symbols of the 2nd kind for spherical coordinates using ($I.5.2.6$)

(**I.5.E.3**) Given the squared line element $ds^2 = dx^2 + e^{2x}\, dy^2$, compute the Christoffel symbols of the 2nd kind.

I.6 Tensors

A tensor is a mathematical object that generalizes scalars, vectors, and matrices to higher-dimensional spaces. Tensors are used to describe various physical quantities, including quantities that have both magnitude and direction in multiple dimensions. They are fundamental in the field of theoretical physics, particularly in areas like relativity, electromagnetism, and fluid dynamics.

Generalisation from scalars to tensors

1. Scalar:
    - A scalar is a single value representing magnitude without direction.
2. Vector:
    - A vector generalizes a scalar. It consists of multiple scalars and has both magnitude and direction.
    - A vector can be seen as a one-dimensional array of scalars.
    - A one-component vector has no direction, only a value.
3. Matrix:
    - A matrix generalizes a vector. It is a two-dimensional array of scalars.
    - A matrix with a single column (or row) can be considered a vector.
    - Matrices are used to represent linear transformations and systems of linear equations.
4. Tensor:
    - A tensor generalizes matrices to higher dimensions. It can have n dimensions, where n is any positive integer.
    - A tensor with two dimensions (rank-2 tensor) is equivalent to a matrix.
    - A tensor with three dimensions can be visualized as a cube of matrices, each layer being a matrix.
    - Example: A_{ij}^{k} represents a tensor with three dimensions where k indicates the layer, and i and j represent row and column indices within each layer.

Here is a visual representation of a rank-3 tensor with three layers of $2x2$ matrices:

$$A_{ij}^k = \left(\underbrace{\begin{pmatrix} a_{11}^1 & a_{12}^1 \\ a_{21}^1 & a_{22}^1 \end{pmatrix}}_{k=1} \; \underbrace{\begin{pmatrix} a_{11}^2 & a_{12}^2 \\ a_{21}^2 & a_{22}^2 \end{pmatrix}}_{k=2} \; \underbrace{\begin{pmatrix} a_{11}^3 & a_{12}^3 \\ a_{21}^3 & a_{22}^3 \end{pmatrix}}_{k=3} \right)$$

Key Points about Tensors:

- Components: A tensor can have components that are scalars, vectors, matrices, or even other tensors.
- Rank: The rank of a tensor is the number of dimensions or indices it has. For example, a scalar is a rank-0 tensor, a vector is a rank-1 tensor, a matrix is a rank-2 tensor, and so on.
- Operations: Tensors follow specific rules for operations like addition, multiplication, and contraction, which are extensions of the corresponding operations for scalars, vectors, and matrices.
- Importance in Physics:
    - Coordinate Independence: Physical laws expressed as tensor equations remain valid across all coordinate systems. This invariance makes tensors extremely powerful for describing physical phenomena in a way that is independent of the choice of coordinates.
    - Applications: Tensors are used to describe stress, strain, and other properties in materials science, the curvature of spacetime in general relativity, and electromagnetic fields, among other things.
    - Definitions of Tensors: In mathematics, a tensor is often defined as a multi-linear map that takes several vectors and returns a scalar. - In physics, tensors are defined based on their transformation properties under changes of coordinates.

Now we will present several definitions of a tensor. We will begin with a definition used in (pure) mathematics, then we have a look at different definitions in physics.

In pure mathematics, a tensor is nothing else than a multilinear mapping: let V be a n-dimensional vector space over a field K, with $K = \mathbb{R}$ or $K = \mathbb{C}$, then a mapping $\mathfrak{I}: V \times V \times \ldots \times V \to W$, where W is also a vector space, is called **multilinear**, if the mapping is linear in all its components. We give some simple examples of multilinear mappings:

- Let $\mathfrak{I}: \mathbb{R} \times \mathbb{R} \to \mathbb{R}$ be defined as $\mathfrak{I}(v, w) = vw$

  This is simply the product of two numbers. In each argument, $\mathfrak{I}$ is linear, so $\mathfrak{I}$ defines a 2-multilinear (in this case: bilinear) mapping

- Let $\mathfrak{I}: \mathbb{R}^n \times \mathbb{R}^n \to \mathbb{R}$ be defined as $\mathfrak{I}(v, w) = v \cdot w$

  In this example, v and w are n-dimensional vectors. $\mathfrak{I}$ maps two vectors to their scalar product.

The tensor-definition in physics is a bit different, as we will see now:

I.6.1 Definition of tensors

We will provide different definitions of tensors. It will turn out that these definitions are equivalent.

Let V be a n-dimensional vector space over a field K, with $K = \mathbb{R}$ or $\mathbb{C}$, and V^* the associated dual vector space to V. Let $r, s \in \mathbb{N}_0$.

$(I.6.001)$ A **tensor** T_r^s of order (r, s) is a multilinear mapping

$$T_s^r : \underbrace{V^* \times V^* \times \ldots \times V^*}_{s \text{ times}} \times \underbrace{V \times V \times \ldots \times V}_{r \text{ times}} \to K, \text{ where } \text{„}\times\text{“ denotes the}$$
cartesian product.

Just to remind: the cartesian product of $\underbrace{V \times V \times \ldots \times V}_{r \text{ times}}$ consists of all tuples

$(v_1, v_2, \ldots, v_r)$ with $v_i \in V$.

T_s^r maps s covariant vectors and r contravariant vectors to a (real) number. The number s is called the **covariant order**, r is called the **contravariant order**. We say, T_s^r is a tensor of **order** (r, s). Furhtermore, $(r + s)$ is called the **rank** of the tensor.

First, note the difference between what mathematicians understand by tensors and how physicians define what a tensor is.

The set $\mathcal{T}_s^r(V) := \{\, T_s^r : is\ a\ tensor\ of\ rank\ (r,s)\,\}$ of all tensors of rank (r,s) is a vector space of dimension n^{s+r}, using the usual addition and scalar multiplication.

Without proof, we state:

Let $\{\, v_1, v_2, \ldots, v_n \,\}$ be a basis of V and $\{\, v^1, v^2, \ldots, v^n \,\}$ the corresponding dual basis. Then the collection

$$\{\, v_{i_1} \otimes v_{i_2} \otimes \ldots \otimes v_{i_s} \otimes v^{j_1} \otimes v^{j_2} \otimes \ldots \otimes v^{j_r} \,\} \quad \text{with}$$

$$1 \le i_1, \ldots, i_s, j_{1,} \ldots, j_r \le n$$

form a basis for $\mathcal{T}_r^s(V)$.

A tensor T_0^0 is defined to be a scalar, so $\mathcal{T}_0^0(V) = K$.

A tensor T_1^0 is called a covariant vector or covector, T_0^1 a contravariant vector.

We continue to use $\{\, v_1, v_2, \ldots, v_n \,\}$ as a basis and $\{\, v^1, v^2, \ldots, v^n \,\}$ as the corresponding dual basis. Now, we take another basis $\{\, \bar{v}_1, \bar{v}_2, \ldots, \bar{v}_n \,\}$ with the corresponding dual basis $\{\bar{v}^i\}$, as we want to investigate the transformation behavior. Then there exists a nonsingular transformation matrix $A = \left[a^j{}_i\right]$, so that $\bar{v}_i$ can be expressed in the old basis:

(1) $\bar{v}_i = v_j\, a^j{}_i$ with some (still unknown) coefficients $a^j{}_i$.

Now, we calculate the scalar product $v^j \cdot \bar{v}_k$:

$$
\begin{aligned}
(2)\ \ v^i \cdot \bar{v}_k \ \ &= \ v^i \cdot v_j\, a^j{}_k && (1) \\[4pt]
&= \ a^j{}_k\, v^i \cdot v_j && \text{scalar product is bilinear} \\[4pt]
&= \ a^j{}_k\, \delta^i{}_j && v^i \cdot v_j = \delta^i{}_j,\ \text{definition of dual vector} \\[2pt]
&= \ a^i{}_k && \text{exchange of indices}
\end{aligned}
$$

So, we can calculate the coefficients a^i_k by using the scalar product between the dual basis elements (covectors), and the new basis.

Now, let $B = \left[b^i_{\ j}\right]$ the inverse of matrix $A = \left[a^i_{\ j}\right]$. In ESN notation this means:

(3) $a^i_{\ j}\, b^j_{\ k} = \delta^j_{\ k}$ and $b^i_{\ j}\, a^j_{\ k} = \delta^j_{\ k}$

We multiply (2) with $b^j_{\ i}$:

(4) $b^j_{\ i}\, v^i \cdot \bar{v}_k = b^j_{\ i}\, a^i_{\ k} = \delta^j_{\ k}$

But as $\{\bar{v}^i\}$ is dual to $\{\bar{v}_i\}$, we have also $\bar{v}^j \cdot \bar{v}_k = \delta^j_{\ k}$. Comparing this result with (4) means:

(5) $\bar{v}^j = b^j_{\ i}\, v^i$

Compare (5) with (1): in (1) the basis vectors transform covariant, in (5) contravariant!

We now draw our attention to the transformation behavior of tensors. Therefore, we take a tensor $T \in \mathcal{T}^2_1$. We adopt the designations from the previous considerations.

We get:

(6) $T^{ij}_{\ \ k} = T(\,v^i, v^j, v_k\,)$

(7) $\bar{T}^{pq}_{\ \ r} = T(\,\bar{v}^p, \bar{v}^q, \bar{v}_r\,)$

Therefore, using (5) and (1), we get:

$$
\begin{aligned}
(8)\ \bar{T}^{pq}_{\ \ r} \quad &= \quad T(\,\bar{v}^p, \bar{v}^q, \bar{v}_r\,) && (7)\\
&= \quad T(\,b^p_{\ i}\, v^i, b^q_{\ j}\, v^j, a^k_{\ r}\, v_r\,) && (5), (1)\\
&= \quad b^p_{\ i}\, b^q_{\ j}\, a^k_{\ r}\, T(\,v^i, v^j, v_k\,) && T \text{ multilinear}\\
&= \quad b^p_{\ i}\, b^q_{\ j}\, a^k_{\ r}\, T^{ij}_{\ \ k} && (6)
\end{aligned}
$$

What you see here is the **classical law of transformation**. Coming back to the **general law of coordinate transformations** (I.3.7), we can reformulate (8) as:

$$(I.6.002) \quad \bar{T}_r^{pq} = \frac{\partial \bar{x}^p}{\partial x^i} \frac{\partial \bar{x}^q}{\partial x^j} \frac{\partial x^k}{\partial \bar{x}^r} \; T_k^{ij}$$

This is the transformation behavior of a (2,1)-tensor. Once more, we point out that the expressions $\frac{\partial \bar{x}^i}{\partial x^j}$ and $\frac{\partial x^i}{\partial \bar{x}^j}$ are the Jacobi matrices and their inverse.

Now, let us take a tensor $T \in \mathcal{T}_0^1$: then we get:

$$\bar{T}_r = \frac{\partial \bar{x}^k}{\partial x^r} \; T_k$$

These are the equations we wanted to have: beginning with the definition of a tensor as a multilinear mapping, we got a more practical way to define tensors:

$(I.6.003)$ **Definition (contravariant/covariant tensor of order 1)**

Let $V = (V^1, \dots, V^n)$ be a n-dimensional vector space over $\mathbb{R}$, x and $\bar{x}$ two coordinate systems which are linked by an admissible coordinate transformation

$$\mathcal{T}: \bar{x} = \bar{x}(x^1, x^2, \dots, x^n)$$

$V^i = V^i(x)$ a scalar field, let V^i be expressible as n real-valued functions

$T^1, T^2, \dots, T^n$ in the (x^i)-system,

$\bar{T}^1, \bar{T}^2, \dots, \bar{T}^n$ in the $(\bar{x}^i)$-system.

The vector field V is a **contravariant tensor of order one** if its components (T^i) and $(\bar{T}^i)$ relative to the respective coordinate system (x^i) and $(\bar{x}^i)$ obey the law of transformation:

$$\bar{T}^i = T^r \frac{\partial \bar{x}^r}{\partial x^i}$$

The vector field V is a **covariant tensor of order 1**, if (x^i) and $(\bar{x}^i)$ obey the law of transformation:

$$\bar{T}^i = T^r \frac{\partial x^r}{\partial \bar{x}^i}$$

Now, let us take a contravariant tensor T of order 2. This tensor takes two upper indices: $T = T^{ij}$. The transformation law then is:

$$(I.6.004) \quad \bar{T}^{ij} = T^{rk} \frac{\partial \bar{x}^r}{\partial x^i} \frac{\partial \bar{x}^k}{\partial x^j}$$

We can also define mixed tensors: take $T = T^i{}_j$, which is a tensor of rank two and has one contravariant and one covariant part. The tensor law now reads as:

$$(I.6.005) \quad \bar{T}^i{}_j = T^r{}_k \frac{\partial \bar{x}^r}{\partial x^i} \frac{\partial x^j}{\partial \bar{x}^k}$$

The point behind is: we can define a tensor by its behavior to coordinate change.

As a generalization, we can define an arbitrary tensor as follows:

$(I.6.006)$ **Definition (tensor of rank (r, s)) – transformation law**

The generalized vector field V is called **tensor of rank** $m = p + q$, contravariant of order p, covariant of order q, if it's components $T^{i_1\, i_2\, \ldots\, i_p}_{j_1\, j_2 \ldots\, j_q}$ in (x^i) and $\bar{T}^{i_1\, i_2\, \ldots\, i_p}_{j_1\, j_2 \ldots\, j_q}$ in $(\bar{x}^i)$ obey the law of transformation

$$\bar{T}^{i_1\, i_2\, \ldots\, i_p}_{j_1\, j_2 \ldots\, j_q} = T^{r_1\, r_2\, \ldots\, r_r}_{s_1\, s_2\, \ldots\, s_q} \frac{\partial \bar{x}^{i_1}}{\partial x^{r_1}} \frac{\partial \bar{x}^{i_2}}{\partial x^{r_2}} \cdots \frac{\partial \bar{x}^{i_p}}{\partial x^{r_p}} \frac{\partial x^{s_1}}{\partial \bar{x}^{j_1}} \frac{\partial x^{s_2}}{\partial \bar{x}^{j_2}} \cdots \frac{\partial x^{s_q}}{\partial \bar{x}^{j_q}}$$

Note, that r_i and s_i are all summation indices, as they all appear exactly two times on the right side.

We will refer $(I.6.1.6)$ also as **the transformation law for tensors**.

A hint: if you see something like this $T^{ij}{}_k$, then this can be a tensor OR the coefficients of a tensor. $T^{ij}{}_k + C$, where C is a constant, makes only sense if $T^{ij}{}_k$ is considered to be a coefficient of the tensor, as you cannot add a constant to a tensor itself.

I.6.1.1 How to write down the transformation law

Writing down the transformation law for a specific tensor can be a challenge, due to the number of indices you have to use. There is a practical path you can follow. As an example, we want to write down the transformation law of the tensor $T^i{}_j{}^k$:

(1) Write the symbol of the transformed tensor $\bar{T}$ on the left-hand side of the transformation equation and T on the right-hand side:

$$\bar{T} = T$$

(2) Add now the indices of T:
$$\bar{T} = T^i{}_j{}^k$$

(3) Do the same on the left side for $\bar{T}$, but take different name of the indices, observing the right positions:
$$\bar{T}^l{}_m{}^n = T^i{}_j{}^k$$

(4) For each upper index of $\bar{T}^l{}_m{}^n$ add on the right side the term $\left(\dfrac{\partial \bar{x}^p}{\partial x^q}\right)$

where p is the name of the index and q is the name of the index on the right side at the same position as p. In our case this is
$$\left(\frac{\partial \bar{x}^l}{\partial x^i}\right)$$

For each lower index on the left side, add $\left(\dfrac{\partial x^q}{\partial \bar{x}^p}\right)$. Again, p is the name of the lower index on the left side, q is the name of the index on the right side at the same position as p.
So, we get in our example:
$$\left(\frac{\partial x^j}{\partial \bar{x}^m}\right)$$

Do this with every index on the left side and add these terms on the right side:
$$\bar{T}^l{}_m{}^n = \left(\frac{\partial \bar{x}^l}{\partial x^i}\right)\left(\frac{\partial x^j}{\partial \bar{x}^m}\right)\left(\frac{\partial \bar{x}^n}{\partial x^k}\right) T^i{}_j{}^k$$

You are done!

I.6.2 Examples of tensors, construction of tensors

I.6.2.1 Zero-order tensors

Zero-order tensors are defined as scalars. The length of a vector is considered as a 0-tensor, as its length is independent from coordinate changes. Temperature, mass, energy, physical constants as c for the speed of light …

I.6.2.2 First-order tensors

First-order tensors can be considered as vectors: they are contravariant or covariant.

The following theorem shows one of the major points in tensor calculus:

$(I.6.007)$ **Theorem**

Let S^i be the components of a contravariant tensor, T_i the components of a covariant vector. If the inner product $E = S^r T_r$ is defined in every coordinate system, then E is an invariant.

Proof:

$$\bar{E} \quad = \quad \bar{S}^i \bar{T}_i$$

$$= \quad S^r \frac{\partial \bar{x}^i}{\partial x^r} T^s \frac{\partial x^s}{\partial \bar{x}^i} \qquad\qquad (I.6.003)$$

$$= \quad S^r T^s \frac{\partial \bar{x}^i}{\partial x^r} \frac{\partial x^s}{\partial \bar{x}^i} \qquad\qquad \text{reorder elements}$$

$$= \quad S^r T^s \frac{\partial x^s}{\partial x^r} \qquad\qquad \text{chain rule } (I.2.060)$$

$$= \quad S^r T^s \, \delta^s_r \qquad\qquad (I.2.045)$$

$$= \quad S^r T^r \qquad\qquad (I.2.022)$$

$$= \ E$$

■

Think of the general scalar product: In $(I.4.009)$ we defined the scalar product as

$$\vec{v} \cdot \vec{w} \ = \ g_{ij} \, v^i \, w^j \ = \ v_i \, w^i \ = \ v^i \, w_i,$$

using the metric tensor g_{ij}. What happens here is that the multiplication of g_{ij} with a contravariant vector v^i returns v_i, so the covariant vector with respect to the metric tensor $(I.4.009)$. Applying the theorem $(I.6.007)$, we conclude that the generalized scalar product is invariant in all coordinate systems.

What happens when we multiply two contravariant vectors?

$(I.6.008)$ **Theorem**:

Let S^i and T^i be the components of a contravariant vectors in $\mathbb{R}^n$. Then the matrix $[U^{ij}] = [S^i \, T^j]$ is a contravariant tensor of order 2.

Proof

As S^i and T^j are contravariant tensors, they fulfill the equations

$$\bar{S}^i \ = \ S^r \, \frac{\partial \bar{x}^i}{\partial x^r} \ \text{and} \ \bar{T}^j \ = \ T^s \, \frac{\partial \bar{x}^j}{\partial x^s}$$

Now, we multiply them:

$$\bar{U}^{ij} := \ \bar{S}^i \, \bar{T}^j \ = \ S^r \, \frac{\partial \bar{x}^i}{\partial x^r} \, T^s \, \frac{\partial \bar{x}^j}{\partial x^s} \ = \ (S^r T^s) \, \frac{\partial \bar{x}^i}{\partial x^r} \, \frac{\partial \bar{x}^j}{\partial x^s} \ = \ U^{rs} \, \frac{\partial \bar{x}^i}{\partial x^r} \, \frac{\partial \bar{x}^j}{\partial x^s}$$

This is just the tensor law for a covariant tensor of order 2 $(I.6.004)$.

■

It should be evident that the multiplication of two covariant tensors of order 1 results in a covariant tensor:

$(I.6.009)$ **Theorem**:

Let $\lambda, \mu \in \mathbb{R}$, both invariants. Let furthermore S^i and T^i contravariant vectors.

Then $\lambda S^i + \mu T^i$ is a covariant vector.

Proof:

As S^i and T^i are contravariant, we have the transformation law

$$\bar{S}^i = S^r \frac{\partial \bar{x}^i}{\partial x^r} \quad \text{and} \quad \bar{T}^i = T^r \frac{\partial \bar{x}^i}{\partial x^r}.$$

Now, let us consider $\bar{\lambda} \bar{S}^i + \bar{\mu} \bar{T}^i$. First, $\lambda = \bar{\lambda}$ and $\mu = \bar{\mu}$, as they are presumed to be invariants. So, we get:

$$\bar{\lambda} \bar{S}^i + \bar{\mu} \bar{T}^i \quad = \quad \lambda \bar{S}^i + \mu \bar{T}^i$$

$$= \quad \lambda S^r \frac{\partial \bar{x}^i}{\partial x^r} + \mu T^r \frac{\partial \bar{x}^i}{\partial x^r}$$

$$= \quad (\lambda S^r + \mu T^r) \frac{\partial \bar{x}^i}{\partial x^r}$$

Which is just the transformation law for a contravariant vector $(I.6.003)$.

■

It should be no surprise that the theorem works also for contravariant vectors so that the result is a contravariant vector.

I.6.2.2.1 Tangent field as a contravariant vector

Let $\mathcal{C}$ be a parameterized curve in the (x^i)-system, $x^i = x^i(t)$. The tangent field $T = T^i$ is then defined by the usual differentiation formula

$$T^i = \frac{dx^i}{dt}$$

Now, consider another system $(\bar{x}^i)$, so that:

$$\bar{x}^i = \bar{x}^i(\, x^1(t),\ x^2(t), \ldots, x^n(t)\,)$$

The tangent field $\bar{T} = \bar{T}^i$ in this new system is given by:

$$\bar{T}^i = \frac{d\bar{x}^i}{dt}$$

As $\bar{x}^i$ is dependent from x^i, we use the multivariate chain rule to obtain:

$$\bar{T}^i = \frac{d\bar{x}^i}{dt} = \frac{\partial \bar{x}^i}{\partial x^r}\frac{dx^r}{dt} = \frac{\partial \bar{x}^i}{\partial x^r} T^r$$

Therefore, the tangent field transforms as

$$\bar{T}^i = \frac{\partial \bar{x}^i}{\partial x^r} T^r$$

This shows that the tangent field is a contravariant tensor of order 1.

I.6.2.2.2 Gradient of a scalar field as covariant tensor

Let f be a differentiable scalar field on a (x^i)-system. Then we can define the gradient ∇f as

$$\nabla f = \left(\frac{\partial f}{\partial x^1}, \frac{\partial f}{\partial x^2}, \ldots, \frac{\partial f}{\partial x^n} \right)$$

We define $T_i = \dfrac{\partial f}{\partial x^i}$

Now consider another system $(\bar{x}^i)$ with $\bar{x}^i = \bar{x}^i(\, x^1,\ x^2, \ldots, x^n\,)$

In $(\bar{x}^i)$, the gradient is given by

$$\nabla \bar{f} = \left(\frac{\partial f}{\partial \bar{x}^1}, \frac{\partial f}{\partial \bar{x}^2}, \ldots, \frac{\partial f}{\partial \bar{x}^n} \right)$$

We define $\bar{T}_i = \dfrac{\partial f}{\partial \bar{x}^i}$

As $\bar{x}^i$ is dependent from x^i, we get with the multivariate chain rule:

$$\bar{T}_i = \frac{\partial f}{\partial \bar{x}^i} = \frac{\partial f}{\partial x^r}\frac{\partial x^r}{\partial \bar{x}^i} = T_i \frac{\partial x^r}{\partial \bar{x}^i}$$

So, the gradient is a covariant tensor of rank 1.

I.6.2.2.3 Example in $\mathbb{R}^2$

Let $T = (T^i)$ be a contravariant vector in $\mathbb{R}^2$. Let (x^i) be a system with $(T^i) = (x^2, x^1)$, so

(1) $T^1 = x^2, \ \ T^2 = x^1$

Define another system $(\bar{x}^i)$ by:

(2) $\bar{x}^1 = (x^2)^2 \neq 0$

(3) $\bar{x}^2 = x^1 x^2$

We want to calculate $(\bar{T}^i)$ in the $(\bar{x}^i)$-system.

As (T^i) is contravariant, we get $(I.6.003)$:

(4) $\bar{T}^i = T^r \dfrac{\partial \bar{x}^i}{\partial x^r} = T^1 \dfrac{\partial \bar{x}^i}{\partial x^1} + T^2 \dfrac{\partial \bar{x}^i}{\partial x^2}$

We build up the Jacobi-matrix

(5) $J = \begin{bmatrix} \dfrac{\partial \bar{x}^1}{\partial x^1} & \dfrac{\partial \bar{x}^1}{\partial x^2} \\[2ex] \dfrac{\partial \bar{x}^2}{\partial x^1} & \dfrac{\partial \bar{x}^2}{\partial x^2} \end{bmatrix} = \begin{bmatrix} 0 & 2\,x^2 \\ x^2 & x^1 \end{bmatrix}$

Replacing the partial derivates from (5) in (4) yields:

(6) $\bar{T}^1 = T^1 \dfrac{\partial \bar{x}^1}{\partial x^1} + T^2 \dfrac{\partial \bar{x}^1}{\partial x^2} = \underbrace{T^1}_{=\,x^2}\,(0) + \underbrace{T^2}_{=\,x^1}\,(2\,x^2) = 2x^1 x^2$

(7) $\bar{T}^2 = T^1 \dfrac{\partial \bar{x}^2}{\partial x^1} + T^2 \dfrac{\partial \bar{x}^2}{\partial x^2} = \underbrace{T^1}_{=\,x^2}\,(x^2) + \underbrace{T^2}_{=\,x^1}\,(x^1) = (x^2)^2 + (x^1)^2$

Looking at (3), we see that

$\bar{T}^1 = 2x^1 x^2 = 2\bar{x}^2.$

Likewise, considering (2), (3) we get:

$$\bar{T}^2 = \underbrace{(x^2)^2}_{=\,\bar{x}^1} + \left(\underbrace{x^1}_{=\,\frac{\bar{x}^2}{x^2}} \right)^2 = \bar{x}^1 + \left(\frac{\bar{x}^2}{x^2} \right)^2 = \bar{x}^1 + \frac{(\bar{x}^2)^2}{\underbrace{(x^2)^2}_{=\,\bar{x}^1}} = \bar{x}^1 + \frac{(\bar{x}^2)^2}{\bar{x}^1}$$

I.6.2.3 Second-order tensors

Second-order tensors are very important in physics. That is why we will focus on these. The advantage is that second-order tensors can be represented as matrices, and we will see that – instead of calculating components via ESN – we can use matrix calculus to work with these tensors.

Examples are:

1. **Stress Tensor in Mechanics**: In solid mechanics, the stress tensor is a second-order tensor that describes the distribution of internal forces within a material. It relates surface forces to areas on a material's surface. The stress tensor includes components such as normal stress and shear stress.
2. **Strain Tensor in Mechanics**: The strain tensor is another second-order tensor in mechanics, describing the deformation of a material in response to applied loads. It relates changes in length and angles between material elements.
3. **Inertia Tensor in Mechanics**: In classical mechanics, the inertia tensor describes how mass is distributed within a rigid body. It is used to calculate moments of inertia and angular momentum.
4. **Dielectric Tensor in Electromagnetism**: In the study of electromagnetic materials, the dielectric tensor describes how the electrical polarization responds to an electric field. It relates the electric displacement field to the electric field.
5. **Permeability Tensor in Electromagnetism**: In the context of electromagnetic materials, the permeability tensor describes how the magnetic field responds to a magnetic field strength. It relates the magnetic induction field to the magnetic field strength.
6. **Diffusion Tensor in Diffusion Processes**: In the field of diffusion, the diffusion tensor characterizes the anisotropic diffusion of particles or substances in a medium. It describes how particles move in different directions due to concentration gradients.

7. **Stress-Strain Tensor in Materials Science**: This tensor combines the stress and strain tensors to relate the stress and strain components in a material. It is commonly used to analyze material behavior under various loads.

8. **Tensor of Intrinsic Curvature in Differential Geometry**: In differential geometry, the Riemann curvature tensor is a second-order tensor that characterizes the intrinsic curvature of a Riemannian manifold. It is used in general relativity to describe the curvature of spacetime.

9. **Second-Order Moment of a Probability Distribution**: In statistics and probability theory, the second-order moment of a probability distribution is represented as a covariance matrix. It describes how two random variables covary.

10. **Diffusion Tensor Imaging (DTI) in Medical Imaging**: In medical imaging, DTI is used to study the diffusion of water molecules in biological tissues, such as the brain. It involves a second-order diffusion tensor that characterizes the diffusion properties in different directions.

These are just a few examples of second-order tensors in various fields.

Second order tensors take two indices. There are three types of tensors:

- Contravariant: T^{ij}

- Covariant: T_{ij}

- Mixed: $T^i{}_j$

Before we have a look at some examples, we need to rewrite the tensor law from ESN into matrix-notation. This will make calculations much easier.

I.6.2.3.1 Rewriting the tensor law

We want to write the tensor law for a tensor of order 2 in matrix notation. We do this for contravariant and covariant tensor:

I.6.2.3.1.1 Tensor law for contravariant tensor of order two

By definition $(I.6.006)$, we have:

$$\bar{T}^{ij} = T^{rs}\frac{\partial \bar{x}^i}{\partial x^r}\frac{\partial \bar{x}^j}{\partial x^s} = \frac{\partial \bar{x}^i}{\partial x^r}\,T^{rs}\frac{\partial \bar{x}^j}{\partial x^s}$$

Define the Jacobi-matrix

$$J := J^i{}_j = \left[\frac{\partial \bar{x}^i}{\partial x^j}\right] = \begin{bmatrix} \dfrac{\partial \bar{x}^1}{\partial x^1} & \dfrac{\partial \bar{x}^1}{\partial x^2} & \cdots & \dfrac{\partial \bar{x}^1}{\partial x^n} \\ \dfrac{\partial \bar{x}^2}{\partial x^1} & \dfrac{\partial \bar{x}^2}{\partial x^2} & \cdots & \dfrac{\partial \bar{x}^2}{\partial x^n} \\ \cdots & \cdots & \cdots & \cdots \\ \dfrac{\partial \bar{x}^n}{\partial x^1} & \dfrac{\partial \bar{x}^n}{\partial x^2} & \cdots & \dfrac{\partial \bar{x}^n}{\partial x^n} \end{bmatrix}$$

Then the tensor transformation law above can be rewritten as

$$(I.6.010)\ \bar{T} = \bar{T}^{ij} = J^i{}_r\, T_{rs}\, J^j{}_s$$

In matrix notation this is

$$(I.6.011)\ \bar{T} = J\,T\,J^T$$

The advantage of rewriting the tensor transformation law as matrices is obvious: it is much easier to work with matrices than to work with ESN. Although ESN implicitly determines the order of operations through the placement of indices, matrix notation makes this order explicit and often simpler to manage.

Take the tensor law from above: $T^{rk}\dfrac{\partial \bar{x}^r}{\partial x^i}\dfrac{\partial \bar{x}^k}{\partial x^j}$. The ESN doesn't tell you directly in which order you have to perform the matrix multiplication.

I.6.2.3.1.2 Tensor law for covariant tensor of order two

By definition $(I.6.006)$ we have:

$$\bar{T}_{ij} = T_{rs}\, \frac{\partial x^r}{\partial \bar{x}^i} \frac{\partial x^s}{\partial \bar{x}^j} = \frac{\partial x^r}{\partial \bar{x}^i}\, T_{rs}\, \frac{\partial x^s}{\partial \bar{x}^j}$$

Defining $\bar{J} := \bar{J}^i_{\ j} = \left[\dfrac{\partial x^i}{\partial \bar{x}^j}\right]$, the equation can be rewritten as

$(\boldsymbol{I.6.012})\ \bar{T} = \bar{T}_{ij} = \bar{J}^r_{\ i}\, T_{rs}\, \bar{J}^s_{\ j}$

In matrix notation this is:

$(\boldsymbol{I.6.013})\ \bar{T} = \bar{J}^T\, T\, \bar{J}$

Let us pause here for a minute: the matrix representation for a second-covariant tensor looks – nearly - the same as the representation of the second-contravariant tensor. *But they differ in their corresponding Jacobi-matrix! They differ also in that the <u>first</u> matrix is the transposed one!*

For the contravariant tensor, we have $J := J_{ij} = \left[\dfrac{\partial \bar{x}^i}{\partial x^j}\right]$, for the covariant tensor we have $\bar{J} := \bar{J}_{ij} = \left[\dfrac{\partial x^i}{\partial \bar{x}^j}\right]$. There is a general link between these two Jacobi-matrices: J is the inverse of $\bar{J}$ and, of course, $\bar{J}$ is the inverse of J.

So, you can rewrite the matrix representation also in the following way:

$(\boldsymbol{I.6.014})\ \bar{T} = J^{-1^T}\, T\, J^{-1}$

It is a matter of preference which approach to follow: calculating the inverse or calculating $\bar{J}$. If you want to directly calculate $\bar{J}$, you might face the problem of having to express (x^i) in terms of $(\bar{x}^i)$. An example are *3D-hypershpere*, defined by:

$x^1 = r \sin \alpha \sin \theta \sin \varphi$

$x^2 = r \sin \alpha \sin \theta \cos \varphi$

$x^3 = r \sin \alpha \cos \theta$

$x^4 = r \cos \alpha$

with $0 \leq \alpha \leq \pi, \ 0 \leq \varphi \leq 2\pi$ and $0 \leq \theta \leq 2\pi$

Try to find the inverse functions! Good luck.

I.6.2.3.3 Examples for second order contravariant tensor

Example in $\mathbb{R}^2$

Let $T = T^{ij}$ be a contravariant tensor of order 2 in a coordinate system (x^i) over $\mathbb{R}^2$.

Let $[T^{ij}] = \begin{bmatrix} T^{11} & T^{12} \\ T^{21} & T^{22} \end{bmatrix} = \begin{bmatrix} 1 & 1 \\ -1 & 2 \end{bmatrix}$

Define another system $(\bar{x}^i)$ by:

$\quad$ (8) $\bar{x}^1 = (x^1)^2 \neq 0$

$\quad$ (9) $\bar{x}^2 = x^1 x^2$

We want to calculate $(\bar{T}^i)$ in the $(\bar{x}^i)$-system.

First, we build up the Jacobi-matrix

$$(1) \ J = J^i{}_r = \begin{bmatrix} \dfrac{\partial \bar{x}^1}{\partial x^1} & \dfrac{\partial \bar{x}^1}{\partial x^2} \\ \dfrac{\partial \bar{x}^2}{\partial x^1} & \dfrac{\partial \bar{x}^2}{\partial x^2} \end{bmatrix} = \begin{bmatrix} 2\,x^1 & 0 \\ x^2 & x^1 \end{bmatrix}$$

We apply $(I.6.012)$:

$$[\bar{T}^{ij}] = \underbrace{\begin{bmatrix} 2x^1 & 0 \\ x^2 & x^1 \end{bmatrix}}_{=J} \underbrace{\begin{bmatrix} 1 & 1 \\ -1 & 2 \end{bmatrix}}_{=T} \underbrace{\begin{bmatrix} 2x^1 & x^2 \\ 0 & x^1 \end{bmatrix}}_{=J^T}$$

$$= \begin{bmatrix} 4(x^1)^2 & 2x^1 x^2 + 2(x^1)^2 \\ 2x^1 x^2 - 2(x^1)^2 & 2(x^1)^2 + (x^2)^2 \end{bmatrix}$$

I.6.2.3.1.3 Kronecker symbol $\delta^i{}_j$ as a mixed tensor of order 2

The Kronecker symbol $\delta^i{}_j$ is a mixed tensor of order 2. Therefore, we demonstrate that the transformation law is fulfilled.

So, if $\delta^i{}_j$ is a mixed tensor of order 2, then the transformation law

$$\bar\delta^p{}_q = \delta^i{}_j \, \frac{\partial \bar x^p}{\partial x^i} \, \frac{\partial x^i}{\partial \bar x^q} \text{ must be valid.}$$

Indeed, we have:

$$
\begin{aligned}
\bar\delta^p{}_q \;&=\; \delta^i{}_j \, \frac{\partial \bar x^p}{\partial x^i} \, \frac{\partial x^j}{\partial \bar x^q} && \text{transformation law}\\[2em]
&=\; \frac{\partial \bar x^p}{\partial x^i} \left(\delta^i{}_j \, \frac{\partial x^j}{\partial \bar x^q} \right) && \text{reorder}\\[2em]
&=\; \frac{\partial \bar x^p}{\partial x^i} \left(\frac{\partial x^i}{\partial \bar x^q} \right) && (I.\,2.022)\\[2em]
&=\; \frac{\partial \bar x^p}{\partial \bar x^q} && \text{multivariate chain rule}\\[2em]
&=\; \delta^p{}_q && (I.\,2.045)
\end{aligned}
$$

We not only showed that $\delta^i{}_j$ is a mixed tensor of order two, but also that $\delta^i{}_j$ always takes the same values in any coordinate system.

∎

Calculation example

Let $U = U_{ij}$ be a covariant tensor of order 2 in (x^i),

$$[U_{ij}] = \begin{bmatrix} U_{11} & U_{12} \\ U_{21} & U_{22} \end{bmatrix} = \begin{bmatrix} x^2 & 0 \\ 0 & x^1 \end{bmatrix}$$

Let the system $(\bar{x}^i)$ be defined as

$$\bar{x}^1 = (x^1)^2 \neq 0$$

$$\bar{x}^1 = x^1\, x^2$$

We calculate $\bar{U}_{ij}$: using $(I.6.013)\ \bar{T} = \bar{J}^T\, T\, \bar{J}$

We have:

$$\bar{J} = \begin{bmatrix} 2\,x^1 & 0 \\ x^2 & x^1 \end{bmatrix}^{-1} = \begin{bmatrix} \dfrac{1}{2\,x^1} & 0 \\[2ex] -\dfrac{x^2}{2\,(x^1)^2} & \dfrac{1}{x^1} \end{bmatrix}$$

Now,

$$\bar{T} \;=\; \bar{J}^T\, T\, \bar{J}$$

$$= \begin{bmatrix} \dfrac{1}{2\,x^1} & -\dfrac{x^2}{2\,(x^1)^2} \\[2ex] 0 & \dfrac{1}{x^1} \end{bmatrix} \begin{bmatrix} x^2 & 0 \\ 0 & x^1 \end{bmatrix} \begin{bmatrix} \dfrac{1}{2\,x^1} & 0 \\[2ex] -\dfrac{x^2}{2\,(x^1)^2} & \dfrac{1}{x^1} \end{bmatrix}$$

$$= \begin{bmatrix} \dfrac{1}{2\,x^1} & -\dfrac{x^2}{2\,(x^1)^2} \\[2ex] 0 & \dfrac{1}{x^1} \end{bmatrix} \begin{bmatrix} \dfrac{x^2}{2\,x^1} & 0 \\[2ex] -\dfrac{x^1 x^2}{2\,(x^1)^2} & 1 \end{bmatrix}$$

$$= \begin{bmatrix} \dfrac{x^1\, x^2 + (x^2)^2}{4\,(x^1)^3} & -\dfrac{x^2}{2\,(x^1)^2} \\[2ex] -\dfrac{x^2}{2\,(x^1)^2} & \dfrac{1}{x^1} \end{bmatrix}$$

∎

I.6.3 Operations on tensors

Now we describe operations on tensors. While addition and multiplication of two tensors does not cause any problems, the tensor product $\otimes$ is a bit more difficult to manage.

I.6.3.1 Addition of tensors

If $A^{i_1\ldots i_p}_{j_1\ldots j_q}$ and $B^{i_1\ldots i_p}_{j_1\ldots j_q}$ are components of (p,q) tensors, then:

$$A^{i_1\ldots i_p}_{j_1\ldots j_q} \pm B^{i_1\ldots i_p}_{j_1\ldots j_q} = C^{i_1\ldots i_p}_{j_1\ldots j_q}$$ are components of a (p,q) tensor.

Thus, the addition is "element"-wise.

I.6.3.2 Scalar multiplication of tensors

If $A^{i_1\ldots i_p}_{j_1\ldots j_q}$ are the components of a (p,q) tensor and ϕ a scalar function, then

$\phi\, A^{i_1\ldots i_p}_{j_1\ldots j_q}$ are the components of a (p,q) tensor.

So, the multiplication of a tensor with a scalar function is also "element"-wise.

I.6.3.3 Equal tensors

If $A^{i_1\ldots i_p}_{j_1\ldots j_q}$ and $B^{i_1\ldots i_p}_{j_1\ldots j_q}$ are components of (p,q) tensors, then the tensor are said to be equal if they are equal in all their components:

$$A^{i_1\ldots i_p}_{j_1\ldots j_q} = B^{i_1\ldots i_p}_{j_1\ldots j_q}$$

One important remark: if two tensors are equal in a coordinate system, then they are equal in every coordinate system.

I.6.3.4 Raising and lowering indices for tensors

Raising or lowering indices is a common operation on tensors. In many cases, it is necessary to ensure that indices match appropriately in mathematical expressions. Raising or lowering indices allows for consistent notation and simplifies calculations. When performing tensor contractions (summations over paired indices), indices must match to perform the operation. In differential geometry, derivative operators such as the covariant derivative require consistent index placement for the components of tensors. Raising or lowering indices ensures that derivatives are applied appropriately.

For lowering an index you will use the metric tensor, for raising you have to use the inverse metric tensor.

We explain the procedures step by step:

1. **Tensor of rank 1**:

We start with a covariant tensor T_i. The only possibility for raising the index is to convert T_i to the covariant version T^i:

$$T^m = g^{im}\, T_i$$

Starting with a contravariant tensor T^i, we get the contravariant version by

$$T_m = g_{im}\, T^i$$

We already encountered these formulas in $(I.4.2.1)$.

2. **Tensor of rank 2**

Let us begin with the covariant tensor $T = T_{ij}$

We raise the first index i:

$$(\boldsymbol{I.6.015})\ T' = T^r_{*j} = g^{ri}T_{ij} \underset{g\ symmetric}{=} g^{ir}\, T_{ij}$$

The star $(*)$ is a placeholder to indicate which index has been lowered or raised.

We can express $(I.6.015)$ as a matrix multiplication. Coming back to our discussion in section $(I.2.1.4.4)$, we see that $(I.6.015)$ can be rewritten as:

$$(\boldsymbol{I.6.016})\ [T'] = [g]^{-1}\, [T]$$

Now, we raise the second index:

$(\boldsymbol{I.6.017})\ T_i^{*r} = g^{rj}\,T_{ij} = g^{jr}\,T_{ij}$, in matrix notation:

$(\boldsymbol{I.6.018})\ [T'] = [T]\,[g]^{-1}$

We can raise both indices "at once":
$(\boldsymbol{I.6.019})\ \ T^{rs} = g^{ri}\,g^{sj}\,T_{ij}$, in matrix notation:
$(\boldsymbol{I.6.020})\ \ \ [T'] = [g]^{-1}[T]\,[g]^{-1}$

Starting now with a contravariant tensor $T = T^{ij}$, we can lower the first, or the second and finally both indices. The procedure is the same, except that we need the metric tensor itself and not its inverse.

Lowering the first index:
$(\boldsymbol{I.6.021})\ T' = T_r^{*j} = g_{ri}T^{ij} \underset{g\ symmetric}{=} g_{ri}\,T^{ij}$

$(\boldsymbol{I.6.021})\ T' = [g]\,[T]$

Lowering the second index:

$(\boldsymbol{I.6.022})\ T_{*r}^{i} = g_{rj}\,T^{ij} = g_{jr}\,T^{ij}$, in matrix notation:

$(\boldsymbol{I.6.023})\ [T'] = [T]\,[g]$

Lowering both indices:

$(\boldsymbol{I.6.024})\ T_{rs} = g_{ri}\,g_{sj}\,T^{ij}$, in matrix notation:
$(\boldsymbol{I.6.025})\ [T'] = [g][T]\,[g]$

3. In a general tensor, you can raise or lower any index. Take, for example, a contravariant tensor or rank 3: T_{ijk}:

$T_{*jk}^{r} = g^{ri}T_{ijk}$ (raise first index)

$T_{i*k}^{*r} = g^{rj}\,T_{ijk}$ (raise 2nd index)

$T_{ij}^{**r} = g^{rk}\,T_{ijk}$ (raise 3rd index)

$$T^{rs}_{**k} = g^{ri}\, g^{sj}\, T_{ijk} \quad \text{(raise 1st and 2nd index)}$$

$$T^{rst} = g^{ri}\, g^{sj}\, g^{tk}\, T_{ijk} \quad \text{(raise all indices at once)}$$

<u>A warning</u>: *dummy indices cannot be raised or lowered across derivatives:*

Let us consider the expression $\dfrac{\partial S_i}{\partial t}\, T^i$ in a curvilinear system.

We then have:

$$
\begin{aligned}
\frac{\partial S_i}{\partial t}\, T^i \;&=\; \frac{\partial\left(S^j g_{ji}\right)}{\partial t}\, T^i && \text{raise index of } S_i \\[2mm]
&=\; \left(\frac{\partial S^j}{\partial t} g_{ji} + S^j \frac{\partial g_{ji}}{\partial t}\right) T^i && \text{apply product rule} \\[2mm]
&=\; \frac{\partial S^j}{\partial t} g_{ji} T^i + S^j \frac{\partial g_{ji}}{\partial t}\, T^i && \text{multiply out} \\[2mm]
&=\; \frac{\partial S^j}{\partial t} T_j + S^j \frac{\partial g_{ji}}{\partial t}\, T^i && \text{lowering index of } T^i \\[2mm]
&=\; \frac{\partial S^i}{\partial t} T_i + S^j \frac{\partial g_{ji}}{\partial t}\, T^i && \text{rename in 1}^{st}\text{term } j \text{ to } i
\end{aligned}
$$

Since the metric tensor varies from one point to another – as we are now in a curvilinear system -, the derivative $\dfrac{d\left(g_{ji}\right)}{dt}$ is non-zero. This means, we got an additional term in the equation. Thus, comparing both sides, we found out:

$$
\frac{\partial S_i}{\partial t}\, T^i = \underbrace{\frac{\partial S^i}{\partial t} T_i + S^j \underbrace{\frac{\partial g_{ji}}{\partial t}}_{\neq 0}\, T^i}_{\neq 0} \neq \frac{\partial S^i}{\partial t} T_i
$$

In tensor identities, we cannot lower and raise dummy indices across derivatives.

∎

I.6.3.4 Contraction of tensors

Tensor contraction is an operation which is only available for mixed tensors.

Suppose $S = \left(S_{j_1\ j_2,\dots\ j_q}^{i_1\ i_2\ \dots\ i_p}\right)$ is a (p,q) tensor. Then the **contraction of S** with respect to a contravariant index i_f with $f \in \{1,p\}$ and a covariant index j_g with $g \in \{1,q\}$ is given by setting $i_f = j_g = u$ and evaluation $S' = \left(S_{j_1\ \dots\ u\ \dots\ j_q}^{i_1\ \dots\ u,\dots\ i_p}\right)$. Summing up over u will give us a $(p-1, q-1)$ tensor, as the index u will disappear after summation.

If we have a mixed tensor T_j^i, we can consider this tensor as a matrix. In this case, the contraction of T_j^i is the trace of the matrix $\left[T_j^i\right]$.

Sometimes, two tensors are contracted using an upper index of one tensor and a lower of the other tensor. In this context, contraction occurs after tensor multiplication.

I.6.3.5 Outer product of tensors (tensor product $\otimes$)

The **outer product** of $S = S_{j_1\ \dots\ j_q}^{i_1\ \dots\ i_p}$ and $T = T_{l_1\ \dots\ l_s}^{k_1\ \dots\ k_r}$, also called **tensor product**, is defined as

$$S \otimes T = S_{j_1\ \dots\ j_q}^{i_1\ \dots\ i_p}\ T_{l_1\ \dots\ l_s}^{k_1\ \dots\ k_r},$$

So, every component of S is multiplied with every component of T.

$S \otimes T$ is a $(p+r, q+s)$ tensor.

An example:

Let us take a contravariant tensor A with components A^i (so it is a vector) and a covariant tensor B with components B_{jk} (a matrix).

Then $C := A \otimes B$ has the components $C_{jk}^i = A^i B_{jk}$. The result is a mixed tensor of rank 3.

I.6.3.6 Inner product of tensors

The **inner product** of $S = S^{i_1 \, \dots \, i_p}_{j_1 \, \dots \, j_q}$ and $T = T^{k_1 \, \dots \, k_r}_{l_1 \, \dots \, l_s}$ with respect to a

contravariant index i_f (from S) and a covariant index l_g (from T)

replaced by a (new) summation index u is defined by:

$$S \cdot T = S^{i_1 \, \dots u \dots \, i_p}_{j_1 \, \dots \, j_q} \, T^{k_1 \, \dots \, k_r}_{l_1 \, \dots u \dots \, l_s}$$

Let us take the example for the outer product, so A is a contravariant tensor with components A^i and a covariant tensor B with components B_{jk}.

Then $C := A \cdot B$ has the components $C_j = A^i \, B_{ij}$. In this example, the result is a covariant tensor of rank 1. It corresponds to the matrix multiplication of B with the vector A.

I.6.3.7 Covariant derivation of tensors

Let $T = T_i(x(t))$ be a covariant tensor, where the (x^i)-functions are parameterized by t. So, we have the transformation law

$$(I.6.026) \quad \bar{T}_i = T_r \, \frac{\partial x^r}{\partial \bar{x}^i}$$

Taking the derivative with respect to $\bar{x}^k$ results in

$$(I.6.027) \quad \frac{\partial \bar{T}_i}{\partial \bar{x}^k} = \frac{\partial T_r}{\partial \bar{x}^k} \frac{\partial x^r}{\partial \bar{x}^i} + T_r \, \frac{\partial^2 x^r}{\partial \bar{x}^k \, \partial \bar{x}^i}$$

As you see, $\dfrac{\partial \bar{T}_i}{\partial \bar{x}^k}$ does not follow the tranformation law in general, the second term spoils it all. Only if $(\bar{x}^i)$ are linear connected to (x^i), the second term will disappear, and only then $\dfrac{\partial \bar{T}_i}{\partial \bar{x}^k}$ is a tensor.

If we want to define a general derivation of a tensor so that the derivative itself is also a tensor, then we must "get rid of" the second term.

Using $(I.5.019)$, we can rewrite $(I.6.027)$

$$(1) \quad \frac{\partial \bar{T}_i}{\partial \bar{x}^k} = \frac{\partial T_r}{\partial \bar{x}^k} \frac{\partial x^r}{\partial \bar{x}^i} + T_r \left(\bar{\Gamma}^s_{ik} \frac{\partial x^r}{\partial \bar{x}^s} - \Gamma^r_{st} \frac{\partial x^s}{\partial \bar{x}^i} \frac{\partial x^t}{\partial \bar{x}^k} \right)$$

With the chain rule, we can express the first term on the right side as:

$$(2) \quad \frac{\partial T_r}{\partial \bar{x}^k} \frac{\partial x^r}{\partial \bar{x}^i} = \frac{\partial T_r}{\partial x^s} \frac{\partial x^r}{\partial \bar{x}^i} \frac{\partial x^s}{\partial \bar{x}^k}$$

Inserting (2) into (1) and outmultiplying the second term in (1), we get:

$$(3) \quad \frac{\partial \bar{T}_i}{\partial \bar{x}^k} = \frac{\partial T_r}{\partial x^s} \frac{\partial x^r}{\partial \bar{x}^i} \frac{\partial x^s}{\partial \bar{x}^k} + T_r \, \bar{\Gamma}^s_{ik} \frac{\partial x^r}{\partial \bar{x}^s} - T_r \, \Gamma^r_{st} \frac{\partial x^s}{\partial \bar{x}^i} \frac{\partial x^t}{\partial \bar{x}^k}$$

Now, we replace $T_r \dfrac{\partial x^r}{\partial \bar{x}^s}$ by $\bar{T}_s$, using $(I.6.022)$. Following our discussion of renaming bound indices, we rename in the *second term* the index s to t, resulting in:

$$(4) \quad \frac{\partial \bar{T}_i}{\partial \bar{x}^k} = \frac{\partial T_r}{\partial x^s} \frac{\partial x^r}{\partial \bar{x}^i} \frac{\partial x^s}{\partial \bar{x}^k} + \bar{T}_t \, \bar{\Gamma}^t_{ik} - T_r \, \Gamma^r_{st} \frac{\partial x^s}{\partial \bar{x}^i} \frac{\partial x^t}{\partial \bar{x}^k}$$

For the third term, we also rename indices: t to r and r to t, so to say at the same time.

$$(5) \quad \frac{\partial \bar{T}_i}{\partial \bar{x}^k} = \frac{\partial T_r}{\partial x^s} \frac{\partial x^r}{\partial \bar{x}^i} \frac{\partial x^s}{\partial \bar{x}^k} + \bar{T}_t \, \bar{\Gamma}^t_{ik} - T_t \, \Gamma^t_{sr} \frac{\partial x^s}{\partial \bar{x}^i} \frac{\partial x^r}{\partial \bar{x}^k}$$

Rearranging the order of the third term and using the Theorem of Schwartz and, we get:

$$(6) \quad \frac{\partial \bar{T}_i}{\partial \bar{x}^k} = \frac{\partial T_r}{\partial x^s} \frac{\partial x^r}{\partial \bar{x}^i} \frac{\partial x^s}{\partial \bar{x}^k} + \bar{T}_t \, \bar{\Gamma}^t_{ik} - T_t \, \Gamma^t_{rs} \frac{\partial x^r}{\partial \bar{x}^i} \frac{\partial x^s}{\partial \bar{x}^k}$$

We arrange now (6) as:

$$(7) \quad \frac{\partial \bar{T}_i}{\partial \bar{x}^k} - \bar{T}_t \, \bar{\Gamma}^t_{ik} = \left(\frac{\partial T_r}{\partial x^s} - T_t \, \Gamma^t_{rs} \right) \frac{\partial x^r}{\partial \bar{x}^i} \frac{\partial x^s}{\partial \bar{x}^k}$$

Now, have a closer look at what you see on the right side: this is the transformation law of a covariant tensor of order 2! So, if we subtract $\bar{T}_t \, \bar{\Gamma}^t_{ik}$

from the derivation of $\dfrac{\partial \bar{T}_i}{\partial \bar{x}^{k}}$, then we have what we want: a tensor. The Christoffel symbols represent the change in basis vectors in a curved space. When imagining transporting a vector along a curved surface or space curve, this vector is influenced not only by the local change of basis vectors but also by the curvature of the space itself. The covariant derivative takes these changes into account and "subtracts" the contributions of the Christoffel symbols to ensure that the derivative remains tensorial.

Intuitively, one can imagine that the Christoffel symbols are the "correction terms" necessary to account for the change of the vector due to the curvature of the space. By subtracting these correction terms, the covariant derivative remains tensorial and respects the underlying geometric properties of the space.

(I.6.028) (Covariant derivative of a covariant tensor rank 1)

In any coordinate system (x^i) the covariant derivative with respect to x^k of a covariant tensor $T = (T_i)$ is the tensor

$$T_{;k} := \left(T_{i;k}\right) = \frac{\partial T_i}{\partial x^k} - \Gamma_{ik}^{t}\, T_t$$

Note that we use the *semicolon* to denote the covariant derivative with respect to k.

In the same way, we can define

(I.6.029) (Covariant derivative of a contravariant tensor rank 1)

In any coordinate system (x^i) the covariant derivative with respect to x^k of a contravariant tensor $T = (T^i)$ is the tensor

$$T_{;k} := \left(T^{i}{}_{;k}\right) = \frac{\partial T^i}{\partial x^k} + \Gamma_{tk}^{i}\, T^t$$

Notice the indices and that here, we have to **add** the Christoffel symbols!

Finally, we can define the covariant derivative of any tensor by:

$$\boxed{\begin{array}{l}
(\boldsymbol{I.6.030})\ (\textit{Covariant derivative of an arbitrary tensor})\\[4pt]
\text{In any coordinate system } (x^i) \text{ the covariant derivative with respect to } x^k \text{ of}\\[2pt]
\text{a } (p,q) \text{ tensor } T = T^{i_1\,i_2\,\dots\,i_p}_{j_1\,j_2\,\dots\,j_q} \text{ is the tensor}\\[8pt]
\end{array}}$$

$$
\begin{aligned}
T^{i_1\,i_2\,\dots\,i_p}_{\;j_1\,j_2\,\dots\,j_q;\,k} =\; & \frac{\partial T^{i_1\,i_2\,\dots\,i_p}_{\;j_1\,j_2\,\dots\,j_q}}{\partial x^k}
+ \Gamma^{i_1}_{tk}T^{t\,i_2\,\dots\,i_p}_{\;j_1\,j_2\,\dots\,j_q}
+ \Gamma^{i_2}_{tk}T^{i_1\,t\,\dots\,i_p}_{\;j_1\,j_2\,\dots\,j_q}
+ \dots + \Gamma^{i_p}_{tk}T^{i_1\,i_2\,\dots\,t}_{\;j_1\,j_2\,\dots\,j_q}\\[6pt]
& - \Gamma^{t}_{j_1 k}T^{i_1\,i_2\,\dots\,i_p}_{\;t\,j_2\,\dots\,j_q}
- \Gamma^{t}_{j_2 k}T^{i_1\,i_2\,\dots\,i_p}_{\;j_1\,t\,\dots\,j_q}
- \Gamma^{t}_{j_q k}T^{i_1\,i_2\,\dots\,i_p}_{\;j_1\,j_2\,\dots\,t}
\end{aligned}
$$

Some notes:

- In rectangle coordinates the metric tensor consists only of constants. So, all Christoffel symbols calculate to zero and the covariant derivative coincides with the partial derivative.
- The covariant derivative of an arbitrary (p,q) tensor is a $(p,q+1)$ tensor.

Where do you need the covariant derivation?

Let us take an example:

In mechanics, tensors are used to describe physical quantities such as stress, strain, and momentum. For example, stress is described by a second-rank tensor, and strain by a fourth-rank tensor. The covariant derivative is used to describe how these quantities change as one moves through space and time. In the context of Newtonian mechanics, the equations of motion for a particle moving under the influence of forces can be expressed as tensor equations using the concept of the covariant derivative.

Covariant derivation plays an immense role in different parts in physics, not only in GR! Later, we will consider an example in electrodynamics.

A warning:

For partial derivatives of a sufficiently continuous function, we already mentioned the **Theorem of Schwartz:**

$$\frac{\partial f}{\partial x}\frac{\partial f}{\partial y} = \frac{\partial f}{\partial y}\frac{\partial f}{\partial x}$$

You have the choice of taking first the derivative w.r.t ∂x and then with respect to ∂y, or the other way round.

This equation does not hold for the covariant derivative!

I.6.3.7.1 Differentiation rules for tensors

Let T and S be tensors. Then the following differentiation rules are valid:

$(\boldsymbol{I.6.031})\ (T+S)_{;k} = T_{;k} + S_{;k}$

$(\boldsymbol{I.6.032})\ [T \otimes S]_{;k} = \left(T_{;k} \otimes S\right) + \left(T \otimes S_{;k}\right)$

$(\boldsymbol{I.6.033})\ (T\,S)_{;k} = \left(T_{;k}\,S\right) + \left(T\,S_{;k}\right)$

I.6.3.7.2 Example on the use of covariant derivation

With the following example, we want to highlight some important points:

- Christoffel symbols are not only used in GR, but also in other areas in physics.
- The covariant derivation *must be observed* when working in curvilinear coordinates.

Let us consider the divergence of a point charge: a point charge is considered as a mathematical singularity, where the electric charge is concentrated at a single point. Due to this property, the divergence of the electric field $\vec{E}$ of a point charge behaves differently compared to continuous charge distributions.

The divergence $\nabla \cdot \vec{E}$ of an electric field $\vec{E}$ around a point charge Q is mathematically defined as:

$$\nabla \cdot \vec{E} = \frac{\rho}{\varepsilon_0}$$

where ρ is the charge density and ε_0 is the electric constant (permittivity of vacuum). For a point charge Q, the charge density ρ becomes $Q\,\delta(\vec{r})$, where $\delta(\vec{r})$ is the Dirac delta function.

The Dirac delta function is a special function that behaves like a point charge. It is zero everywhere except at the origin, where it is infinitely large, so the product of Q and $\delta(\vec{r})$ creates a delta peak at the position of the charge.

Since the Dirac delta function is infinitely large in all directions, the divergence of the electric field vanishes everywhere except at the location of the point charge. At the position of the point charge itself, the divergence is not well-defined, but in integral form, it can be interpreted as the total charge divided by the volume it "occupies," which in this case is infinitely large.

We expect that $\nabla \cdot \vec{E} = 0$.

First, we notice that

$$div \, \vec{v} = \nabla_i \, v^i = v^i_{\;;i} = \frac{\partial v^i}{\partial x^i} + \Gamma^i_{ij} \, v^j$$

We use the covariant derivation.

As the point charge is radial, it makes sense to switch to spherical coordinates. We need the Christoffel symbols (see $I.\,A.\,1.3$). The only non-zero Christoffel symbols are:

$$\Gamma^r_{\theta\theta} = -r$$

$$\Gamma^r_{\phi\phi} = -r \, sin^2(\theta)$$

$$\Gamma^\theta_{r\theta} = {}^1\!/_r$$

$$\Gamma^\theta_{\phi\phi} = -\cos(\theta)\sin(\theta)$$

$$\Gamma^\phi_{r\phi} = {}^1\!/_r$$

$$\Gamma^\phi_{\theta\phi} = \cos(\theta)/\sin(\theta)$$

As $\vec{E}$ is radial, $|\vec{E}| \sim \frac{1}{r^2}$. Therefore, we can presume:

$$E = E^i = \begin{pmatrix} E^1 \\ E^2 \\ E^3 \end{pmatrix} = \begin{pmatrix} \frac{c}{r^2} \\ 0 \\ 0 \end{pmatrix} \text{ for a constant } c \text{ and } r \neq 0$$

Now, we calculate the divergence:

$$div \, \vec{E} = \nabla_i \, E^i = E^i_{\;;i} = \frac{\partial E^i}{\partial x^i} + \Gamma^i_{ij} \, E^j$$

We calculate every $E^i_{\;;i}$:

$$E^1_{\;;1} = \underbrace{\frac{\partial E^1}{\partial r}}_{=-\frac{2c}{r^3}} + \underbrace{\Gamma^1_{11}}_{=0} E^1 + \underbrace{\Gamma^1_{12}}_{=0} E^2 + \underbrace{\Gamma^1_{13}}_{=0} E^3 = -\frac{2c}{r^3}$$

$$E^2{}_{;2} = \underbrace{\frac{\partial E^2}{\partial \theta}}_{=0} + \underbrace{\Gamma^2_{21}}_{=1/r}\ \underbrace{E^1}_{c/r^2} + \underbrace{\Gamma^2_{22}\ E^2}_{=0} + \underbrace{\Gamma^2_{23}\ E^3}_{=0} = \frac{c}{r^3}$$

$$E^3{}_{;3} = \underbrace{\frac{\partial E^3}{\partial \varphi}}_{=0} + \underbrace{\Gamma^3_{31}}_{=1/r}\ \underbrace{E^1}_{c/r^2} + \underbrace{\Gamma^3_{32}\ E^2}_{=0} + \underbrace{\Gamma^3_{33}\ E^3}_{=0} = \frac{c}{r^3}$$

Summing up $E^i{}_{;i}$, we get 0. So $div\,\vec{E} = 0$.

If we had taken only the partial derivatives, ignoring the correcting factors given by the Christoffel symbols, we would have gotten a non-zero result in contrary to our expectation.

I.6.4 Quotient rule

The following theorem, called "Quotient Rule" can be used to determine the tensor character of an object T without verifying the tensor transformation law. The general form of the Quotient Rule is:

$(I.6.034)$ **Quotient Rule** (**extended form**)

If $T^{i_1 i_2 \dots i_p}_{j_1 j_2 \dots j_q k}\, V^k \equiv S^{i_1 i_2 \dots i_p}_{j_1 j_2 \dots j_q}$ are components of a tensor for an arbitrary contravariant vector V^k, then $T^{i_1 i_2 \dots i_p}_{j_1 j_2 \dots j_q j_{q+1}}$ is a tensor.

In a simpler form, we can formulate:

$(I.6.035)$ **Quotient Rule** (**simple form**)

If A and B are tensors and if the expression $A = BT$ is invariant under coordinate transformation, then T is a tensor.

The quotient rule gives us a good way to determine if an object is a tensor or not:

$(I.6.036)$ If $T_i V^i \equiv E$ is invariant for all contravariant tensors V^i, then T_i is a covariant tensor.

($I.6.037$) If $T_{ij} V^i \equiv U_j$ are components of a covariant for all contravariant vectors V^i, then T_{ij} is a covariant tensor of rank 2.

($I.6.038$) If $T_{ij} U^i V^j \equiv E$ is invariant for all contravariant tensors U^i, V^i, then T_{ij} is a covariant tensor of rank 2.

An example of how the quotient rule can be used:

Let us say we have a tensor T_{ij} and a tensor U_{ij} defined explicitly as:

$$T_{ij} = \begin{bmatrix} 1 & 2 \\ 3 & 4 \end{bmatrix}, U_{ij} = \begin{bmatrix} 2 & 4 \\ 6 & 8 \end{bmatrix}$$

satisfying the equation

$$T_{ij} S_{ij} = U_{ij} \text{ for all } i, j.$$

This is not to be considered as an ESN equation, so there is no summation. It states that

$$T_{11} S_{11} = U_{11}$$

$$T_{12} S_{12} = U_{12} \quad \text{etc.}$$

As all elements T_{ij} are non-zero, we can solve for S_{ij}:

$$S_{11} = \frac{U_{11}}{T_{11}} = \frac{2}{1} = 2$$

$$S_{12} = \frac{U_{12}}{T_{12}} = \frac{4}{2} = 2$$

$$S_{21} = \frac{U_{21}}{T_{21}} = \frac{6}{3} = 2$$

$$S_{22} = \frac{U_{22}}{T_{22}} = \frac{8}{4} = 2$$

So $S_{ij} = 2$ for all i,j. As S_{ij} is constant in all elements, S_{ij} is a tensor.

I.6.E Exercises

$(\boldsymbol{I.6.E.1})$ Let $[g_{ij}] = \begin{bmatrix} 0 & 1 \\ 1 & 1 \end{bmatrix}$ be the metric tensor for a 2-dimensional vectorspace over $\mathbb{R}$. Let $T_{ij} = \begin{bmatrix} 0 & 2 \\ 0 & 0 \end{bmatrix}$ be a tensor.

Determine $T^i{}_j$, $T_j{}^i$ and T^{ij}

$(\boldsymbol{I.6.E.2})$ Let $[g_{ij}] = \begin{bmatrix} 2 & -1 \\ -1 & 2 \end{bmatrix}$ be a metric tensor and T^i_{*jkl} a 4-rank-tensor. Let us assume that the only non-zero element of T^i_{*jkl} is

$$T^1_{*212} = \pi$$

Calculate T_{1212} and T_{2212}

$(\boldsymbol{I.6.E.3})$ Let A_{ij} be a symmetric tensor and B^{ij} be an anti-symmetric tensor. Show that $A_{ij} B^{ij} = 0$

$(\boldsymbol{I.6.E.4})$ Determine if the following tensor expressions are legitimate and explain your answer.

(a) $T^k{}_i + S_i$

(b) $T^k_k - S^n_m$

(c) $T^{ijm}{}_{ijk} - T^{lnm}{}_{lnk}$

$(\boldsymbol{I.6.E.5})$ Write down the transformation laws for (a) covariant and contravariant vectors and (b) for covariant, contravariant, and mixed rank-2 tensors between different coordinate systems.

$(\boldsymbol{I.6.E.6})$ Let $T_{\alpha\beta}$ be a rank-2 tensor and $g_{\mu\nu}$ a metric tensor.

Calculate (1) $T^\alpha{}_\beta$ and (2) $T^{\alpha\beta}$.

Suppose $T_{\alpha\beta}$ has a symmetric part $S_{\alpha\beta}$ and an antisymmetric part $A_{\alpha\beta}$. Write down the raised versions of (3) $S^{\alpha\beta}$ and (4) $A^{\alpha\beta}$.

($I.6.E.7$) Let the metric tensor be given by:

$$[g_{ij}] = \begin{bmatrix} 1 & 0 \\ 0 & 4 \end{bmatrix}$$

And a tensor T_{ij} given by:

$$[T_{ij}] = \begin{bmatrix} 2 & 1 \\ 3 & 5 \end{bmatrix}$$

Calculate $T^i{}_j$ and T^{ij}.

($I.6.E.8$) Let g_{ij} be the metric tensor for spherical coordinates and let $A_{\mu\nu}$ be a tensor with the following non-zero components:

$$A_{rr} = r^2$$

$$A_{r\theta} = \theta$$

$$A_{\theta\theta} = sin(\theta)$$

$$A_{\theta\phi} = \phi$$

$$A_{\phi\phi} = 1$$

Determine $A^\mu{}_\nu$.

($I.6.E.9$) Let (x^i) be a coordinate system defined by

$$x = x^1(u, v) = u + v$$

$$y = x^2(u, v) = u^2 - v^2$$

Let $T_{\alpha\beta}$ be a tensor in the (x, y)-base.

(1) *Calculate the metric tensor* $g_{\mu\nu}$ in the (u, v)-base.

(2) Express $T_{\alpha\beta}$ in the (u, v)-base.

I.7 Scale Factors (Lamé coefficients)

In curvilinear coordinate systems, **scale factors** - also called **Lamé coefficients** - play an important role in understanding the geometry of the coordinate system. Scale factors, also known as metric coefficients, quantify how lengths, areas, and volumes change when moving along the coordinate directions.

Let us consider a coordinate transformation from a Cartesian coordinate system to a curvilinear coordinate system described by coordinates $x^i = x^i(u_1, u_2, \ldots, u_m)$, $i = 1, \ldots, n$. The scale factors associated with this transformation are denoted as h_i or h^i, depending on whether they are covariant or contravariant. They are defined as the length of the partial derivatives of the new coordinates with respect to the Cartesian coordinates:

$$h_i = \sqrt{\left(\frac{\partial x^1}{\partial u^i}\right)^2 + \left(\frac{\partial x^2}{\partial u^i}\right)^2 + \ldots + \left(\frac{\partial x^n}{\partial u^i}\right)^2}$$

These scale factors provide information about how the coordinate grid deforms as we move around in space.

In particular:

- Covariant scale factors h_i describe how the spacing between lines of constant x^i changes with respect to the Cartesian coordinates. They affect the magnitude of differentials and can be used to calculate surface areas and volumes.
- Contravariant scale factors h^i describe how the spacing between grid lines of constant u_i changes with respect to the curvilinear coordinates. They are often used in the transformation of vectors and tensors between coordinate systems.

They capture the geometric distortion of the coordinate grid, providing insight into how lengths, areas, and volumes change as we move between different coordinate directions.

$(\boldsymbol{I.7.001})$If the metric tensor g_{ij} for (x^i) is a diagonal matrix, then the scale factors are simply the root of the entries along the diagonal, so we get in this case:

$$h_i = \sqrt{g_{ii}}$$

<u>A warning</u>: if you intend to calculate the scale factors as the roots of the entries of the (diagonal) metric tensor, then you should pay special attention on HOW you solve the square. An example:

In *Inverse Prolate Coordinates* (see I.A.2.8) we get as the metric tensor:

$$[g_{ij}] = diag \left\{ \begin{array}{c} \dfrac{a^2(\cosh^2(\eta) - \sin^2(\theta))}{(\boldsymbol{\cosh^2(\eta) - \cos^2(\theta)})^2} \\[2ex] \dfrac{a^2(\cosh^2(\eta) - \sin^2(\theta))}{(\boldsymbol{\cosh^2(\eta) - \cos^2(\theta)})^2} \\[2ex] \dfrac{a^2 \cosh^2(\eta) \sin^2(\theta)}{(\boldsymbol{\cosh^2(\eta) - \cos^2(\theta)})^2} \end{array} \right.$$

Now, $(\cosh^2(\eta) - \cos^2(\theta))^2 = (\cos^2(\theta) - \cosh^2(\eta))^2$, so we can write:

$$[g_{ij}] = diag \left\{ \begin{array}{c} \dfrac{a^2(\cosh^2(\eta) - \sin^2(\theta))}{(\boldsymbol{\cos^2(\theta) - \cosh^2(\eta)})^2} \\[2ex] \dfrac{a^2(\cosh^2(\eta) - \sin^2(\theta))}{(\boldsymbol{\cos^2(\theta) - \cosh^2(\eta)})^2} \\[2ex] \dfrac{a^2 \cosh^2(\eta) \sin^2(\theta)}{(\boldsymbol{\cos^2(\theta) - \cosh^2(\eta)})^2} \end{array} \right.$$

This is the same metric tensor for our coordinate system. They differ in what we put in bold. Let us take the first entry and calculate the root. We get:

$$\sqrt{\frac{a^2(\cosh^2(\eta) - \sin^2(\theta))}{(\cosh^2(\eta) - \cos^2(\theta))^2}} = \frac{a\sqrt{(\cosh^2(\eta) - \sin^2(\theta))}}{\sqrt{(\cosh^2(\eta) - \cos^2(\theta))^2}}$$

$$= \frac{a\sqrt{(\cosh^2(\eta) - \sin^2(\theta))}}{\boldsymbol{\cosh^2(\eta) - \cos^2(\theta)}}$$

And for the second matrix, first entry, we get:

$$\sqrt{\frac{a^2(\cosh^2(\eta) - \sin^2(\theta))}{(\cos^2(\theta) - \cosh^2(\eta))^2}} = \frac{a\sqrt{(\cosh^2(\eta) - \sin^2(\theta))}}{\sqrt{(\cos^2(\theta) - \cosh^2(\eta))^2}}$$

$$= \frac{a\sqrt{(\cosh^2(\eta) - \sin^2(\theta))}}{\cos^2(\theta) - \cosh^2(\eta)}$$

But $\cosh^2(\eta) - \cos^2(\theta) \neq \cos^2(\theta) - \cosh^2(\eta)$. Which one to choose? As we are dealing with vector spaces over $\mathbb{R}$, we can only take a root from a real positive(!) number. In our case it is easy to show that $\cosh^2(\eta) - \cos^2(\theta) \geq 0$ as the definition of the coordinate systems states that $\eta \geq 0, \theta \in [0, \pi]$.

Especially in curvilinear orthogonal coordinate systems (and most of the important systems used in physics are orthogonal), the metric tensor is always a diagonal matrix, so the calculation of the scale factors therefore is simple. The inverse also holds: if the metric tensor is diagonal, the coordinate system is orthogonal.

With the scale factors we can define the orthonormal basis vectors for the curvilinear functions, and we can define the gradient, divergence, and rotations.

Before we get on, let us take the time for an example on how to calculate the scale factors for spherical coordinates:

The set of functions for spherical coordinates are:

$$x^1(r, \theta, \phi) = r \sin \theta \cos \phi$$

$$x^2(r, \theta, \phi) = r \sin \theta \sin \phi$$

$$x^3(r, \theta, \phi) = r \cos \theta$$

With $r \geq 0, \quad 0 \leq \theta \leq \pi, \quad 0 \leq \phi \leq 2\pi$

Step 1:

As we will need all differentials of x^i, we build up the Jacobi-matrix:

$$\mathcal{J}_{f(r,\theta,\varphi)} = \begin{bmatrix} \partial x^1/\partial r & \partial x^1/\partial\theta & \partial x^1/\partial\phi \\ \partial x^2/\partial r & \partial x^2/\partial\theta & \partial x^2/\partial\phi \\ \partial x^3/\partial r & \partial x^3/\partial\theta & \partial x^3/\partial\phi \end{bmatrix} =$$

$$\begin{bmatrix} \cos(\phi)\sin(\theta) & r\cos(\theta)\cos(\phi) & -r\sin(\theta)\sin(\phi) \\ \sin(\theta)\sin(\phi) & r\cos\theta\,\sin(\phi) & r\cos(\phi)\sin(\theta) \\ \cos(\theta) & -r\sin(\theta) & 0 \end{bmatrix}$$

Step 2:

For each column sum up the squared entries, then take the square root:

$$h_r \;=\; \sqrt{(\cos(\phi)\sin(\theta))^2 + (\sin(\theta)\sin(\phi))^2 + (\cos(\theta))^2}$$

$$=\; \sqrt{\sin^2(\theta)\underbrace{(\cos^2(\phi) + \sin^2(\phi))}_{=1} + \cos^2(\theta)}$$
$$\underbrace{\phantom{\sqrt{\sin^2(\theta)(\cos^2(\phi) + \sin^2(\phi)) + \cos^2(\theta)}}}_{=1}$$

$$=\; 1$$

$$h_\theta \;=\; \sqrt{(r\cos(\theta)\cos(\phi))^2 + (r\cos\theta\,\sin(\phi))^2 + (-r\sin(\theta))^2}$$

$$=\; \sqrt{r^2\cos^2(\theta)\left(\underbrace{\cos^2(\phi) + \sin^2(\phi)}_{=1}\right) + r^2\sin^2(\theta)}$$

$$=\; \sqrt{r^2\left(\underbrace{\cos^2(\theta) + \sin^2(\theta)}_{=1}\right)}$$

$$=\; r$$

$$h_\phi \;=\; \sqrt{(-r\sin(\theta)\sin(\phi))^2 + (r\cos(\phi)\sin(\theta))^2 + 0^2}$$

$$= \sqrt{r^2 \, sin^2(\theta) \, \underbrace{(sin^2(\phi) + \, cos^2(\phi))}_{=1}}$$

$$= \quad r \, sin(\theta)$$

Another, much simpler way to calculate the scale factors is to have a look at the metric tensor $[g_{ij}]$. As you have the Jacobi matrix, you can create the metric tensor as $[g_{ij}] = \mathcal{J}_{f(r,\theta,\varphi)}^T \, \mathcal{J}_{f(r,\theta,\varphi)}$. In spherical coordinates we get:

$$[g_{ij}] = \begin{bmatrix} 1 & 0 & 0 \\ 0 & r^2 & 0 \\ 0 & 0 & r^2 sin^2(\theta) \end{bmatrix}$$

The matrix is diagonal, so you can take the root values of the diagonal entries. By naming the scale factors (h_r, h_θ, h_ϕ) you must follow the order in which you build up the Jacobi matrix.

I.7.1 Application of scale vectors

In this chapter, we want to explore where the scale factors can be used. We do this only in three dimensions.

I.7.1.1 Element of arc length

The element of arc length in three dimensions can be generalized as follows:

$$(I.7.001) \; ds = \sqrt{(h_1 \, x^1)^2 + (h_2 \, x^2)^2 + (h_3 \, x^3)^2}$$

I.7.1.2 Element of surface area

The element of surface in three dimensions area can be generalized as:

$$(I.7.002) \; d\sigma = \sqrt{(h_1 h_2 \, x^1 x^2)^2 + (h_1 h_3 \, x^1 x^3)^2 + (h_2 h_3 \, x^2 x^3)^2}$$

I.7.1.3 Volume element

If (x^i) is a coordinate system and $J = J_{(x_1, x_2, \dots, x_n)}$ is the corresponding Jacobi-matrix, then the volume element dV can be expressed by the formula:

$$(\textbf{\textit{I}.7.003})\ dV = |\det(J)| \prod_i dx^i$$

Especially if (x^i) is an orthogonal coordinate system and $h_1, h_2, \ldots, h_n$ the corresponding scale factors, then

$$dV = \prod_i h_i \prod_i dx^i \quad \text{(no summation here)}$$

Why? The orthogonality of (x^i) implies that the metric tensor g_{ij} is a diagonal matrix and it follows that $\det([g_{ij}]) = \prod_j g_{jj}$. Now $[g_{ij}] = J^T J$. Furthermore $\det(J^T) = \det(J)$, so $\det([g_{ij}]) = \det(J)^2$. We conclude that

$$\det(J) = \sqrt{\det([g_{ij}])} = \sqrt{\prod_j g_{jj}} = \sqrt{g_{11}\, g_{22} \cdots g_{nn}} = \sqrt{g_{11}}\,\sqrt{g_{22}}\,\cdots\sqrt{g_{nn}}.$$

On the other hand, we already know that for an orthogonal coordinate system the scale factors h_i correspond to the square roots of g_{ii}.

I.7.1.4 Normalized basis vectors

If (e_i) is a basis, then you can create a normalized basis by:

$$(\textbf{\textit{I}.7.004})\ \hat{e}_i = \frac{e_i}{h_i}$$

I.8 Derivatives and their generalizations, part 2

In this chapter we go back to the derivatives of part 1 and we see, how gradient, divergence, curl and Laplacian can be expressed in general coordinate systems. We will use the scale factors h_i and the normalized unit vectors $\hat{e}_i$.

I.8.1 Gradient in general coordinates

The gradient in a general coordinate system (x^i) takes the form

$$(\boldsymbol{I.8.001}) \quad \nabla S = g^{ik} \frac{\partial S}{\partial x^k} \frac{\partial}{\partial x^i}$$

Note that we use the inverse metric tensor here.

Let us consider an example:

We have a coordinate system given by

$$x(u, v) = u + \alpha v$$

$$y(u, v) = v$$

with $\alpha \neq 0$.

Here, α is a shear parameter (a constant that defines how much the x-coordinate is "skewed" by the v-coordinate).

To calculate the gradient, we need the basis vectors $\boldsymbol{g}_1$ and $\boldsymbol{g}_2$ associated with the coordinates u and v. These are obtained as partial derivatives of the position vector $\boldsymbol{r}(u, v)$ with respect to u and v.

The position vector $\boldsymbol{r}$ in Cartesian coordinates is:

$$r(u, v) = x\,\boldsymbol{i} + y\,\boldsymbol{j}$$

Using the relationships $x = u + \alpha v$ and $y = v$, we have:

$$r(u, v) = (u + \alpha v)\,\boldsymbol{i} + v\,\boldsymbol{j}$$

Partial derivative with respect to u:

$$\boldsymbol{g}_1 = \frac{\partial \boldsymbol{r}}{\partial u} = \frac{\partial}{\partial u}(u + \alpha v)\,\boldsymbol{i} + v\boldsymbol{j} = \boldsymbol{i}$$

Partial derivative with respect to v:

$$\boldsymbol{g}_2 = \frac{\partial \boldsymbol{r}}{\partial v} = \frac{\partial}{\partial v}(u + \alpha v)\,\boldsymbol{i} + v\boldsymbol{j} = \alpha\boldsymbol{i} + \boldsymbol{j}$$

The metric tensor calculates as:

$$[g_{ij}] = \begin{bmatrix} 1 & \alpha \\ \alpha & \alpha^2 + 1 \end{bmatrix}$$

and the inverse metric tensor as:

$$[g^{ij}] = \begin{bmatrix} \alpha^2 + 1 & -\alpha \\ -\alpha & 1 \end{bmatrix}$$

Finally, we can compute:

$$
\begin{aligned}
\nabla S = g^{ik}\frac{\partial S}{\partial x^k}\frac{\partial}{\partial x^i} &= g^{ik}\frac{\partial S}{\partial x^k}\boldsymbol{g}_i \\
&= \left(g^{11}\frac{\partial S}{\partial u} + g^{12}\frac{\partial S}{\partial v}\right)\boldsymbol{g}_1 + \left(g^{21}\frac{\partial S}{\partial u} + g^{22}\frac{\partial S}{\partial v}\right)\boldsymbol{g}_2 \\
&= \left((\alpha^2 + 1)\frac{\partial S}{\partial u} - \alpha\frac{\partial S}{\partial v}\right)\boldsymbol{i} + \left(-\alpha\frac{\partial S}{\partial u} + \frac{\partial S}{\partial v}\right)(\alpha\boldsymbol{i} + \boldsymbol{j}) \\
&= \left(\alpha^2\frac{\partial S}{\partial u}\right)\boldsymbol{i} + \left(\frac{\partial S}{\partial v} - \alpha\frac{\partial S}{\partial u}\right)\boldsymbol{j}
\end{aligned}
$$

∎

If (x^i) is orthogonal, the metric tensor $[g_{ij}]$ and its inverse $[g^{ij}]$ are diagonal matrices. The equation above reduces therefore to:

$$\nabla S = g^{ii}\frac{\partial S}{\partial x^i}\frac{\partial}{\partial x^i}$$

As $[g^{ij}]$ is diagonal, the component g^{ii} calculates as:

$$g^{ii} = \frac{1}{g_{ii}} = \frac{1}{h_i{}^2}$$

where h_i are the scale factors.

Next, we relate the partial derivative $\dfrac{\partial}{\partial x^i}$ to the unit vector $\hat{e}_i$. Since the basis vector in the direction of x^i has a length h_i, the unit vector is:

$$\frac{\partial}{\partial x^i} = \frac{\hat{e}_i}{h_i}$$

Putting all pieces together, we can write the gradient in an orthogonal system as:

$$(\boldsymbol{I.8.002})\ \nabla S = \sum_{i=1}^{n} \frac{1}{h_i} \frac{\partial S}{\partial x^i}\, \hat{e}_i$$

We cannot use ESN notation here, as the right side would contain three times the summation index i, which in ESN is not allowed.

Note that the equation above uses **normalized** basis vectors. There is another way to write the gradient of a scalar field, this time with <u>unnormalized</u> basis vectors:

$$(\boldsymbol{I.8.003})\ \nabla S = \sum \frac{\partial S}{\partial x^i}\, g^{ij}\, e_j$$

In general, you will use the representation with normalized basis vectors.

For the standard Euclidian Space, we get the unit matrix as the metric tensor, so all scale factors calculate to $\sqrt{1} = 1$. It is easy to see that we get the well-known formular of the gradient.

Let us see what we get in spherical coordinates: in $(I.A.1.3)$ we calculated the scale factors as

$$h_r = 1$$

$$h_\theta = r$$

$$h_\phi = r \sin(\theta)$$

Using $(I.8.001)$ the gradient for spherical coordinates calculates as:

$$\nabla S = \sum_{i=1}^{n} \frac{1}{h_i} \frac{\partial S}{\partial x^i} \, \hat{e}_i = \frac{\partial S}{\partial r} \hat{e}_r + \frac{1}{r} \frac{\partial S}{\partial \theta} \hat{e}_\theta + \frac{1}{r \sin(\theta)} \frac{\partial S}{\partial \varphi} \hat{e}_\varphi$$

Now, let us see the result when we use $(I.8.003)$:

First, we note that the metric tensor is given by

$$[g_{ij}] = \begin{bmatrix} 1 & 0 & 0 \\ 0 & r^2 & 0 \\ 0 & 0 & r^2 \sin^2(\theta) \end{bmatrix}$$

The inverse is:

$$[g^{ij}] = \begin{bmatrix} 1 & 0 & 0 \\ 0 & \dfrac{1}{r^2} & 0 \\ 0 & 0 & \dfrac{1}{r^2 \sin^2(\theta)} \end{bmatrix}$$

$$\nabla S = \frac{\partial f}{\partial x^i} g^{ij} e_j$$

$$= \frac{\partial S}{\partial r} \underbrace{g^{rr}}_{=1} e_r + \frac{\partial S}{\partial r} \underbrace{g^{r\theta}}_{=0} e_\theta + \frac{\partial S}{\partial r} \underbrace{g^{r\varphi}}_{=0} e_\varphi +$$

$$\frac{\partial S}{\partial \theta} \underbrace{g^{\theta r}}_{=0} e_r + \frac{\partial S}{\partial \theta} \underbrace{g^{\theta\theta}}_{=1/r^2} e_\theta + \frac{\partial S}{\partial \theta} \underbrace{g^{\varphi\theta}}_{=0} e_\varphi +$$

$$\frac{\partial S}{\partial \varphi} \underbrace{g^{\varphi r}}_{=0} e_r + \frac{\partial S}{\varphi} \underbrace{g^{\varphi\theta}}_{=0} e_\theta + \frac{\partial S}{\partial \varphi} \underbrace{g^{\varphi\varphi}}_{=1/r^2\sin^2(\theta)} e_\varphi$$

$$= \frac{\partial S}{\partial r} e_r + \frac{1}{r^2} \frac{\partial S}{\partial \theta} e_\theta + \frac{1}{r^2 \sin^2(\theta)} \frac{\partial S}{\partial \varphi} e_\varphi$$

Notice the difference between using the unnormalized and normalized unit vectors.

I.8.2 Divergence in general coordinates

In n-dimensional space with a general coordinate system (x^i) and metric tensor $g = g_{ij}$, the divergence of a vector field V with **contravariant** components V^i can be written as

$$(\boldsymbol{I.8.004})\; div\, V(V^1, V^2, \ldots, V^n) \underset{(I.6.025)}{=} \frac{\partial V^i}{\partial x^i} + V^k\, \Gamma^i_{ki}$$

$$= \frac{1}{\sqrt{|det([g])|}} \frac{\partial}{\partial x^i}\left(\sqrt{|det([g])|}\, V^i\right)$$

where $det([g])$ is the determinant of g_{ij}. In the first line you will find the covariant derivative of a contravariant tensor of rank 1.

The last equal sign holds as

$$\Gamma^i_{ki} = \frac{\partial ln(\,|det([g])|\,)}{\partial x^i} = \frac{1}{\sqrt{|det([g])|}} \frac{\partial}{\partial x^i}\left(\sqrt{|det([g])|}\right)$$

Throughout, the Einstein summation convention (summing over repeated indices) is used. Note that V^i here are **contravariant** components in the **unnormalized** coordinate basis $\dfrac{\partial}{\partial x^i}$.

This formula represents the divergence with respect to the local coordinate basis, which is generally not orthonormal unless the coordinate system is Cartesian. While mathematically correct and often convenient for certain calculations, some physical applications prefer using an orthonormal basis, because orthonormal basis vectors have a clearer physical interpretation (each basis vector has unit length and is mutually orthogonal).

In orthogonal coordinates (such as cylindrical or spherical), the metric tensor is diagonal, $[g_{ij}] = diag((h_1)^2, (h_2)^2, \ldots, (h_n)^2)$. One can construct local orthonormal vectors by dividing each coordinate basis vector by its corresponding scale factor $h_i = \sqrt{g_{ii}}$. If one prefers to express the divergence in terms of these **normalized** basis vectors, an alternative form arises:

$$(\boldsymbol{I.8.005})\ div\ V(V^1, V^2, \ldots, V^n) = \sum_{i=1}^{n} \frac{1}{\sqrt{|det([g])|}} \frac{\partial}{\partial x^i}\left(\sqrt{\frac{|det([g])|}{g_{ii}}}\ V^i\right)$$

Here again, V^i are taken to be the contravariant components of V. The factor $\sqrt{|det([g])|}$ effectively captures how the basis vectors are rescaled to become orthonormal, which can simplify certain physical interpretations.

Let us apply this formula to find the divergence in cylindrical coordinates:

The (covariant) metric tensor for cylindrical coordinates is given by:

$$[g_{ij}] = \begin{bmatrix} 1 & 0 & 0 \\ 0 & r^2 & 0 \\ 0 & 0 & 1 \end{bmatrix}$$

It follows that:

$$det([g_{ij}]) = r^2$$

Now, we apply $(I.8.005)$:

$$div\ V(V^r, V^\theta, V^h) = \underbrace{\frac{1}{r}}_{1/\sqrt{|det(g)|}} \frac{\partial}{\partial x^r}\left(\sqrt{\underbrace{\frac{r^2}{1}}_{\frac{|det(g)|}{g_{rr}}}}\ V^r\right) +$$

$$\underbrace{\frac{1}{r}}_{1/\sqrt{|det(g)|}} \frac{\partial}{\partial x^\theta}\left(\sqrt{\underbrace{\frac{r^2}{r^2}}_{\frac{|det(g)|}{g_\theta}}}\ V^\theta\right) + \underbrace{\frac{1}{r}}_{1/\sqrt{|det(g)|}} \frac{\partial}{\partial x^h}\left(\sqrt{\underbrace{\frac{r^2}{1}}_{\frac{|det(g)|}{g_{hh}}}}\ V^h\right) =$$

$$\frac{1}{r}\frac{\partial}{\partial x^r}(r\,V^r) + \frac{1}{r}\frac{\partial}{\partial x^\theta}(V^\theta) + \frac{\partial}{\partial x^h}(V^h)$$

You should recognize the form you find normally in physic textbooks.

Special case: orthogonal system

In case that (x^i) is an **orthogonal** system, the square root of the determinant of the metric tensor is just the product of the Lamé-coefficients h_i, furthermore we can express $\sqrt{\dfrac{|det([g])|}{g_{ii}}}$ as $\prod_{k \neq i} h_k$, so we have:

$$(\boldsymbol{I.8.006})\ div\ V = \sum_i \frac{1}{\prod_j h_j} \frac{\partial}{\partial x^i}\left(V^i \prod_{k \neq i} h_k\right)$$

Note that in the second product we exclude h_i.

In two dimensions you get:

$$div\ V = \frac{1}{h_1 h_2}\left[\frac{\partial}{\partial u^1}(h_2 V^1) + \frac{\partial}{\partial u^2}(h_1 V^2)\right]$$

In three dimensions you get:

$$div\ V = \frac{1}{h_1 h_2 h_3}\left[\frac{\partial}{\partial u^1}(h_2 h_3 V^1) + \frac{\partial}{\partial u^2}(h_1 h_3 V^2) + \frac{\partial}{\partial u^3}(h_1 h_2 V^3)\right]$$

I.8.2.E Exercise

$(\boldsymbol{I.8.2.E.1})$ Given the vector field $V = \begin{pmatrix} V_\sigma \\ V_\tau \\ V_z \end{pmatrix} = \begin{pmatrix} \frac{1}{\sigma^2} \\ 0 \\ 0 \end{pmatrix}$, calculate the divergence in parabolic 3D coordinates given by the set of functions

$$x^1(\sigma,\tau,z) = \sigma\tau\ cos(\phi)$$
$$x^2(\sigma,\tau,z) = \sigma\tau\ sin(\phi)$$
$$x^3(\sigma,\tau,z) = \frac{1}{2}(\tau^2 - \sigma^2)$$

$(\boldsymbol{I.8.2.E.2})$ Given the following coordinate system:

$$u(x,y,z) = x^2 + y^2$$

$v(x, y, z) = xy$

$w(x, y, z) = z$

Calculate curl of the following vector field:

$V = (V_1, V_2, V_3) = u\hat{e}_u + v^2 \hat{e}_v + w\hat{e}_w$

I.8.3 Curl in general coordinates

Strictly speaking, the **curl** of a vector field as a vector is defined in three-dimensional spaces. This is because the cross-product A×B is a 3D-only construct.

In three dimensions, let V be a vector field whose components V_i are **covariant** (index down). Also let $\{e_i\}$ be the (generally non-orthonormal) coordinate basis vectors corresponding to $\partial/\partial x^i$. The metric determinant is $det([g_{ij}]) \neq 0$, and ε^{ijk} is the (contravariant) Levi-Civita symbol in 3D.

$$\underbrace{}_{=g}$$

Then a fully covariant derivative approach to the curl in these coordinates reads:

$$(I.8.007) \; \boldsymbol{curl}\, V(V_1, V_2, V_3) = \frac{1}{\sqrt{|det([g])|}} \; \varepsilon^{ijk} \; e_i \left(\frac{\partial V_k}{\partial x^j} - \Gamma_{jk}^m \, V_m \right)$$

$$= \frac{1}{\sqrt{|det([g])|}} \; \varepsilon^{ijk} \; e_i \; \frac{\partial V_k}{\partial x^j}$$

We need to explain the last equal sign: ε^{ijk} is anti-symmetric and the Christoffel symbol is symmetric in the lower indices. Therefore, we get:

$$\varepsilon^{ijk} \Gamma_{jk}^m = -\varepsilon^{ijk} \Gamma_{kj}^m$$

So, summing up results that the Christoffel symbols cancel out. Good news, we don't need them for curl!

This formula is used in contexts where:

1. The coordinate system may be curved or nontrivial (e.g., a curved 3D manifold).

2. The basis vectors e_i are not "fixed" but vary with position.

3. A full covariant derivative (with Christoffel symbols) is necessary, as in general relativity or other geometric settings where you want coordinate-independent definitions.

In many familiar applications (e.g., standard cylindrical, spherical, or other "static" coordinate systems in a flat 3D space), one often sees the simpler classical curl expression:

$$(I.\,8.\,008)\ \boldsymbol{curl}\ V(V_1, V_2, V_3) = \frac{1}{\sqrt{|det([g])|}}\ \varepsilon^{ijk}\ e_i \left(\frac{\partial V_k}{\partial x^j} - \frac{\partial V_j}{\partial x^k} \right)$$

$$= \frac{1}{\sqrt{|det(g)|}} \left[e_1 \left(\frac{\partial V_3}{\partial x^2} - \frac{\partial V_2}{\partial x^3} \right) + e_2 \left(\frac{\partial V_1}{\partial x^3} - \frac{\partial V_3}{\partial x^1} \right) \right.$$

$$\left. + e_3 \left(\frac{\partial V_2}{\partial x^1} - \frac{\partial V_1}{\partial x^2} \right) \right]$$

Here, the Christoffel symbols do not explicitly appear. This form is valid when:

- We are working in standard 3D space (flat) with known coordinate systems (e.g., cylindrical or spherical).

- The dependence of the basis vectors on position can effectively be captured by the metric determinant factor and the partial derivatives themselves—no further "connection terms" are needed.

- **Important**: V_i here are still covariant components, and e_i are the coordinate basis vectors (not necessarily orthonormal). If you want an orthonormal basis $\hat{e}_i$, you must insert additional scale-factor terms.

- When the coordinate system is fixed (e.g., cylindrical or spherical coordinates) and the basis vectors are explicitly defined.

- When corrections due to changing, basis vectors aren't necessary.

Because $\{e_i\}$ in a general (or even orthogonal) coordinate system are not necessarily unit length, some applications prefer to express the curl in terms of orthonormal basis vectors $\hat{e}_i$. In an orthogonal system, each basis vector can be normalized by dividing by the respective scale factor $h_i = \sqrt{g_{ii}}$. Then the curl formula picks up extra $1/h_i$ factors:

$(I.8.009)\ \boldsymbol{curl\ V}(V_1, V_2, V_3)$

$$= \frac{1}{\sqrt{|det(g)|}}\left[e_1 \frac{1}{h_1}\left(\frac{\partial V_3}{\partial x^2} - \frac{\partial V_2}{\partial x^3}\right) + e_2 \frac{1}{h_2}\left(\frac{\partial V_1}{\partial x^3} - \frac{\partial V_3}{\partial x^1}\right)\right.$$
$$\left. + e_3 \frac{1}{h_3}\left(\frac{\partial V_2}{\partial x^1} - \frac{\partial V_1}{\partial x^2}\right)\right]$$

For an orthogonal system (x^i), $\sqrt{|det([g])|}$ equals to $h_1 h_2 h_3$, where $h_i = \sqrt{g_{ii}}$ (scale factors, cf. (I.7.1)). Explicitly calculating then $(I.8.3.1b)$ results in:

$(I.8.010)\ \boldsymbol{curl\ in\ orthogonal\ systems}$:

$curl\ V(V^1, V^2, V^3)$

$$= \frac{1}{h_2 h_3}\left(\frac{\partial}{\partial x^2}(h_3 V^3) - \frac{\partial}{\partial x^3}(h_2 V^2)\right)\hat{e}_1$$
$$+ \frac{1}{h_3 h_1}\left(\frac{\partial}{\partial x^3}(h_1 V^1) - \frac{\partial}{\partial x^1}(h_3 V^3)\right)\hat{e}_2$$
$$+ \frac{1}{h_1 h_2}\left(\frac{\partial}{\partial x^1}(h_2 V^2) - \frac{\partial}{\partial x^2}(h_1 V^1)\right)\hat{e}_3$$

Important: Do not prematurely "cancel" or "shorten" the h_i factors under a derivative if they depend on x^i.

Notice, that we used **contravariant** V^i as − in orthogonal systems only - $V^i = g^{ii} V_i = \frac{V_i}{(h_i)^2}$ (see $(I.4.010)$):

Example:

If we take (x^i) as spherical coordinates, we already got earlier the scale factors

$$h_r = 1, \qquad h_\theta = r, \qquad h_\phi = r \sin(\theta)$$

For a vector field $V = (V^r, V^\theta, V^\varphi)$ we get:

$$curl_{spherical}\, V = \frac{1}{r^2 \sin(\theta)} \left(\frac{\partial}{\partial \theta} (r \sin(\theta)\, V^\varphi) - \frac{\partial}{\partial \phi} (r\, V^\theta) \right) \hat{e}_r$$

$$+ \frac{1}{r^2 \sin(\theta)} \left(\frac{\partial}{\partial \phi} (r\, V^r) - \frac{\partial}{\partial r} (r \sin(\theta)\, V^\varphi) \right) \hat{e}_\theta$$

$$+ \frac{1}{r^2 \sin(\theta)} \left(\frac{\partial}{\partial r} (r \sin(\theta)\, V^\theta) - \frac{\partial}{\partial \theta} (\sin(\theta)\, V^r) \right) \hat{e}_\varphi$$

We write it now as a vector and we shorten the factors where possible:

$$curl_{spherical}\, V = \begin{pmatrix} \dfrac{1}{r \sin(\theta)} \left(\dfrac{\partial}{\partial \theta} (\sin(\theta)\, V^\varphi) - \dfrac{\partial}{\partial \phi} \left(r\, V^2_\theta \right) \right) \\[2em] \dfrac{1}{r} \left(\dfrac{1}{\sin(\theta)} \dfrac{\partial}{\partial \phi} (r\, V^r) - \dfrac{\partial}{\partial r} (r\, V^\varphi) \right) \\[2em] \dfrac{1}{r} \left(\dfrac{\partial}{\partial r} (r\, V^\theta) - \dfrac{\partial}{\partial \theta} (V^r) \right) \end{pmatrix}$$

∎

Now, let V_i be a covariant vector. Then, we have:

$$V_{i;j} = \frac{\partial V_i}{\partial x^j} - V_k\, \Gamma^k_{ij}$$

$$V_{j;i} = \frac{\partial V_j}{\partial x^i} - V_k\, \Gamma^k_{ji}$$

As Christoffel symbols are symmetric in the lower indices, we get:

$$V_{i;j} - V_{j;i} = \frac{\partial V_i}{\partial x^j} - \frac{\partial V_j}{\partial x^i}$$

This is a covariant tensor of second order, so

$$curl\, V_i = V_{i;j} - V_{j;i}$$

I.8.3.E Exercise

$(\boldsymbol{I.8.3.E.1})$ Show that the following vector field is irrotational:

$$V(r,\theta,\phi) = r^2 \sin^2(\theta)\big(\, 3\sin(\theta)\cos(\phi)\hat{e}_r + 3\cos(\theta)\cos(\phi)\hat{e}_\theta - \sin(\phi)\hat{e}_\phi\big)$$

I.8.4 Laplacian in general coordinates

The Laplacian-operator $\nabla^2\varphi$, also notated as $\Delta\varphi$, of a scalar field φ in general coordinates can be written as:

$$(\boldsymbol{I.8.011}) \quad \nabla^2\varphi = \frac{1}{\sqrt{\det([g])}}\,\frac{\partial}{\partial x^i}\left(\sqrt{\det([g])}\; g^{ij}\,\frac{\partial\varphi}{\partial x^j}\right)$$

Note, please, that you must sum up over i and j. Note also, that we use the determinant of the covariant tensor $[g] = [g_{ij}]$ and the contravariant metric tensor g^{ij}.

In case that (x^i) is an orthogonal system, the square root of the determinant is just the product of the Lamé-coefficients, the component g^{ij} equals to $\frac{1}{g_{ij}}$.

Therefore, we can formulate:

> If (x^i) is orthogonal, then:
>
> $$(\boldsymbol{I.8.012}) \quad \nabla^2\varphi = \sum_i \frac{1}{\prod_j h_j}\,\frac{\partial}{\partial x^i}\left(\frac{\prod_j h_j}{h_i^{\,2}}\,\frac{\partial\varphi}{\partial x^i}\right)$$

We note that $\dfrac{\prod_j h_j}{h_i^{\,2}} = \dfrac{1}{h_i}\,\prod_{j,j\neq i} h_j$, so we can also write

$$(\boldsymbol{I.8.013}) \quad \nabla^2\varphi = \frac{1}{\prod_j h_j}\sum_i \frac{\partial}{\partial x^i}\left(\frac{1}{h_i}\prod_{j,j\neq i} h_j\,\frac{\partial\varphi}{\partial x^i}\right)$$

Expanding the equation to two dimensions gives the formula:

$$(I.8.014)\quad \nabla^2\varphi \;=\; \frac{1}{h_1 h_2}\frac{\partial}{\partial x^1}\left(\frac{h_2}{h_1}\frac{\partial\varphi}{\partial x^1}\right) + \frac{1}{h_1 h_2}\frac{\partial}{\partial x^2}\left(\frac{h_1}{h_2}\frac{\partial\varphi}{\partial x^2}\right)$$

Expanding the equation to three dimensions gives the formula:

$$\nabla^2\varphi \;=\; \frac{1}{h_1 h_2 h_3}\frac{\partial}{\partial x^1}\left(\frac{h_2 h_3}{h_1}\frac{\partial\varphi}{\partial x^1}\right)$$

$$+\; \frac{1}{h_1 h_2 h_3}\frac{\partial}{\partial x^2}\left(\frac{h_1 h_3}{h_2}\frac{\partial\varphi}{\partial x^2}\right)$$

$$+\; \frac{1}{h_1 h_2 h_3}\frac{\partial}{\partial x^3}\left(\frac{h_1 h_2}{h_3}\frac{\partial\varphi}{\partial x^3}\right)$$

<u>Example</u>: Laplacian in polar coordinates

We note that polar coordinates are orthogonal.

As before, we use the scale factors $h_r = 1$ and $h_\theta = r$

Using $(I.8.014)$, we get for a scalar field φ:

$$\nabla^2\varphi \;=\; \frac{1}{h_r h_\theta}\frac{\partial}{\partial r}\left(\frac{h_\theta}{h_r}\frac{\partial\varphi}{\partial r}\right) + \frac{1}{h_r h_\theta}\frac{\partial}{\partial \theta}\left(\frac{h_r}{h_\theta}\frac{\partial\varphi}{\partial \theta}\right)$$

$$=\; \frac{1}{1\,r}\frac{\partial}{\partial r}\left(\frac{r}{1}\frac{\partial\varphi}{\partial r}\right) + \frac{1}{1\,r}\frac{\partial}{\partial \theta}\left(\frac{1}{r}\frac{\partial\varphi}{\partial \theta}\right)$$

$$=\; \frac{1}{r}\frac{\partial}{\partial r}\left(r\frac{\partial\varphi}{\partial r}\right) + \frac{1}{r^2}\frac{\partial^2\varphi}{\partial \theta^2}$$

$$=\; \frac{\partial^2\varphi}{\partial r^2} + \frac{1}{r}\frac{\partial\varphi}{\partial r} + \frac{1}{r^2}\frac{\partial^2\varphi}{\partial \theta^2}$$

I.8.4.E Exercise

($I.8.4.E.1$) Let $\Theta(r,\theta,h) = r\,cos(\theta) + h^2$. Calculate $\nabla^2\Theta(r,\theta,h)$ in cylindrical coordinates.

I.9 Coordinate transformations and integrals

Sometimes, it is difficult to integrate a function. It might be easier to change the coordinate system from, let us say, cartesian to polar coordinates, cylindrical coordinates, or sphere coordinates. We need a clear understanding of what is going on.

Consider an "normal" (Riemann-) integral:

$$\int_a^b f(x)\, dx$$

How can we interpret this integral? We know that dx is an abbreviation for infinitesimal distance of point x. Calculating $f(x)\, dx$ is *the area of a rectangle with $f(x)$ as the length of the one side and dx the length of the second side.* So, the integral $\int_a^b f(x)\, dx$ is nothing else than summing up infinitesimal small areas of rectangles between a and b. This form of integral is called Riemann-integral.

Now, let us take the integral

$$\int\int_A f(x^1, x^2)\, dx^1\, dx^2$$

Here, we are in $\mathbb{R}^2$. The expression $f(x^1, x^2)\, dx^1 dx^2$ calculates the volume of the cube of the area A given by the sides $f(x^1, x^2)$, dx^1 and dx^2. So, the integral above sums up infinitesimal small cubes.

We are coming now to a point where we consider an integral over a function f and a transformation.

We have the following general theorem:

Theorem (Transformation of integrals):

Let $U \subseteq \mathbb{R}^n$ be an open set, $\phi: U \rightarrow \phi(U) \subseteq \mathbb{R}^n$ a diffeomorphism. Then a function f can be integrated if and only if the function

$x \rightarrow f\big(\phi(x)\big)\, \big|\det\big(J_{\phi(x)}\big)\big|$ can be integrated. In this case the following equation holds:

$$\int\limits_{\phi(U)} f(y)\,dy = \int\limits_{U} f(\phi(x))\,\left|\det\left(J_{\phi(x)}\right)\right|$$

A function ϕ is called a **diffeomorphism** if ϕ is bijective, ϕ is continiously differentible and ϕ^{-1} is also continiously differentible.

When transforming coordinates in multiple dimensions, especially in integrals over multiple variables, we often use the Jacobian determinant to enable integration from one coordinate representation to another. The Jacobian determinant is a scaling factor that describes the change in volume in coordinate space due to the transformation. The Jacobian determinant shows how much the volume in the original space changes when moving into the new space. If the Jacobian determinant is constant, the volume is conserved.

In a sense, you can think of the Jacobian determinant as the analogous concept to the derivative g'(x) in the one-dimensional substitution rule:

$$\int f(x)\,dx = \int f(\varphi(u)) * \varphi'(u)\,du$$

It takes into account the change in variables and their influence on integration. Both concepts are crucial for performing integral calculus in different coordinate systems and with different sets of variables.

Let us see some examples:

I.9.1 Integration in polar coordinates

Let $U \subseteq \mathbb{R}^2$. Using polar coordinates, we get:

$$(\textit{I.9.001})\int\limits_{U} f(x,y)\,dx\,dy = \int_{r}\int_{\varphi} f(r\cos(\varphi), r\sin(\varphi))\,r\,dr\,d\varphi$$

Notice, that $r > 0$ is the determinant of the corresponding Jacobi-matrix for polar coordinates.

I.9.2 Integration in cylindrical coordinates

Let $U \subseteq \mathbb{R}^3$. Using cylindrical coordinates, we get:

$$(\boldsymbol{I.9.002}) \iiint_U f(x,y,z)\,dx\,dy\,dz$$
$$= \int_r \int_\varphi \int_z f(\,r\cos(\varphi)\,,r\sin(\varphi)\,,z)\,r\,dz\,d\varphi\,dr$$

I.9.3 Integration in spherical coordinates

Let $U \subseteq \mathbb{R}^3$. Using spherical coordinates, we get:

$$(\boldsymbol{I.9.003}) \iiint_U f(x,y,z)\ dx\,dy\,dz =$$

$$\int_r \int_\varphi \int_\vartheta f(r\cos(\varphi)\sin(\vartheta)\,,r\sin(\varphi)\sin(\vartheta)\,,r\cos(\vartheta))\,r^2\sin(\vartheta)\,d\vartheta\,d\varphi\,dr$$

I.9.4 Example: diamond shape

The diamond shape A, as shown below, is bound by the functions

$$x + 2y = 2,$$

$$x - 2y = 2,$$

$$x + 2y = -2,$$

$$x - 2y = -2$$

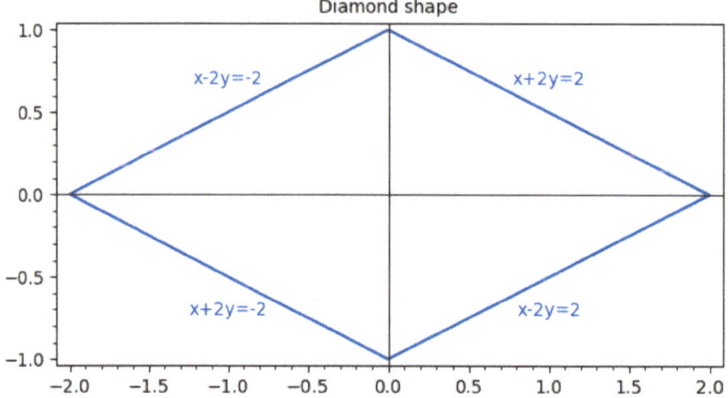

We want to calculate the integral

$$I = \iint_A (3x + 6y)^2 \, dA$$

It is not easy to integrate over the area A, as A is neither vertical nor horizontal. We would have to split up the area into two triangular regions. So, we will use a coordinate transformation that transforms the diamond shape into a rectangle, denoted by $\bar{A}$.

This can be done by using

(1) $u = x + 2y$ and

(2) $v = x - 2y$

The boundary lines are now $u = \pm 2$ and $v = \pm 2$

We need to find the inverse function so that we can express x and y in terms of u and v.

Adding (1) and (2) gives: $u + v = x$. Substracting (1) and (2) results in $u - v = 4y$. Thus, we get:

$$x = \frac{1}{2}(u + v)$$

$$y = \frac{1}{4}(u - v)$$

Now, we substitute x and y in the integrand. We get:

$$(3x + 6y)^2 = \left(\frac{3}{2}(u + v) + \frac{6}{4}(u - v)\right)^2 = 9(u)^2$$

As the next step, we need the Jacobi matrix:

$$J = \begin{bmatrix} \dfrac{\partial x}{\partial u} & \dfrac{\partial x}{\partial v} \\ \dfrac{\partial y}{\partial u} & \dfrac{\partial y}{\partial v} \end{bmatrix} = \begin{bmatrix} \dfrac{1}{2} & \dfrac{1}{2} \\ \dfrac{1}{4} & -\dfrac{1}{4} \end{bmatrix}$$

The determinant than calculates as

$$\det(J_{(u,v)}) = \frac{1}{2}\left(-\frac{1}{4}\right) - \frac{1}{4}\frac{1}{2} = -\frac{1}{4}$$

So, we have:

$$d\bar{A} = \left|\det(J_{(u,v)})\right| dA = \left|-\frac{1}{4}\right| dA = \frac{1}{4} dA$$

Using all together, we get:

$$\begin{aligned} I &= \int_{-2}^{2}\int_{-2}^{2} 9\,(u)^2\,\frac{1}{4}\,du\,dv \\ &= \int_{-2}^{2}\left[\frac{3}{4}\,(u)^3\right]_{-2}^{2}\,dv \\ &= \int_{-2}^{2} 6 - (-6)\,dv \\ &= 2 * 12 - (-2 * 12) = 48 \end{aligned}$$

I.9.5 Example

We are given a region in $\mathbb{R}^3$ defined by:

$$D = \{\, (x, y, z) \in \mathbb{R}^3, \qquad 0 \le x \le y$$
$$0 <= z <= 3$$
$$1 \le (x)^2 + (y)^2 \le 4 \}$$

Let us first decipher the conditions to see what kind of object the set D defines. We are dealing with a 3D-object, but as the z variable has a condition without any connection to x and y, we may begin with a 2-dimensional analysis.

The first condition tells us, that the points of D are uniquely in the first quadrant. Observe that $x \le y$, so in the first step we get:

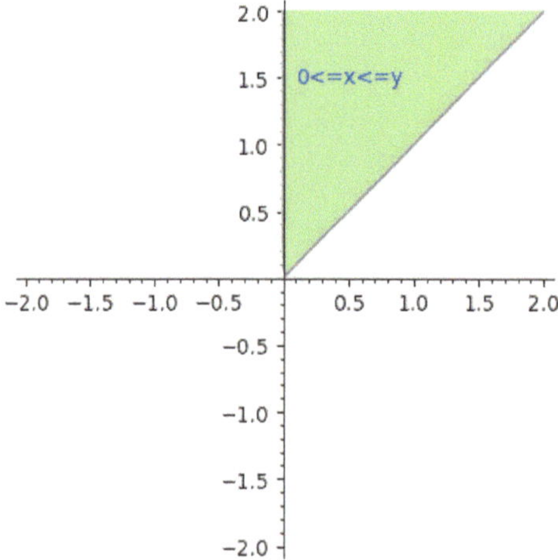

The condition $1 \le (x)^2 + (y)^2 \le 4$ describes two circles, one with radius 1 and the other with radius 2. Remember, $(x)^2 + (y)^2 = R^2$ describes a circle of radius R. In our example you have to take the square root, then! So, we have to take the inner of the circle with radius 2 minus the circle of radius 1. Putting this information in 2D, we get:

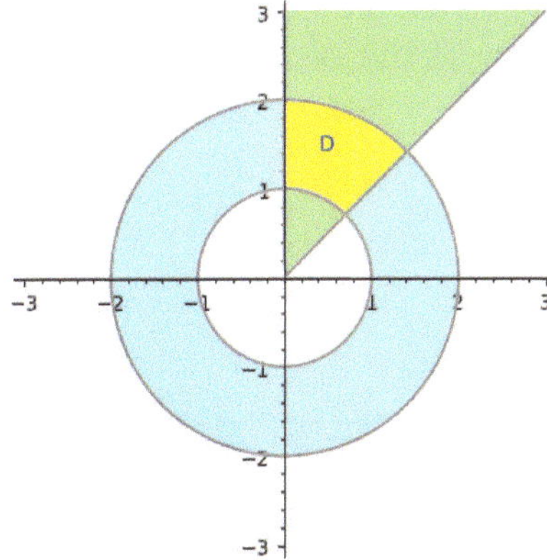

The yellow region is what we have so far. The condition $0 <= z <= 3$ tells us that the object defined so far has a thickness of 3, so, finally, we get something like this:

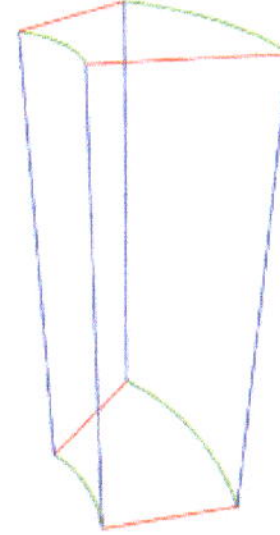

Let us calculate the volume V_D. In cartesian coordinates, we need to solve the following equation:

$$
\begin{aligned}
V_D &= \iiint_V 1 \, dV \\
&= \iint_{1<=(x)^2+(y)^2<=4} \int_0^3 1 \, dz \, d(x,y) \\
&= 3 \iint_{1<=(x)^2+(y)^2<=4} 1 \, d(x,y)
\end{aligned}
$$

The inner integral with dz was easy to manage, but the remaining integral is more difficult because of the boundaries.

Whenever you see something like $(x)^2 + (y)^2$, we are dealing with circles. Then you should change to polar coordinates, as the circle with radius R is simply represented as $r(t) = R$.

So, let us write down the polar transformation:

$$x = r \cos(\varphi)$$

$$y = r \sin(\varphi)$$

Let us talk now about r: we know that r is the radius of the circle, so we conclude that $r \in [1, 2]$.

Seeing at picture (2), we can conclude that $\varphi \in \left[{}^{\pi}/_{4}, {}^{\pi}/_{2}\right]$.

Now, we apply $(I.9.001)$:

$$
\begin{aligned}
V_D &= 3 \iint_{1 <= (x)^2 + (y)^2 <= 4} 1\, d(x, y) \\
&= 3 \int_{{}^{\pi}/_{4}}^{{}^{\pi}/_{2}} \int_{1}^{2} r\, dr\, d\varphi = \frac{9}{8}\pi
\end{aligned}
$$

I.9.E Exercises

$(I.9.E.1)$ Sphere inside of a cylinder.

Let $A = \{(x, y, z) \in \mathbb{R}^3 : x^2 + y^2 + z^2 \leq 1\}$,

$$B = \{(x, y, z) \in \mathbb{R}^3 : x^2 + y^2 \leq \frac{1}{2}\}$$

Calculate the volume of $V_{A \cap B} = A \cap B$.

$(I.9.E.2)$ Polar transformation

Calculate the integral

$$\iint_A 3x + 4(y)^2 \, dx\, dy$$

with $A = \{(x, y) \in \mathbb{R}^2, y \geq 0,\ 1 \leq (x)^2 + (y)^2 <= 4\}$

$(I.9.E.3)$ Cylindrical transformation

Let $B = \{(x, y, z) \in \mathbb{R}^3,\ (x)^2 + (y)^2 \leq 1,\ 0 \leq z \leq 1\}$.

Calculate

$$I = \int_B y\sqrt{(x)^2 + (y)^2} + z\ dx\ dy\ dz$$

$(I.9.E.4)$ Cardioid

Let $B = \{(x, y) \in \mathbb{R}^2, (x^2 + y^2)^2 - 2x(x^2 + y^2) - y^2 = 0\}$

And let $f(x, y) = 1$.

Calculate the area of f above the region B.

$(I.9.E.5)$ Part of a sphere

Calculate the volume of the section of the sphere

$x^2 + y^2 + z^2 \leq R^2$ and the half-space $z \leq R/2$.

I.10 Solution of exercises

($I.2.1.E.1$) Determine for the following equations if these are valid ESN equations:

1. $a_i + b_i = c_i$

   The expression is valid. It represents component-wise addition of two vector components, a_i and b_i, to form a new vector component c_i.

2. $a_i\, b_i = c_i\, d_i$

   This is **not** a valid ESN equation. In Einstein notation, you cannot multiply two vector components this way. It would require a summation over an index to make it valid, such as $a_i\, b_i = c_i$, which represents the dot product of two vectors.

3. $a_i b_i = c_i$

   This equation is **not** valid in ESN. In ESN, when you have repeated indices, they must be summed over. Here, the indices i in both a_i and b_i are repeated, but they are not summed over. It would be valid if written as:

   $$a_i b_i = \sum_{i=1}^{n} c_i$$

   This equation represents the dot product of two vectors a and b, resulting in a scalar c, where the sum over i represents the component-wise multiplication and addition.

4. $t_{ij}\, v^i = w^j$

   This equation is valid. We have a summation index i and a bounded index j on both sides.

It represents the contraction of a tensor t_{ij} with a vector v^i to produce another vector w^j. This is a common operation in tensor calculus, and we will see this equation later.

5. $a_{ij}\, b^{jk}\, c_k = d_i$

This is a valid ESN equation. It represents a contraction of the tensor a_{ij} with the tensor b_{jk} followed by a contraction with the vector c_k to produce a new vector d_i. This kind of operation is commonly used in tensor calculus. We will see it again.

6. $a_{ijk}\, b^{ij}\, c^k = d_i$

This equation is invalid as i is a summation index on the left side but a free index on the right side. Rule 4 $(I.2.006)$ is violated.

7. $m_{ij}\, v^j = w_i$

This equation is valid. It represents a multiplication of a matrix M with components m_{ij} with vector v. The result is a vector w with components w_i.

8. $v_i\, w^i = x$

This equation is valid. It represents the scalar product of the vector $v = (v_i)$ and $w = (w^i)$

9. $\varepsilon_{ijk}\, a^i\, b^j = c_k$

This expression is valid. It represents the cross product of two vectors a_i and b_j to produce another vector c_k, where ε_{ijk} is the Levi-Civita symbol (see chapter $I.2.1.3$) that accounts for the direction of the resulting vector. We will come back to the Levi-Civita symbol.

(**$I.2.1.2.E.1$**)

1. $\delta_{ij}\, a^i b^j = x$

   This expression is valid. Referring to ($I.2.022$) , we see that $\delta_{ij}\, a^i\, b_j = a_j\, b_j$. The Kronecker delta δ_{ij} ensures that the indices i and j are aligned for summation, resulting in a scalar x.

2. $\delta_{ij} + \delta_{ij} = 2$

   The expression is not valid: if $i = j$, then the equation above is true, but not in the other case. The equation $\delta_{ij} + \delta_{ij} = 2\,\delta_{ij}$ is correct.

3. $a_{ijk}\, \delta^{il} = b_{jk}$

This expression is invalid, as the bounded index l does not appear on the right side. Rule 4 ($I.2.006$) is violated.

(**$I.2.1.3.E.1$**) Determine if the following expressions are valid in ESN:

1. $\epsilon_{ilm}\, T_{jkl} = S_{ijm}$

   This expression is valid. It represents the contraction of the tensor T_{jkl} with the Levi-Civita symbol ϵ_{ilm} for the indices i, l and m. The Levi-Civita symbol performs a permutation of indices, and this operation results in the new tensor S_{ijm}.

2. $\epsilon_{lmn}\, a_{ijl}\, a_{kmn} = b_{ik}$

   This expression is valid. It represents a combination of operations involving the Levi-Civita symbol, the tensor a_{ijl}, and index contractions. The Levi-Civita symbol effectively performs a permutation of indices, and the equation results in a new tensor b_{ik} formed by contracting indices l and m from the tensor a_{ijl} and indices n from the tensor a_{kmn}.

3. $\epsilon_{ijk}\, v_j\, \delta_{kl}\, w_l = x_i$

   This equation is valid in ESN. It represents a cross-product operation between vectors v_j and w_l, with the result being a new vector x_i. The Kronecker delta δ_{kl} ensures that the indices k and l are aligned for summation.

4. $\epsilon_{ijk}\, \delta_{il}\, M_{lk} = N_{ij}$

   This equation is valid in ESN. It represents the contraction of the matrix M_{lk} with the Levi-Civita symbol ϵ_{ijk}, followed by a contraction with the Kronecker delta δ_{il}. The Kronecker delta ensures that the indices j and l are aligned for summation.

$(I.2.1.4.4.E.1)$ For a matrix A the trace of A – noted as $Tr(A)$ - is defined as the sum of its diagonal elements. Express the trace in ESN notation.

Solution:

$$Tr(A) = a_{ii}$$

$(I.2.1.4.4.E.2)$ Let A, B and C be matrices, so that you can perform the product ABC. Express this multiplication in ESN.

First, we note that $(AB)C = A(BC)$. So, we do not need to pay attention what multiplication we have to do first. However, we need to pay attention to the order of multiplication.

We define $\tilde{D} = AB$. Then we get: $\tilde{D}^i{}_j = A^i{}_r\, B^r{}_j$.
Now, we define $D = \tilde{D}C$, resulting in $D^k{}_l = \tilde{D}^k{}_s\, C^s{}_l$. Replacing $\tilde{D}^k{}_s$, (see $(I.2.014)$) we get:
$$D^k{}_l = A^k{}_r\, B^r{}_s\, C^s{}_l$$

$(I.2.1.4.4.E.3)$ Let A, B and C be matrices, so that the operation

$A(B + C)$ can be calculated. Express this term in ESN.

Solution:

The ESN of $A(B + C)$ can be expressed by:

$A_{ij} = A_{ik}\, B_{kj} + A_{ik}\, C_{kj}$

$(\boldsymbol{I.2.1.5.E.1})$ Given a scalar field $\phi(x^1, x^2, x^3)$, express $\dfrac{\partial \phi}{\partial x^i}$ and $\dfrac{\partial \phi}{\partial x^i \partial x^j}$ in ESN, where $i = 1,2,3$.

Solution:

$$\frac{\partial \phi}{\partial x^i} = \partial_i \phi$$

$$\frac{\partial \phi}{\partial x^i \partial x^j} = \partial_i \partial_j \phi$$

∎

$(\boldsymbol{I.2.1.5.E.2})$ Given a vector field $A^i(x^1, x^2, x^3)$, express $\dfrac{\partial A^i}{\partial x^j}$ in ESN, where $i = 1,2,3$.

Solution:

$$\frac{\partial A^i}{\partial x^j} = \partial_j A^i$$

∎

$(\boldsymbol{I.2.2.E.1})$ Let $r(t)$ be defined as:

$$r(t) = \begin{pmatrix} \cos 2\pi t \\ \sin 2\pi t \\ k\,t \end{pmatrix} \text{ with } k, t \in \mathbb{R}, k \text{ a fixed value} \neq 0.$$

Parameterize $r(t)$ by arc length.

Solution:

Let us first have a view on the function:

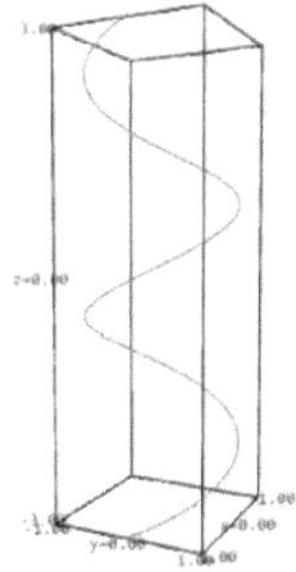

$r(t)$ represents a spiral.

We calculate $\dot{r}$ as:

$$\dot{r} = \frac{dr}{dt} = \begin{pmatrix} -2\pi\,sin(\,2\pi t\,) \\ 2\pi\,cos(\,2\pi t\,) \\ k \end{pmatrix}$$

It follows that

$$|\dot{r}| = \sqrt[2]{4\pi^2\,sin^2(\,2\pi t\,) + 4\pi^2\,cos^2(\,2\pi t\,) + k^2} = \sqrt[2]{4\pi^2 + k^2}$$

Now, we calculate the arc length

$$s(t) = \int_0^t \sqrt[2]{4\pi^2 + k^2}\; du = \sqrt[2]{4\pi^2 + k^2}\; t$$

The inverse function, then, is

$$t(s) = \frac{s}{\sqrt[2]{4\pi^2 + k^2}}$$

Replacing the last term in the parameterization, we get:

$$
r(s) = \begin{pmatrix} \cos\left(2\pi \dfrac{s}{\sqrt[2]{4\pi^2 + k^2}}\right) \\[1.5em] \sin\left(2\pi \dfrac{s}{\sqrt[2]{4\pi^2 + k^2}}\right) \\[1.5em] k \dfrac{s}{\sqrt[2]{4\pi^2 + k^2}} \end{pmatrix}
$$

($\boldsymbol{I.2.2.E.2}$) Let $r(t)$ be defined as:

$$
r(t) = \begin{pmatrix} \sin(2t)\,\cos(t) \\ \sin(2t)\,\sin(t) \end{pmatrix} \text{ with } t \in [0, 2\pi]
$$

Determine if $r(t)$ is parameterized by arc length.

Solution:

We need to check if $|\dot{r}(t)| = 1$

Using the product rule and chain rule, we get:

$$
\dot{r}(t) = \begin{pmatrix} 2\cos(2t)\,\cos(t) - \sin(2t)\,\sin(t) \\ \sin(t)\,\cos(t) - 2\cos(2t)\,\sin(t) \end{pmatrix}
$$

As for $|\dot{r}(t)|$, we get (using c for cos and s for sin):

$$
|\dot{r}(t)| = \sqrt{(2c\,(2t)\,c(t) - s(2t)\,s\,(t))^2 + (s(t)\,c\,(t) - 2c(2t)\,s(t))^2}
$$

Now, $|\dot{r}(0)| = 2$, so the equation $|\dot{r}(t)| = 1$ does not hold for all $t \in [0, 2\pi]$. That is why $r(t)$ is not parameterized by arc length.

■

$I.2.3.2.E.1$ Given the equation $(x^1)^2 \, (x^2)^4 - 3 = \sin(x^1 \, x^2)$.

Calculate $\dfrac{dy}{dx}$.

Solution:

This is obviously an implicit function of x^1 and x^2. We have to rewrite the function as:

$$(x^1)^2 \, (x^2)^4 - 3 - \sin(x^1 \, x^2) = 0$$

Defining

$$f(x^1, x^2) = (x^1)^2 \, (x^2)^4 - 3 - \sin(x^1 \, x^2),$$

we can calculate now by using $(I.2.050)$:

$$\frac{dx^2}{dx^1} = -\frac{\partial f_{x^1}}{\partial f_{x^2}} = -\frac{2\,x^1\,(x^2)^4 - x^2\cos(x^1\,x^2)}{4\,(x^2)^3\,(x^1)^2 - x^1\cos(x^1\,x^2)}$$
$$= \frac{x^2\cos(x^1\,x^2) - 2\,x^1\,(x^2)^4}{x^1\cos(x^1\,x^2) - 4\,(x^2)^3\,(x^1)^2}$$

∎

$I.2.3.4.E.1$

Define $T(x,y,z) = T_0/(1 + x^2 + y^2 + z^2)$ with $T_0 \in \mathbb{R}$.

Calculate the gradient.

Solution:

We calculate the partial derivatives as follows:

$$\frac{\partial T}{\partial x} = \frac{\partial}{\partial x}\left(T_0/(1 + x^2 + y^2 + z^2)\right) = -\frac{2x\,T_0}{(1 + x^2 + y^2 + z^2)^2}$$

$$\frac{\partial T}{\partial y} = \frac{\partial}{\partial y}(T_0/(1 + x^2 + y^2 + z^2)) = -\frac{2y\,T_0}{(1 + x^2 + y^2 + z^2)^2}$$

$$\frac{\partial T}{\partial z} = \frac{\partial}{\partial z}(T_0/(1 + x^2 + y^2 + z^2)) = -\frac{2z\,T_0}{(1 + x^2 + y^2 + z^2)^2}$$

$$\Delta T = \begin{pmatrix} -\dfrac{2x\,T_0}{(1 + x^2 + y^2 + z^2)^2} \\[2ex] -\dfrac{2y\,T_0}{(1 + x^2 + y^2 + z^2)^2} \\[2ex] -\dfrac{2z\,T_0}{(1 + x^2 + y^2 + z^2)^2} \end{pmatrix}$$

∎

$I.2.3.8.E.1$: Given the function $f(r,\theta) = (\,e^{-r}\sin(\theta)\,,\,e^{-r}\cos(\theta)\,)$. Calculate the Jacobi-matrix $J(r,\theta)$

Solution:

$$J(r,\theta) = \begin{pmatrix} -e^{-r}\sin(\theta) & e^{-r}\cos(\theta) \\ e^{r}\cos(\theta) & -e^{r}\sin(\theta) \end{pmatrix}$$

∎

$I.2.4.E.1$ Given the function $z = \cos(\,x^2\,(x^1)^2\,)$ with $x^1(t) = t^4 - 2t$, $x^2(t) = 1 - t^6$, calculate $\partial z/\partial t$.

Solution:

$$\frac{\partial z(\,x^1(t),\,x^2(t))}{\partial t} = \frac{\partial z}{\partial x^1}\frac{\partial x^1}{\partial t} + \frac{\partial z}{\partial x^2}\frac{\partial x^2}{\partial t}$$

$$= \quad \underbrace{-2\, x^1\, x^2 \sin(\, x^2\, (x^1)^2\,)}_{=\frac{\partial z}{\partial x^1}}\,\underbrace{(4t^3 - 2)}_{=\frac{\partial x^1}{\partial t}}$$

$$+ \left[\underbrace{-x^2 \sin(\, x^2\, (x^1)^2\,)}_{=\frac{\partial z}{\partial x^2}}\,\underbrace{(-6t^5)}_{=\frac{\partial x^2}{\partial t}} \right]$$

$$= \quad -2\left[\underbrace{(t^4 - 2t)}_{=x^1}\,\underbrace{(1 - t^6)}_{=x^2}\right] \sin\left(\left[\underbrace{(1 - t^6)}_{=x^2}\,\underbrace{(t^4 - 2t)^2}_{=(x^1)^2}\right]\right)(\,4t^3$$
$$- 2)$$

$$+ \left[-\,\underbrace{(t^4 - 2t)^2}_{=(x^1)^2}\, \sin\left(\left[\underbrace{(1 - t^6)}_{=x^2}\,\underbrace{(t^4 - 2t)^2}_{=(x^1)^2}\right]\right)(-6t^5)\right]$$

∎

$$(\boldsymbol{I.2.4.E.2})$$

Given the function

$z = 4\, x^2 \sin(\, 2\, x^1)$ with
$x^1(u, p) = 3u - p, \quad x^2(u, p) = p^2 u, \quad u(t) = t^2 + 1$

Calculate $\dfrac{dz}{dt}$ and $\dfrac{dz}{dv}$

Solution:

The problem with this exercise is that the dependencies are concatenated. Concatenation in partial differentials means *multiplication*. Therefore, we get:

$$\frac{dz}{dt} \quad = \quad \frac{dz}{dx^1}\frac{dx^1}{du}\frac{du}{dt} + \frac{dz}{dx^2}\frac{dx^2}{du}\frac{du}{dt}$$

$$= \quad \underbrace{8\,x^2 \cos(\,2\,x^1)}_{=\frac{dz}{dx^1}}\;\underbrace{(3)}_{=\frac{dx^1}{du}}\;\underbrace{(2t)}_{=\frac{du}{dt}} + \underbrace{(4\sin\,(\,2\,x^1)\,)}_{=\frac{dz}{dx^2}}\,\underbrace{(p^2)}_{=\frac{dx^2}{du}}\;\underbrace{(2t)}_{=\frac{du}{dt}}$$

$$= \quad 48\,x^2 \cos(\,2\,x^1) + 8\,t\,(p^2)\,\sin\,(\,2\,x^1)$$

$$\frac{dz}{dp} \;=\; \frac{dz}{dx^1}\frac{dx^1}{dp} + \frac{dz}{dx^2}\frac{dx^2}{dp}$$

$$= \quad \underbrace{8\,x^2 \cos(\,2\,x^1)}_{=\frac{dz}{dx^1}}\,\underbrace{(-1)}_{=\frac{dx^1}{dp}} + \underbrace{(4\sin\,(\,2\,x^1)\,)}_{=\frac{dz}{dx^2}}\;\underbrace{(2pu)}_{=\frac{dx^2}{dp}}$$

$$= \quad -8\,x^2 \cos(\,2\,x^1) + 8pu\,\sin\,(\,2\,x^1)$$

We do not use "back-substitution" with these derivatives – it would become unreadable.

∎

($I.2.4.E.3$) Given the equations

$$w = f(\,x^1, x^2, x^3\,)$$

$$x^1 = x^1(t)$$

$$x^2 = x^2(u, v, p\,)$$

$$x^3 = x^3(\,v, p\,)$$

$$v = v(r, u\,)$$

$$p = p(\,t, u\,)$$

Determine $\dfrac{\partial w}{\partial t}$ and $\dfrac{\partial w}{\partial u}$.

Solution:

As in the previous exercise we have to deal with concatenated chain rules.

We get:

$$(1) \quad \frac{\partial w}{\partial t} = \frac{\partial w}{\partial x^1}\frac{\partial x^1}{\partial t} + \frac{\partial w}{\partial x^2}\frac{\partial x^2}{\partial t} + \frac{\partial w}{\partial x^3}\frac{\partial x^3}{\partial t}$$

That is the chain rule in its basic form. Now, we calculate $\dfrac{\partial x^i}{\partial t}$ for $i = 1,2,3$:

- $$\frac{\partial x^1}{\partial t} = \frac{\partial x^1}{\partial t}$$
  Here, there is nothing to do as $x^1 = x^1(t)$ is only dependant from t.

- $$\frac{\partial x^2}{\partial t} = \frac{\partial x^2}{\partial u}\frac{\partial u}{\partial t} + \frac{\partial x^2}{\partial v}\frac{\partial v}{\partial t} + \frac{\partial x^2}{\partial p}\frac{\partial p}{\partial t}$$

  $$\frac{\partial u}{\partial t} = 0 \text{ as it is not dependant from } t$$

  $$\frac{\partial v}{\partial t} = 0 \text{ , with the same argument.}$$

  $$\frac{\partial p}{\partial t} = \frac{\partial p}{\partial t}, \text{ as } p \text{ is dependant from t}$$

We get:

$$\frac{\partial x^2}{\partial t} = \frac{\partial x^2}{\partial u}\underbrace{\frac{\partial u}{\partial t}}_{=0} + \frac{\partial x^2}{\partial v}\underbrace{\frac{\partial v}{\partial t}}_{=0} + \frac{\partial x^2}{\partial p}\frac{\partial p}{\partial t} = \frac{\partial x^2}{\partial p}\frac{\partial p}{\partial t}$$

$$\frac{\partial x^3}{\partial t} = \frac{\partial x^3}{\partial v}\underbrace{\frac{\partial v}{\partial t}}_{=0} + \frac{\partial x^3}{\partial p}\frac{\partial p}{\partial t} = \frac{\partial x^3}{\partial p}\frac{\partial p}{\partial t}$$

Replacing $\dfrac{\partial x^i}{\partial t}$ in equation (1) gives us:

$$\frac{\partial w}{\partial t} = \frac{\partial w}{\partial x^1}\frac{\partial x^1}{\partial t} + \frac{\partial w}{\partial x^2}\frac{\partial x^2}{\partial p}\frac{\partial p}{\partial t} + \frac{\partial w}{\partial x^3}\frac{\partial x^3}{\partial p}\frac{\partial p}{\partial t}$$

Now, we calculate the second part of the exercise, and again we begin with the basic formula:

$$\frac{\partial w}{\partial u} = \frac{\partial w}{\partial x^1}\frac{\partial x^1}{\partial u} + \frac{\partial w}{\partial x^2}\frac{\partial x^2}{\partial u} + \frac{\partial w}{\partial x^3}\frac{\partial x^3}{\partial u}$$

Again, we calculate the partial derivatives:

- $\dfrac{\partial x^1}{\partial u} = 0$

- $\dfrac{\partial x^2}{\partial u} = \dfrac{\partial x^2}{\partial u}\underbrace{\dfrac{\partial u}{\partial u}}_{=1} + \dfrac{\partial x^2}{\partial v}\dfrac{\partial v}{\partial u} + \dfrac{\partial x^2}{\partial p}\dfrac{\partial p}{\partial u}$

- $\dfrac{\partial x^3}{\partial u} = \dfrac{\partial x^3}{\partial v}\dfrac{\partial v}{\partial u} + \dfrac{\partial x^2}{\partial p}\dfrac{\partial p}{\partial u}$

Putting all together, we get:

$$\frac{\partial w}{\partial u} = \frac{\partial w}{\partial x^2}\frac{\partial x^2}{\partial u} + \frac{\partial w}{\partial x^2}\frac{\partial x^2}{\partial v}\frac{\partial v}{\partial u} + \frac{\partial w}{\partial x^2}\frac{\partial x^2}{\partial p}\frac{\partial p}{\partial u} + \frac{\partial w}{\partial x^3}\frac{\partial x^3}{\partial v}\frac{\partial v}{\partial u} + \frac{\partial w}{\partial x^3}\frac{\partial x^3}{\partial p}\frac{\partial p}{\partial u}$$

∎

$(I.2.6.E.1)$:

Calculate the dual basis of the basis vectors $\left\{ \begin{pmatrix} -5 \\ 2 \\ 4 \end{pmatrix}, \begin{pmatrix} -5 \\ 1 \\ 3 \end{pmatrix}, \begin{pmatrix} 3 \\ -2 \\ -3 \end{pmatrix} \right\}$

Solution:

We apply $(I.2.064)$:

$$\vec{g}_1 = \begin{pmatrix} -5 \\ 2 \\ 4 \end{pmatrix}, \ \vec{g}_2 = \begin{pmatrix} -5 \\ 1 \\ 3 \end{pmatrix}, \ \vec{g}_3 = \begin{pmatrix} 3 \\ -2 \\ -3 \end{pmatrix} \quad \text{in } \mathbb{R}^3.$$

Step 1: build up the metric tensor $g_{ij} = \vec{g}_i \cdot \vec{g}_j$

We get:

$$[g_{ij}] = \begin{bmatrix} 45 & 39 & -31 \\ 39 & 35 & -26 \\ -31 & -26 & 22 \end{bmatrix}$$

Calculating the inverse of $[g_{ij}]$ gives us:

$$[g^{ij}] = [g_{ij}]^{-1} = \begin{pmatrix} 45 & 39 & -31 \\ 39 & 35 & -26 \\ -31 & -26 & 22 \end{pmatrix}^{-1} = \begin{pmatrix} 94 & -52 & 71 \\ -52 & 29 & -39 \\ 71 & -39 & 54 \end{pmatrix}$$

Take now the basis vectors as row-vectors:

$$\vec{g_1}^{\,T} = (\,-5, \quad 2, \quad 4\,)$$

$$\vec{g_2}^{\,T} = (\,-5, \quad 1, \quad 3\,)$$

$$\vec{g_3}^{\,T} = (\,3, \quad -2, \quad -3\,)$$

Calculate

$$\left[\vec{g}^{iT}\right] = [g_{ij}]^{-1} [\,\vec{g_1}\ \vec{g_2}\ \cdots\ \vec{g_n}\,]^T$$

$$= \begin{pmatrix} 94 & -52 & 71 \\ -52 & 29 & -39 \\ 71 & -39 & 54 \end{pmatrix} \begin{pmatrix} -5 & 2 & 4 \\ -5 & 1 & 3 \\ 3 & -2 & -3 \end{pmatrix}$$

$$= \begin{pmatrix} 3 & -6 & 7 \\ -2 & 3 & -4 \\ 2 & -5 & 5 \end{pmatrix}$$

We get:

$$\vec{g}^1 = (3\ -6\ 7\,)^T = \begin{pmatrix} 3 \\ -6 \\ 7 \end{pmatrix}$$

$$\vec{g}^2 = (2\ \ 3\ \ 4\,)^T = \begin{pmatrix} -2 \\ 3 \\ -4 \end{pmatrix}$$

$$\vec{g}^3 = (2\ -5\ \ 5\,)^T = \begin{pmatrix} 2 \\ -5 \\ 5 \end{pmatrix}$$

∎

I.2.7.E.1

Proof the following identities:

(a) $(a \otimes b)^T = b \otimes a$

(b) $(a \otimes b)(c \otimes d) = (b \cdot c)\, a \otimes d$

Solution:

(a) $(a \otimes b)^T \underset{(I.2.066)}{=} (ab^T)^T = b^{T^T} a^T = ba^T \underset{(I.2.066)}{=} b \otimes a$

(b)

$$(a \otimes b)(c \otimes d)\, x \quad = \quad (a \otimes b)\,(d \cdot x)\, c \qquad (I.2.069)$$

$$= \quad (d \cdot x)\,(b \cdot c)\, a$$

$$= \quad (b \cdot c)\,(d \cdot x)\, a$$

$$= \quad (b \cdot c)\,(a \otimes d)\, x \qquad (I.2.069)$$

∎

$(I.4.E.1)$ Given the following ESN expressions: simplify them in a way that the metric coefficients g_{ij} vanish.

(a) $a^m g_{lm}$

(b) $g^{pk} u_p$

(c) $g^{ij}\, b^k\, g_{kj}$

(d) $g_{ij} g^{jk}$

Solution:

$$(a) \quad a^m g_{lm} \underset{(I.4.009)}{=} a_l$$

$$(b) \quad g^{pk} u_p \underset{(I.4.010)}{=} u^k$$

$$(c) \quad g^{ij} b^k g_{kj} \underset{(I.4.009)}{=} g^{ij} b_j \underset{(I.4.010)}{=} b^i$$

$$(d) \quad g_{ij} g^{jk} \underset{(1*)}{=} \delta_{ik}$$

(1^*) is due to the definition of the metric tensor

■

$(I.4.E.2)$ Let V be a 2-dimensional vector space and e^1, e^2 a basis. Let g be the metric tensor associated to e^1 and e^2. Suppose we know that

$$g(e^1, e^1) = 3, \ g(e^1, e^2) = 1 \text{ and } g(e^2, e^2) = 2,$$

where $g(v, w) := v^T g \, v$

Calculate

    (d) $|e^1|$ and $|e^2|$
    (e) $|e^1 + e^2|$
    (f) $|e^1 - 2e^2|$

Solution:

As the metric tensor is symmetric, we conclude that $g(e^1, e^2) = g(e^2, e^1)$ so:

$$[g_{ij}] = \begin{bmatrix} g(e^1, e^1) & g(e^1, e^2) \\ g(e^2, e^1) & g(e^2, e^2) \end{bmatrix} = \begin{bmatrix} 3 & 1 \\ 1 & 2 \end{bmatrix}$$

We use $(I.4.014)$:

(a) $|e^1| = \sqrt{g(e^1,e^1)} = \sqrt{3}, \quad |e^2| = \sqrt{g(e^2,e^2)} = \sqrt{2}$

(b) $|e^1 + e^2| = \sqrt{g(e^1 + e^2, e^1 + e^2)} =$

$\sqrt{g(e^1,e^1) + g(e^2,e^2) + 2g(e^1,e^2)} = \sqrt{7}$

(c) $|e^1 - 2e^2|^2 \underset{(I.4.014)}{=} (1 \;\; -2) \begin{bmatrix} 3 & 1 \\ 1 & 2 \end{bmatrix} \begin{pmatrix} 1 \\ -2 \end{pmatrix} = (1 \;\; -2) \begin{pmatrix} 1 \\ -3 \end{pmatrix} = 7,$

hence $|e^1 - 2e^2| = \sqrt{7}$

■

$(I.4.E.3)$ Given the basis vectors

$$\vec{g}_1 = \begin{pmatrix} -5 \\ 2 \\ 4 \end{pmatrix}, \qquad \vec{g}_2 = \begin{pmatrix} -5 \\ 1 \\ 3 \end{pmatrix}, \qquad \vec{g}_3 = \begin{pmatrix} 3 \\ -2 \\ -3 \end{pmatrix},$$

compute the metric tensor.

Solution:

Following $(I.4.007)$, we calculate:

$$[g_{ij}] = [\vec{g}_1 \; \vec{g}_2 \; \cdots \; \vec{g}_n]^T [\vec{g}_1 \; \vec{g}_2 \; \cdots \; \vec{g}_n]$$

$$= \begin{pmatrix} -5 & 2 & 4 \\ -5 & 1 & 3 \\ 3 & -2 & 1 \end{pmatrix} \begin{pmatrix} -5 & -5 & 3 \\ 2 & 1 & -2 \\ 4 & 3 & -1 \end{pmatrix}$$

$$= \begin{pmatrix} 45 & 39 & -23 \\ 39 & 35 & 20 \\ -23 & -20 & 14 \end{pmatrix}$$

■

$(\boldsymbol{I.4.E.4})$ Show that the vectors $\vec{v} = \begin{pmatrix} 0 \\ 1 \\ 2\,b\,r\,sin(\theta) \end{pmatrix}$ and $\vec{w} =$
$\begin{pmatrix} 0 \\ -2\,b\,r\,sin(\theta) \\ r^2\,cos^2(\theta) \end{pmatrix}$ are orthogonal under cylindrical coordinates (see also
$I.A.1.2$) defined as

$$x^1(r,\theta,h\,) = r\,cos\,\theta$$

$$x^2(r,\theta,h\,) = r\,sin\,\theta$$

$$x^3(r,\theta,h) = h$$

with $r \geq 0,\ 0 \leq \theta \leq 2\pi,\ -\infty < h < \infty$.

Solution:

Step 1: calculate the Jacobi matrix.

The Jacobi-matrix for cylindrical coordinates is:

$$
\mathcal{J}_{(r,\theta,h)} \;=\;
\begin{bmatrix}
\dfrac{\partial x^1}{\partial r} & \dfrac{\partial x^1}{\partial \theta} & \dfrac{\partial x^1}{\partial h} \\[2mm]
\dfrac{\partial x^2}{\partial r} & \dfrac{\partial x^2}{\partial \theta} & \dfrac{\partial x^2}{\partial h} \\[2mm]
\dfrac{\partial x^3}{\partial r} & \dfrac{\partial x^3}{\partial \theta} & \dfrac{\partial x^3}{\partial h}
\end{bmatrix}
=
\begin{bmatrix}
cos\,\theta & -r\,sin\,\theta & 0 \\
sin\,\theta & r\,cos\,\theta & 0 \\
0 & 0 & 1
\end{bmatrix}
$$

Step 2: calculate the metric tensor.

Using $(I.4.008)$ we get the metric tensor $\left[g_{ij}\right]$ as

$$
\left[g_{ij}\right] \;=\;
\begin{bmatrix}
1 & 0 & 0 \\
0 & r^2 & 0 \\
0 & 0 & 1
\end{bmatrix}
$$

Step 3: check orthogonality by using the metric tensor.

We defined the general scalar product of two vectors as $v \cdot w = g_{ij}\,v^i w^j$. So v and w are orthogonal if and only if $g_{ij}\,v^i w^j = 0$.

We translate the equation $g_{ij}\, v^i w^j$ in matrix term:

$$[0, 1, 2\, b\, r\, sin(\theta)] \underbrace{\begin{bmatrix} 1 & 0 & 0 \\ 0 & r^2 & 0 \\ 0 & 0 & 1 \end{bmatrix} \begin{pmatrix} 0 \\ -2\, b\, r\, sin(\theta) \\ r^2\, cos^2(\theta) \end{pmatrix}}_{caculate\ this\ first!}$$

$$= \quad (0, 1, 2\, b\, r\, sin(\theta)) \begin{pmatrix} 0 \\ -2\, b\, rsin(\theta)\, r^2\, cos^2(\theta) \\ r^2\, cos^2(\theta) \end{pmatrix}$$

$$= \quad 0 - 2\, b\, r\, sin(\theta)\, r^2\, cos^2(\theta) + 2\, b\, r\, sin(\theta)\, r^2\, cos^2(\theta)$$

$$= \quad 0$$

$\blacksquare$

($I.4.E.5$) Show that the contravariant vectors $\vec{v} = \begin{pmatrix} -x^1/\,x^2 \\ 1 \\ 0 \end{pmatrix}$ and $\vec{w} = \begin{pmatrix} 1/x^2 \\ 0 \\ 0 \end{pmatrix}$ in curvilinear coordinates (x^i) are orthogonal if is related to rectangular coordinates $(\bar{x}^i)$ by

$$\bar{x}^1 = x^2, \quad \bar{x}^2 = x^3, \quad \bar{x}^3 = x^1 x^2$$

Solution:

We need to do the same steps as in exercise ($I.4.E.4$):

The Jacobi-matrix calculates as:

$$\mathcal{J}_{(x^1, x^2, x^3)} \quad = \quad \begin{bmatrix} \dfrac{\partial \bar{x}^1}{\partial x^1} & \dfrac{\partial \bar{x}^1}{\partial x^2} & \dfrac{\partial \bar{x}^1}{x^3} \\[2mm] \dfrac{\partial \bar{x}^2}{\partial x^1} & \dfrac{\partial \bar{x}^2}{\partial x^2} & \dfrac{\partial \bar{x}^2}{\partial x^3} \\[2mm] \dfrac{\partial \bar{x}^3}{\partial x^1} & \dfrac{\partial \bar{x}^3}{\partial x^2} & \dfrac{\partial \bar{x}^3}{\partial x^3} \end{bmatrix} = \begin{bmatrix} 0 & 1 & 0 \\ 0 & 0 & 1 \\ x^2 & x^1 & 0 \end{bmatrix}$$

Using ($I.4.008$) we get the metric tensor $\left[g_{ij} \right]$ as

$$[g_{ij}] \quad = \quad \mathcal{J}_{(x^1,x^2,x^3)}{}^T \, \mathcal{J}_{(x^1,x^2,x^3)} = \begin{bmatrix} (x^2)^2 & x^1 x^2 & 0 \\ x^1 x^2 & 1+(x^1)^2 & 0 \\ 0 & 0 & 1 \end{bmatrix}$$

Now we calculate:

$$\left[-\frac{x^1}{x^2}, 1, 0\right] \underbrace{\begin{bmatrix} (x^2)^2 & x^1 x^2 & 0 \\ x^1 x^2 & 1+(x^1)^2 & 0 \\ 0 & 0 & 1 \end{bmatrix} \begin{pmatrix} 1/x^2 \\ 0 \\ 0 \end{pmatrix}}_{calculate\ this\ first!} = \left[-\frac{x^1}{x^2}, 1, 0\right] \begin{pmatrix} (x^2)^2/x^2 \\ x^1 x^2/x^2 \\ 0 \end{pmatrix}$$

$$= \left[-\frac{x^1}{x^2}, 1, 0\right] \begin{pmatrix} x^2 \\ x^1 \\ 0 \end{pmatrix} = -\frac{x^1 x^2}{x^2} + x^1 = 0$$

∎

$(I.4.E.6)$ (building the metric tensor for spherical coordinates):

Let the cartesian coordinates x^1, x^2 and x^3 be expressed by:

$$x^1 = x^1(r, \theta, \phi\ ) = r \sin \theta \cos \phi$$

$$x^2 = x^2(r, \theta, \phi\ ) = r \sin \theta \sin \phi$$

$$x^3 = x^3(r, \theta, \phi) = r \, \cos \theta$$

with $r \geq 0$, $0 \leq \theta \leq \pi$ and $0 \leq \phi \leq 2\pi$.

(c) Calculate the metric tensor.

(d) Represent the squared line element s^2 as a linear combination of tensor products

Solution:

(a) We follow $(I.4.008)$. We calculate the Jacobi matrix:

$$J(r, \theta, \phi) \;=\; \begin{bmatrix} \dfrac{\partial x^1}{\partial r} & \dfrac{\partial x^1}{\partial \theta} & \dfrac{\partial x^1}{\partial \phi} \\[2ex] \dfrac{\partial x^2}{\partial r} & \dfrac{\partial x^2}{\partial \theta} & \dfrac{\partial x^2}{\partial \phi} \\[2ex] \dfrac{\partial x^3}{\partial r} & \dfrac{\partial x^3}{\partial \theta} & \dfrac{\partial x^3}{\partial \phi} \end{bmatrix}$$

$$= \begin{bmatrix} \cos(\phi)\sin(\theta) & r\cos(\theta)\cos(\phi) & -r\sin(\theta)\sin(\phi) \\ \sin(\theta)\sin(\phi) & r\cos(\theta)\sin(\phi) & r\cos(\phi)\sin(\theta) \\ \cos(\theta) & -r\sin(\theta) & 0 \end{bmatrix}$$

Using $c = \cos, s = \sin$:

$$[g_{ij}] \;=\; J^T J =$$

$$\begin{bmatrix} c(\phi)\,s(\theta) & s(\theta)\,s(\phi) & c(\theta) \\ r\,c(\theta)\,c(\phi) & r\,c(\theta)\,s(\phi) & -r\,s(\theta) \\ -r\,s(\theta)\,s(\phi) & r\,c(\phi)\,s(\theta) & 0 \end{bmatrix} \begin{bmatrix} c(\phi)\,s(\theta) & r\,c(\theta)\,c(\phi) & -r\,s(\theta)\,s(\phi) \\ s(\theta)\,s(\phi) & r\,c(\theta)\,s(\phi) & r\,c(\phi)\,s(\theta) \\ c(\theta) & -r\,s(\theta) & 0 \end{bmatrix}$$

$$= \begin{bmatrix} 1 & 0 & 0 \\ 0 & r^2 & 0 \\ 0 & 0 & r^2\sin^2(\theta) \end{bmatrix}$$

We shorten all calculations at this point. The reader may do the calculations and may consider that $\cos^2(\alpha) + \sin^2(\alpha) = 1$.

(b) Represent the squared line element s^2 as a linear combination of tensor products

Following $(I.4.023)$ $ds^2 = g_{ij}\, dx^i \otimes dx^j$ we get:

$$ds^2 \;=\; g_{ij}\, dx^i \otimes dx^j$$

$$\underset{g\ diag}{=} \quad g_{11}\, dx^1 \otimes dx^1 + g_{22}\, dx^2 \otimes dx^2 + g_{33}\, dx^3 \otimes dx^3$$

$$= \quad dr \otimes dr + r^2 \, d\theta \otimes d\theta + r^2 \sin^2(\theta) \, d\phi \otimes d\phi$$

∎

$(I.4.E.7)$ **(Arc length)**

Calculate the arc length of the curve

$$\mathfrak{C}: \begin{cases} x^1 = \ln\left(\sqrt[2]{1+t^2}\right) \\ x^2 = arc\tan t \end{cases}$$

with $0 \le t \le 2$.

Solution:

Graphical representation of the curve:

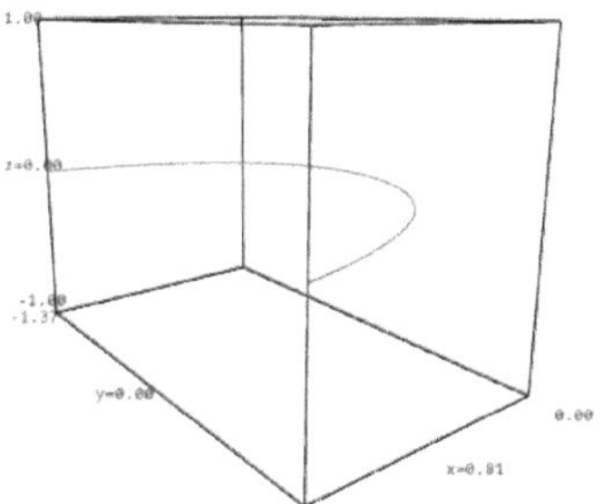

First, we are in Cartesian coordinates, so $g_{ij} = \delta_{ij}$. We want to use $(I.4.020)$, so first we calculate the differentials:

$$\frac{dx^1}{dt} = \frac{t}{t^2+1}$$

$$\frac{dx^2}{dt} = \frac{1}{t^2+1}$$

    (1) We note that both differentials are positive.

Now we get:

$$\left(\frac{ds}{dt}\right)^2 = \left(\frac{dx^i}{dt}\right)^T [g_{ij}] \left(\frac{dx^i}{dt}\right)$$

$$= \begin{bmatrix} \dfrac{t}{t^2+1} & \dfrac{1}{t^2+1} \end{bmatrix} \begin{bmatrix} \dfrac{t}{t^2+1} \\[2mm] \dfrac{1}{t^2+1} \end{bmatrix}$$

$$= \frac{1}{1+t^2}$$

Using $(I.4.020)$, we get:

$$s = \int_0^2 \sqrt{\left|\frac{dx^i}{dt}\frac{dx^j}{dt}\right|}\, dt$$

$$= \int_0^2 \sqrt{\frac{1}{1+t^2}}\, dt$$

$$= \int_0^2 \frac{1}{\sqrt{1+t^2}}\, dt$$

$$= [arcsinh(t)]_0^2$$

$$= arcsinh(2)$$

$(I.4.E.8)$ (arc length of implizit function)

Determine the length of $(2x^1 + (x^1)^2 + (x^2)^2)^2 = 4\,((x^1)^2 + (x^2)^2)$

Solution:

This is an implicit function in two dimensions. In case you encounter an implicit function in two dimensions, it is recommendable to switch over to polar coordinates. That is what we will do now (see $(I.A.1.1)$):

$$x^1 = r \cos \varphi$$

$$x^2 = r \sin \varphi$$

As we intend to use the squared line element for polar coordinates, we need to compute r. But first, we replace x^1 and x^2 in (1):

$$(2r \cos \varphi + (r \cos \varphi)^2 + (r \sin \varphi)^2)^2 = 4 \ ((r \cos \varphi)^2 + (r \sin \varphi)^2)$$

As $(r \cos \varphi)^2 + (r \sin \varphi)^2 = r^2$ we get:

$$(2r \cos \varphi + r^2)^2 = 4 \, r^2$$

Out multiplying the left side, we get:

$$4 \, r^2 \, \cos^2 \varphi + 4r \cos \varphi \ r^2 + r^4 = 4 \, r^2$$

Now, divide both sides by r^2:

$$4 \, \cos^2 \varphi + 4r \cos \varphi \ + r^2 = 4 \text{ , so}$$

$$4 \, \cos^2 \varphi + 4r \cos \varphi \ + r^2 - 4 = 0$$

This is a polynomial of degree 2, so we get two solutions:

$$(2) \, r = \ -2 \cos \varphi \ \pm 2$$

Let us have a look at what we get as a plot:

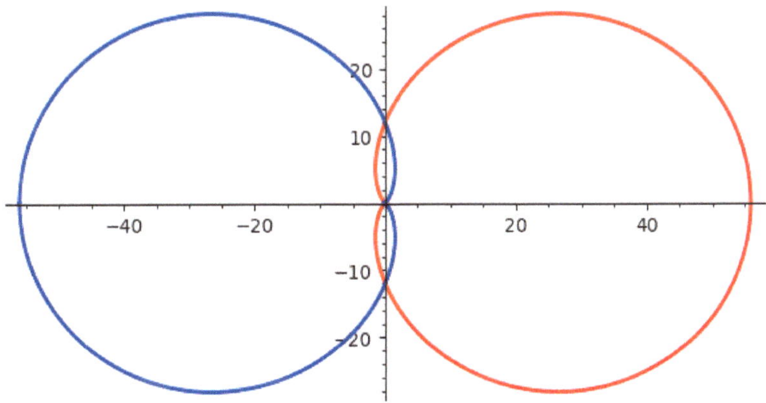

The red cardioid is the one with $r = -2\cos\varphi + 2$, the blue one with $r = -2\cos\varphi - 2$.

Now, we compute the squared line element for polar coordinates:

$$(3)\quad ds^2 = dr^2 + r^2\,d\theta^2.$$

Applying the chain rule, we can rewrite (3) as

$$(4)\quad ds^2 = \left(\frac{dr}{d\varphi}\right)^2 d\varphi^2 + r^2\,d\theta^2$$

We calculate $\frac{dr}{d\varphi}$ as

$$(5)\quad \frac{dr}{d\varphi} = \frac{d(-2\cos\varphi \pm 2)}{d\varphi} = 2\sin(\varphi)$$

Replacing (5) into (4):

$$
\begin{aligned}
ds^2 &= \left(2\sin(\varphi)\right)^2 d\varphi^2 + (-2\cos\varphi \pm 2)^2 d\varphi^2\\
&= 4\left(\sin^2(\varphi) + \cos^2(\varphi) \pm 2\cos(\varphi) + 1\right)d\varphi^2\\
&= 4\left(1 \pm 2\cos(\varphi) + 1\right)d\varphi^2\\
&= 4\left(2 \pm 2\cos(\varphi)\right)d\varphi^2\\
&= 8\left(1 \pm \cos(\varphi)\right)d\varphi^2
\end{aligned}
$$

For integrating, we need ds, so we need to take the square root of ds^2. Integrating trigonometric expressions embedded in square roots can be really very complicated, therefore we need to express $\sqrt{1 \pm \cos(\varphi)}$ in a form that we get rid off the square root.

We claim:

$$(6)\quad \sqrt{1 - \cos(\varphi)} = \sqrt{2}\,\sin(\varphi/2) \text{ with } \varphi \in [0, 2\pi]:$$

We use the identity $2\sin^2(a) = 1 - \cos(2a)$,

setting $a = \varphi/2$

So, we have:

$$2\sin^2(\varphi/2) = 1 - \cos(\varphi)$$

$$\sin^2\left(\varphi/2\right) = \frac{1 - \cos(\varphi)}{2}$$

$$\sin\left(\varphi/2\right) = \pm\sqrt{\frac{1 - \cos(\varphi)}{2}}$$

$$\sin\left(\varphi/2\right) \underset{\varphi \in [0,2\pi]}{=} +\sqrt{\frac{1 - \cos(\varphi)}{2}}$$

$$\sqrt{2}\,\sin\left(\varphi/2\right) = \sqrt{1 - \cos(\varphi)}$$

In a similar way, we can conclude

(7) $\sqrt{1 + \cos(\varphi)} = \sqrt{2}\left|\cos\left(\varphi/2\right)\right|$ with $\varphi \in [0, 2\pi]$

Notice that we need to take the absolute value, as $\sqrt{1 + \cos(\varphi)}$ has a zero in the interval $\varphi \in [0, 2\pi]$

Let us first have a look at the graph of the function $\sqrt[2]{8}\left|\sqrt[2]{2}\,\sin\left(\dfrac{\varphi}{2}\right)\right|$:

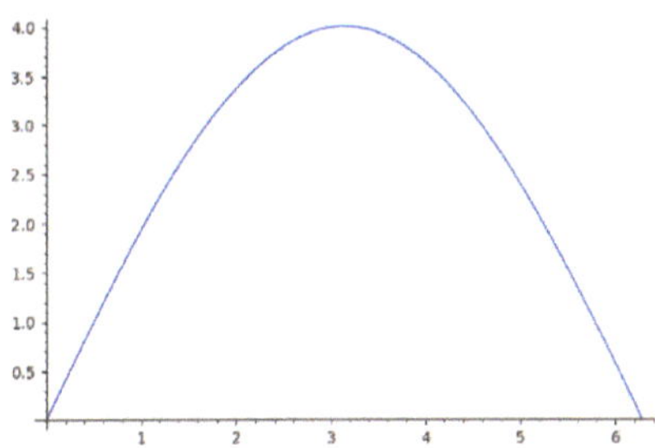

As we can see, all values are positive, so we can integrate the function between 0 and 2π.

Now we look at $\sqrt[2]{8}\left|\sqrt[2]{2}\,\cos\left(\dfrac{\varphi}{2}\right)\right|$:

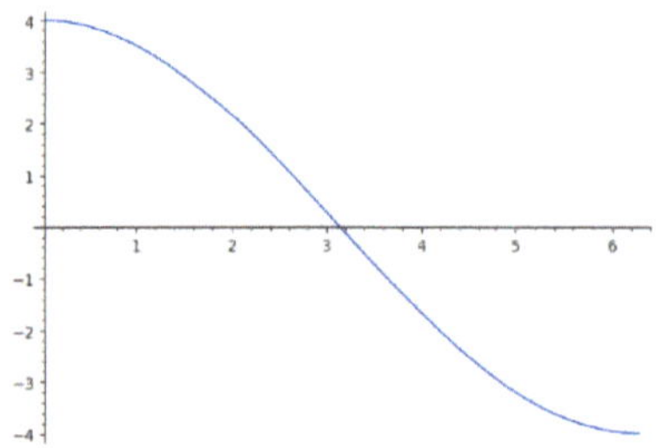

As we see, we need to integrate from 0 to π and then from π to 2π.

$$S_1 = \int_0^{2\pi} \sqrt{8} \left| \sqrt{2} \, sin \left(\frac{\varphi}{2} \right) \right| \, d\varphi = 4 \left[-2 \, cos \left(\frac{\varphi}{2} \right) \right]_0^{2\pi} = 16$$

$$S_2 = \int_0^{2\pi} \sqrt{8} \left| \sqrt{2} \, cos \left(\frac{\varphi}{2} \right) \right| \, d\varphi$$

$$= \int_0^{\pi} \sqrt{8} \left| \sqrt{2} \, cos \left(\frac{\varphi}{2} \right) \right| \, d\varphi + \int_{\pi}^{2\pi} \sqrt{8} \left| \sqrt{2} \, cos \left(\frac{\varphi}{2} \right) \right| \, d\varphi$$

$$= 4 \left[2 \, sin \left(\frac{\varphi}{2} \right) \right]_0^{\pi} + 4 \left[2 \, sin \left(\frac{\varphi}{2} \right) \right]_{\pi}^{2\pi}$$

$$= 16$$

∎

$(I.4.E.9)$:

Verify that the curve

$$\Phi(t) = \begin{pmatrix} \sin(t) \\ \dfrac{\sin^2(t)}{2} \\ \dfrac{1}{2}(t - \sin(t)\cos(t)) \end{pmatrix}$$

is already parameterized by arc length.

Solution:

We only need to check if the norm of the tangent vector is equal to 1:

We have:

$$T = \frac{\partial \Phi}{\partial t} = \begin{pmatrix} \cos(t) \\ \cos(t)\sin(t) \\ \dfrac{1}{2}(1 - \cos^2(t) + \sin^2(t)) \end{pmatrix} = \begin{pmatrix} \cos(t) \\ \cos(t)\sin(t) \\ \sin^2(t) \end{pmatrix}$$

$$\begin{aligned} \|T\|^2 &= \cos^2(t) + \cos^2(t)\sin^2(t) + \sin^4(t) \\ &= \underbrace{\cos^2(t) + \sin^2(t)\underbrace{\left(\cos^2(t) + \sin^2(t)\right)}_{=1}}_{=1} \\ &= 1 \end{aligned}$$

So, the curve is already parameterized by arc length.

∎

I.4.E.10:

Let $x^1 = t$, $x^2 = arc\sin\left(\frac{1}{t}\right)$, $x^3 = \sqrt{t^2 - 1}$ be a curve in spherical coordinates. Determine the length of the arc with

$1 \leq t \leq 2$

Solution:

The metric tensor for spherical coordinates (see $I.A.1.3)$) is given by

$$[g_{ij}] = \begin{bmatrix} 1 & 0 & 0 \\ 0 & (x^1)^2 & 0 \\ 0 & 0 & (x^1 \sin(x^2))^2 \end{bmatrix}$$

Replacing x^1 and x^2 with the definitions given for this excercise results in:

$$[g_{ij}] = \begin{bmatrix} 1 & 0 & 0 \\ 0 & (t)^2 & 0 \\ 0 & 0 & \left(t\sin\left(arc\sin\left(\frac{1}{t}\right)\right)\right)^2 \end{bmatrix} = \begin{bmatrix} 1 & 0 & 0 \\ 0 & t^2 & 0 \\ 0 & 0 & 1 \end{bmatrix}$$

Now, we take the derivatives of x^i with respect to t:

$$\frac{dx^1}{dt} = \frac{dt}{dt} = 1$$

$$\frac{dx^2}{dt} = \frac{d(arc\sin\left(\frac{1}{t}\right))}{dt} = \sqrt{\frac{1}{t^2(t^2-1)}}$$

$$\frac{dx^3}{dt} = \frac{d(\sqrt{t^2-1})}{dt} = \sqrt{\frac{t^2}{t^2-1}}$$

We calculate now the squared line element:

$$\begin{pmatrix} \dfrac{dx^1}{dt} & \dfrac{dx^2}{dt} & \dfrac{dx^3}{dt} \end{pmatrix} [g_{ij}] \begin{pmatrix} \dfrac{dx^1}{dt} \\ \dfrac{dx^2}{dt} \\ \dfrac{dx^3}{dt} \end{pmatrix}$$

Replacing the values we calculated so far, we get:

$$\underbrace{\left(1 \quad \sqrt{\dfrac{1}{t^2(t^2-1)}} \quad \sqrt{\dfrac{t^2}{t^2-1}}\right)}_{=\left(\frac{dx^1}{dx} \; \frac{dx^2}{dx} \; \frac{dx^3}{dx}\right)^T} \underbrace{\begin{bmatrix} 1 & 0 & 0 \\ 0 & t^2 & 0 \\ 0 & 0 & 1 \end{bmatrix}}_{=\,g_{ij}} \underbrace{\begin{pmatrix} \sqrt{\dfrac{1}{t^2(t^2-1)}} \\ \sqrt{\dfrac{t^2}{t^2-1}} \end{pmatrix}}_{=\left(\frac{dx^1}{dx} \; \frac{dx^2}{dx} \; \frac{dx^3}{dx}\right)} =$$

$$\left(1 \quad \sqrt{\dfrac{1}{t^2(t^2-1)}} \quad \sqrt{\dfrac{t^2}{t^2-1}}\right) \begin{pmatrix} t^2\sqrt{\dfrac{1}{t^2(t^2-1)}} \\ \sqrt{\dfrac{t^2}{t^2-1}} \end{pmatrix} =$$

$$1 + \frac{1}{t^2-1} + \frac{t^2}{t^2-1} = \frac{2\,t^2}{t^2-1}$$

Now we use $(I.4.020)$:

$$s = \int_a^b \sqrt{\left(\frac{dx^1}{dt} \quad \frac{dx^2}{dt} \quad \cdots \quad \frac{dx^n}{dt}\right)^T [g_{ij}] \begin{pmatrix} \dfrac{dx^1}{dt} \\ \dfrac{dx^2}{dt} \\ \cdots \\ \dfrac{dx^n}{dt} \end{pmatrix}}\; dt$$

$$= \int_1^2 \sqrt{\frac{2\,t^2}{t^2-1}}\; dt$$

$$= \sqrt{2\,(t^2-1)}\,\Big]_1^2$$

$$= \sqrt{6}$$

$\blacksquare$

($I.5.E.1$) Given the set of functions for *helix coordinate system*

$$x^1(\rho,\psi,\varsigma) = \rho\cos(\psi + a\varsigma)$$

$$x^2(\rho,\psi,\varsigma) = \rho\sin(\psi + a\varsigma)$$

$$x^3(\rho,\psi,\varsigma) = \varsigma$$

With $\rho \geq 0$, $\psi \in [0, 2\pi]$, $-\infty < \varsigma < \infty$

Calculate the Christoffel symbols of the 2nd kind by using

1. ($I.5.015$)

2. ($I.5.008$)

3. ($I.5.013$)

Solution:

Anyway, we need the Jacobi-matrix:

$$J = \begin{bmatrix} \partial x^1/\partial\rho & \partial x^1/\partial\psi & \partial x^1/\partial\varsigma \\ \partial x^2/\partial\rho & \partial x^2/\partial\psi & \partial x^2/\partial\varsigma \\ \partial x^3/\partial\rho & \partial x^3/\partial\psi & \partial x^3/\partial\varsigma \end{bmatrix}$$

$$= \begin{bmatrix} \cos(\psi + a\varsigma) & -\rho\sin(\psi + a\varsigma) & -a\rho\sin(\psi + a\varsigma) \\ \sin(\psi + a\varsigma) & \rho\cos(\psi + a\varsigma) & a\rho\cos(\psi + a\varsigma) \\ 0 & 0 & 1 \end{bmatrix}$$

We get:

$$J^T = \begin{bmatrix} \cos(\psi + a\varsigma) & \sin(\psi + a\varsigma) & 0 \\ -\rho\,\sin(\psi + a\varsigma) & \rho\,\cos(\psi + a\varsigma) & 0 \\ -a\,\rho\,\sin(\psi + a\varsigma) & a\,\rho\,\cos(\psi + a\varsigma) & 1 \end{bmatrix}$$

1. (*1.5.015*):

As J is not orthogonal, we need to calculate J^{T-1}:

$$J^{T-1} = \begin{bmatrix} \cos(\psi + a\varsigma) & -\dfrac{\sin(\psi + a\varsigma)}{\rho} & 0 \\[2ex] \sin(\psi + a\varsigma) & \dfrac{\cos(\psi + a\varsigma)}{\rho} & 0 \\[2ex] 0 & -a & 1 \end{bmatrix}$$

Next, we calculate the differentials of J^T

$$\frac{\partial}{\partial \rho} J^T = \begin{bmatrix} 0 & 0 & 0 \\ -\sin(\psi + a\varsigma) & \cos(\psi + a\varsigma) & 0 \\ -a\,\sin(\psi + a\varsigma) & a\,\cos(\psi + a\varsigma) & 0 \end{bmatrix}$$

$$\frac{\partial}{\partial \psi} J^T = \begin{bmatrix} -\sin(a\varsigma + \psi) & \cos(a\varsigma + \psi) & 0 \\ -\rho\,\cos(a\varsigma + \psi) & -\rho\,\sin(a\varsigma + \psi) & 0 \\ -a\,\rho\,\cos(a\varsigma + \psi) & -a\,\rho\,\sin(a\varsigma + \psi) & 0 \end{bmatrix}$$

$$\frac{\partial}{\partial \varsigma} J^T = \begin{bmatrix} -a\,\sin(a\varsigma + \psi) & a\,\cos(a\varsigma + \psi) & 0 \\ -a\,\rho\,\cos(a\varsigma + \psi) & -a\,\rho\,\sin(a\varsigma + \psi) & 0 \\ -a^2\,\rho\,\cos(a\varsigma + \psi) & -a^2\,\rho\,\sin(a\varsigma + \psi) & 0 \end{bmatrix}$$

Now comes the weird parts:

For simplification, we define:

$$s = \sin(\psi + a\varsigma)$$

$$c = \cos(\psi + a\varsigma)$$

$$[\Gamma^{i}{}_{j\rho}] = \left(\frac{\partial}{\partial \rho} J^T \right) J^{T-1}$$

$$= \begin{bmatrix} 0 & 0 & 0 \\ -s & c & 0 \\ -a\,s & a\,c & 0 \end{bmatrix} \begin{bmatrix} c & -\dfrac{s}{\rho} & 0 \\ s & \dfrac{c}{\rho} & 0 \\ 0 & -a & 1 \end{bmatrix}$$

$$= \begin{bmatrix} 0 & 0 & 0 \\ 0 & \dfrac{1}{\rho} & 0 \\ 0 & \dfrac{a}{\rho} & 0 \end{bmatrix} = \begin{bmatrix} \Gamma^{\rho}_{\rho\rho} & \Gamma^{\psi}_{\rho\rho} & \Gamma^{\varsigma}_{\rho\rho} \\ \Gamma^{\rho}_{\psi\rho} & \Gamma^{\psi}_{\psi\rho} & \Gamma^{\varsigma}_{\psi\rho} \\ \Gamma^{\rho}_{\varsigma\rho} & \Gamma^{\psi}_{\varsigma\rho} & \Gamma^{\varsigma}_{\varsigma\rho} \end{bmatrix}$$

$$\left[\Gamma^{i}_{\ j\psi} \right] \quad = \quad \left(\frac{\partial}{\partial\psi} J^{T} \right) J^{T-1}$$

$$= \begin{bmatrix} -s & c & 0 \\ -\rho\,c & -\rho\,s & 0 \\ -a\,\rho\,c & -a\,\rho\,s & 0 \end{bmatrix} \begin{bmatrix} c & -\dfrac{s}{\rho} & 0 \\ s & \dfrac{c}{\rho} & 0 \\ 0 & -a & 1 \end{bmatrix}$$

$$= \begin{bmatrix} 0 & \dfrac{1}{\rho} & 0 \\ -\rho & 0 & 0 \\ -a\,\rho & 0 & 0 \end{bmatrix} = \begin{bmatrix} \Gamma^{\rho}_{\rho\psi} & \Gamma^{\psi}_{\rho\psi} & \Gamma^{\varsigma}_{\rho\psi} \\ \Gamma^{\rho}_{\psi\psi} & \Gamma^{\psi}_{\psi\psi} & \Gamma^{\varsigma}_{\psi\psi} \\ \Gamma^{\rho}_{\varsigma\psi} & \Gamma^{\psi}_{\varsigma\psi} & \Gamma^{\varsigma}_{\varsigma\psi} \end{bmatrix}$$

$$\left[\Gamma^{i}_{\ j\varsigma} \right] \quad = \quad \left(\frac{\partial}{\partial\varsigma} J^{T} \right) J^{T-1}$$

$$= \begin{bmatrix} -a\,s & a\,c & 0 \\ -a\,\rho\,c & -a\,\rho\,s & 0 \\ -a^{2}\,\rho\,c & -a^{2}\,\rho\,s & 0 \end{bmatrix} \begin{bmatrix} c & -\dfrac{s}{\rho} & 0 \\ s & \dfrac{c}{\rho} & 0 \\ 0 & -a & 1 \end{bmatrix}$$

$$
= \begin{bmatrix} 0 & \dfrac{a}{\rho} & 0 \\[2mm] -a\,\rho & 0 & 0 \\[2mm] -a^2\,\rho & 0 & 0 \end{bmatrix} = \begin{bmatrix} \Gamma^\rho_{\rho\varsigma} & \Gamma^\psi_{\rho\varsigma} & \Gamma^\varsigma_{\rho\varsigma} \\[2mm] \Gamma^\rho_{\psi\varsigma} & \Gamma^\psi_{\psi\varsigma} & \Gamma^\varsigma_{\psi\varsigma} \\[2mm] \Gamma^\rho_{\varsigma\varsigma} & \Gamma^\psi_{\varsigma\varsigma} & \Gamma^\varsigma_{\varsigma\varsigma} \end{bmatrix}
$$

Thus, we get the following non-zero Christoffel symbols:

$$
\Gamma^\psi_{\psi\rho} = \frac{1}{\rho}, \quad \Gamma^\psi_{\varsigma\rho} = \frac{a}{\rho}, \quad \Gamma^\psi_{\rho\psi} = \frac{1}{\rho}, \quad \Gamma^\rho_{\psi\psi} = -\rho, \quad \Gamma^\rho_{\varsigma\psi} = -a\,\rho,
$$

$$
\Gamma^\psi_{\rho\varsigma} = \frac{a}{\rho}, \quad \Gamma^\rho_{\psi\varsigma} = -a\,\rho, \quad \Gamma^\rho_{\varsigma\varsigma} = -a^2\,\rho
$$

2. *using* (*I*. 5.008)

We need the metric tensor. Using the same definitions of s and c as in the previous section, we get:

$$
\left[g_{ij} \right] = J^T J
$$

$$
= \begin{bmatrix} c & s & 0 \\ -\rho\,s & \rho\,c & 0 \\ -a\,\rho\,s & a\,\rho\,c & 1 \end{bmatrix} \begin{bmatrix} c & -\rho\,s & -a\,\rho\,s \\ s & \rho\,c & a\,\rho\,c \\ 0 & 0 & 1 \end{bmatrix}
$$

$$
= \begin{bmatrix} 1 & 0 & 0 \\ 0 & \rho^2 & a\,\rho^2 \\ 0 & a\,\rho^2 & 1+a^2\,\rho^2 \end{bmatrix} = \begin{bmatrix} g_{\rho\rho} & g_{\rho\psi} & g_{\rho\varsigma} \\ g_{\psi\rho} & g_{\psi\psi} & g_{\psi\varsigma} \\ g_{\varsigma\rho} & g_{\varsigma\psi} & g_{\varsigma\varsigma} \end{bmatrix}
$$

We also need the inverse and the differentials of the metric tensor:

$$
[g^{ij}] = \begin{bmatrix} 1 & 0 & 0 \\ 0 & a^2 + \dfrac{1}{\rho^2} & -a \\ 0 & -a & 1 \end{bmatrix}
$$

$$\frac{\partial}{\partial \rho}\left[g_{ij}\right] = \begin{bmatrix} 0 & 0 & 0 \\ 0 & 2\rho & 2a\rho \\ 0 & 2a\rho & 2a^2\rho \end{bmatrix}$$

$$\frac{\partial}{\partial \psi}\left[g_{ij}\right] = \begin{bmatrix} 0 & 0 & 0 \\ 0 & 0 & 0 \\ 0 & 0 & 0 \end{bmatrix}$$

$$\frac{\partial}{\partial \varsigma}\left[g_{ij}\right] = \begin{bmatrix} 0 & 0 & 0 \\ 0 & 0 & 0 \\ 0 & 0 & 0 \end{bmatrix}$$

Now, we can use $(I.5.008)$ for calculating the Christoffel symbols by using the formula:

$$\Gamma^m_{ij} = \frac{1}{2} g^{ml} \left(\frac{\partial g_{il}}{\partial x^j} + \frac{\partial g_{lj}}{\partial x^i} - \frac{\partial g_{ji}}{\partial x^l} \right)$$

Best practice:

In order not to get lost in indices, expand the ESN equation by the variable l, taking the values ρ, ψ and ς:

$$\Gamma^m_{ij} = \frac{1}{2} g^{m\rho} \left(\frac{\partial g_{i\rho}}{\partial x^j} + \frac{\partial g_{\rho j}}{\partial x^i} - \frac{\partial g_{ji}}{\partial \rho} \right) + \frac{1}{2} g^{m\psi} \left(\frac{\partial g_{i\psi}}{\partial x^j} + \frac{\partial g_{\psi j}}{\partial x^i} - \frac{\partial g_{ji}}{\partial \psi} \right)$$
$$+ \frac{1}{2} g^{m\varsigma} \left(\frac{\partial g_{i\varsigma}}{\partial x^j} + \frac{\partial g_{\varsigma j}}{\partial x^i} - \frac{\partial g_{ji}}{\partial \varsigma} \right)$$

You can now replace m, i, j with values of ρ, ψ and ς.

Now, write down the matrices $\frac{\partial}{\partial \rho}\left[g_{ij}\right]$, $\frac{\partial}{\partial \psi}\left[g_{ij}\right]$, $\frac{\partial}{\partial \varsigma}\left[g_{ij}\right]$ on a "second" paper, the same with $\left[g^{ij}\right]$. In our case, $\frac{\partial}{\partial \psi}\left[g_{ij}\right]$ and $\frac{\partial}{\partial \varsigma}\left[g_{ij}\right]$ are always zero-matrices, so whenever in the equation there is a term by $\partial \psi$ or $\partial \varsigma$, it's a zero. Fill in now the values for $\frac{\partial}{\partial \rho}\left[g_{ij}\right]$ and $\left[g^{ij}\right]$. The "second" paper will enable you to look up quickly the values you need.

$$\Gamma^{\rho}{}_{\rho\rho} = \frac{1}{2}\underbrace{g^{\rho\rho}}_{=1}\underbrace{\left(\underbrace{\frac{\partial g_{\rho\rho}}{\partial\rho}}_{=0} + \underbrace{\frac{\partial g_{\rho\rho}}{\partial\rho}}_{=0} - \underbrace{\frac{\partial g_{\rho\rho}}{\partial\rho}}_{=0}\right)}_{=0}$$

$$+ \frac{1}{2}\underbrace{g^{\rho\psi}}_{=0}\underbrace{\left(\underbrace{\frac{\partial g_{\rho\psi}}{\partial\rho}}_{=0} + \underbrace{\frac{\partial g_{\psi\rho}}{\partial\rho}}_{=0} - \underbrace{\frac{\partial g_{\rho\rho}}{\partial\psi}}_{=0}\right)}_{=0}$$

$$+ \frac{1}{2}\underbrace{g^{\rho\varsigma}}_{=0}\left(\underbrace{\frac{\partial g_{\rho\varsigma}}{\partial\rho}}_{=0} + \underbrace{\frac{\partial g_{\varsigma\rho}}{\partial\rho}}_{=0} - \underbrace{\frac{\partial g_{\rho\rho}}{\partial\varsigma}}_{=0}\right) = 0$$

$$\Gamma^{\rho}{}_{\rho\psi} = \frac{1}{2}\underbrace{g^{\rho\rho}}_{=1}\underbrace{\left(\underbrace{\frac{\partial g_{\rho\rho}}{\partial\psi}}_{=0} + \underbrace{\frac{\partial g_{\rho\psi}}{\partial\rho} - \frac{\partial g_{\psi\rho}}{\partial\rho}}_{=0}\right)}_{=0}$$

$$+ \frac{1}{2}\underbrace{g^{\rho\psi}}_{=0}\underbrace{\left(\underbrace{\frac{\partial g_{\rho\psi}}{\partial\psi}}_{=0} + \underbrace{\frac{\partial g_{\psi\psi}}{\partial\rho}}_{=2\rho} - \underbrace{\frac{\partial g_{\psi\rho}}{\partial\psi}}_{=0}\right)}_{=0}$$

$$+ \frac{1}{2}\underbrace{g^{\rho\varsigma}}_{=0}\underbrace{\left(\underbrace{\frac{\partial g_{\rho\varsigma}}{\partial\psi}}_{=0} + \underbrace{\frac{\partial g_{\varsigma\psi}}{\partial\rho}}_{=2a\rho} - \underbrace{\frac{\partial g_{\psi\rho}}{\partial\varsigma}}_{=0}\right)}_{=0} = 0$$

$$\Gamma^{\rho}{}_{\rho\varsigma} = \underbrace{\frac{1}{2}\underbrace{g^{\rho\rho}}_{=1}\left(\underbrace{\frac{\partial g_{\rho\rho}}{\partial\varsigma}}_{=0}+\underbrace{\frac{\partial g_{\rho\varsigma}}{\partial\rho}}_{=0}-\frac{\partial g_{\varsigma\rho}}{\partial\rho}\right)}_{=0}$$

$$+\ \underbrace{\frac{1}{2}\underbrace{g^{\rho\psi}}_{=0}\left(\underbrace{\frac{\partial g_{\rho\psi}}{\partial\varsigma}}_{=0}+\underbrace{\frac{\partial g_{\psi\varsigma}}{\partial\rho}}_{=2a\rho}-\underbrace{\frac{\partial g_{\varsigma\rho}}{\partial\psi}}_{=0}\right)}_{=0}$$

$$+\ \underbrace{\frac{1}{2}\underbrace{g^{\rho\psi}}_{=0}\left(\underbrace{\frac{\partial g_{\rho\psi}}{\partial\varsigma}}_{=0}+\underbrace{\frac{\partial g_{\psi\varsigma}}{\partial\rho}}_{=2a\rho}-\underbrace{\frac{\partial g_{\varsigma\rho}}{\partial\psi}}_{=0}\right)}_{=0}$$

$$\Gamma^{\rho}{}_{\psi\psi} = \underbrace{\frac{1}{2}\underbrace{g^{\rho\rho}}_{=1}\left(\underbrace{\frac{\partial g_{\psi\rho}}{\partial\psi}}_{=0}+\underbrace{\frac{\partial g_{\rho\psi}}{\partial\psi}}_{=0}-\underbrace{\underbrace{\frac{\partial g_{\psi\psi}}{\partial\rho}}_{=2\rho}}_{=-2\rho}\right)}_{-\rho}$$

$$+\ \underbrace{\frac{1}{2}\underbrace{g^{\rho\psi}}_{=0}\left(\underbrace{\frac{\partial g_{\psi\psi}}{\partial\psi}}_{=0}+\underbrace{\frac{\partial g_{\psi\psi}}{\partial\psi}}_{=0}-\underbrace{\frac{\partial g_{\psi\psi}}{\partial\psi}}_{=0}\right)}_{=0}$$

$$+ \quad \frac{1}{2}\underbrace{g^{\rho\varsigma}}_{=0}\left(\underbrace{\frac{\partial g_{\psi\varsigma}}{\partial \psi}}_{=0}+\underbrace{\frac{\partial g_{\varsigma\psi}}{\partial \psi}}_{=0}-\underbrace{\frac{\partial g_{\psi\psi}}{\partial \varsigma}}_{=0}\right)}_{=0} = -\rho$$

$$\Gamma^{\rho}{}_{\psi\varsigma} \quad = \quad \underbrace{\frac{1}{2}\underbrace{g^{\rho\rho}}_{=1}\left(\underbrace{\frac{\partial g_{\psi\rho}}{\partial \varsigma}}_{=0}+\underbrace{\frac{\partial g_{\rho\varsigma}}{\partial \psi}}_{=0}-\underbrace{\frac{\partial g_{\varsigma\psi}}{\partial \rho}}_{=2a\rho}\right)}_{=-a\rho}$$

$$+ \quad \underbrace{\frac{1}{2}\underbrace{g^{\rho\psi}}_{=0}\left(\underbrace{\frac{\partial g_{\psi\psi}}{\partial \varsigma}}_{=0}+\underbrace{\frac{\partial g_{\psi\varsigma}}{\partial \psi}}_{=0}-\underbrace{\frac{\partial g_{\varsigma\psi}}{\partial \psi}}_{=0}\right)}_{=0}$$

$$+ \quad \underbrace{\frac{1}{2}\underbrace{g^{\rho\varsigma}}_{=0}\left(\underbrace{\frac{\partial g_{\psi\varsigma}}{\partial \varsigma}}_{=0}+\underbrace{\frac{\partial g_{\varsigma\varsigma}}{\partial \psi}}_{=0}-\underbrace{\frac{\partial g_{\varsigma\psi}}{\partial \varsigma}}_{=0}\right)}_{=0} = -a\rho$$

$$\Gamma^{\rho}{}_{\varsigma\varsigma} \quad = \quad \underbrace{\frac{1}{2}\underbrace{g^{\rho\rho}}_{=1}\left(\underbrace{\frac{\partial g_{\varsigma\rho}}{\partial \varsigma}}_{=0}+\underbrace{\frac{\partial g_{\rho\varsigma}}{\partial \varsigma}}_{=0}-\underbrace{\underbrace{\frac{\partial g_{\varsigma\varsigma}}{\partial \rho}}_{=2a^2\rho}}_{=-2a^2\rho}\right)}_{=-a^2\rho}$$

$$+ \quad \frac{1}{2} \underbrace{g^{\rho\psi}}_{=0} \left(\underbrace{\frac{\partial g_{\varsigma\psi}}{\partial \varsigma}}_{=0} + \underbrace{\frac{\partial g_{\psi\varsigma}}{\partial \varsigma}}_{=0} - \underbrace{\frac{\partial g_{\varsigma\varsigma}}{\partial \psi}}_{=0} \right)}_{=0}$$

$$+ \quad \frac{1}{2} \underbrace{g^{\rho\varsigma}}_{=0} \left(\underbrace{\frac{\partial g_{\varsigma\varsigma}}{\partial \varsigma}}_{=0} + \underbrace{\frac{\partial g_{\varsigma\varsigma}}{\partial \varsigma}}_{=0} - \underbrace{\frac{\partial g_{\varsigma\varsigma}}{\partial \varsigma}}_{=0} \right)}_{=0} = -a^2\rho$$

$$\Gamma^{\psi}{}_{\rho\rho} \quad = \quad \underbrace{\frac{1}{2} \underbrace{g^{\psi\rho}}_{=0} \left(\underbrace{\frac{\partial g_{\rho\rho}}{\partial \rho}}_{=0} + \underbrace{\frac{\partial g_{\rho\rho}}{\partial \rho}}_{=0} - \underbrace{\frac{\partial g_{\rho\rho}}{\partial \rho}}_{=0} \right)}_{=0}$$

$$+ \quad \frac{1}{2} \underbrace{g^{\psi\psi}}_{=a^2+\frac{1}{\rho^2}} \left(\underbrace{\frac{\partial g_{\rho\psi}}{\partial \rho}}_{=0} + \underbrace{\frac{\partial g_{\psi\rho}}{\partial \rho}}_{=0} - \underbrace{\frac{\partial g_{\rho\rho}}{\partial \psi}}_{=0} \right)}_{=0}$$

$$+ \quad \frac{1}{2} \underbrace{g^{\psi\varsigma}}_{=-a} \left(\underbrace{\frac{\partial g_{\rho\varsigma}}{\partial \rho}}_{=0} + \underbrace{\frac{\partial g_{\varsigma\rho}}{\partial \rho}}_{=0} - \underbrace{\frac{\partial g_{\rho\rho}}{\partial \varsigma}}_{=0} \right)}_{=0} = 0$$

$$\Gamma^{\psi}{}_{\rho\psi} \quad = \quad \underbrace{\frac{1}{2} \underbrace{g^{\psi\rho}}_{=0} \left(\underbrace{\frac{\partial g_{\rho\rho}}{\partial \psi}}_{=0} + \underbrace{\frac{\partial g_{\rho\psi}}{\partial \rho}}_{=0} - \underbrace{\frac{\partial g_{\psi\rho}}{\partial \rho}}_{=0} \right)}_{=0}$$

$$+ \quad \frac{1}{2} \underbrace{g^{\psi\psi}}_{=a^2+\frac{1}{\rho^2}} \underbrace{\left(\underbrace{\frac{\partial g_{\rho\psi}}{\partial \psi}}_{=0} + \underbrace{\frac{\partial g_{\psi\psi}}{\partial \rho}}_{=2\rho} - \underbrace{\frac{\partial g_{\psi\rho}}{\partial \psi}}_{=0} \right)}_{=a^2\rho+\frac{1}{\rho}}$$

$$+ \quad \frac{1}{2} \underbrace{g^{\psi\varsigma}}_{=-a} \underbrace{\left(\underbrace{\frac{\partial g_{\rho\varsigma}}{\partial \psi}}_{=0} + \underbrace{\frac{\partial g_{\varsigma\psi}}{\partial \rho}}_{=2a\rho} - \underbrace{\frac{\partial g_{\psi\rho}}{\partial \varsigma}}_{=0} \right)}_{=-a^2\rho} = \frac{1}{\rho}$$

$$\Gamma^{\psi}_{\rho\varsigma} \quad = \quad \frac{1}{2} \underbrace{g^{\psi\rho}}_{=0} \underbrace{\left(\underbrace{\frac{\partial g_{\rho\rho}}{\partial \varsigma}}_{=0} + \underbrace{\frac{\partial g_{\rho\varsigma}}{\partial \rho}}_{=0} - \underbrace{\frac{\partial g_{\varsigma\rho}}{\partial \rho}}_{=0} \right)}_{=0}$$

$$+ \quad \frac{1}{2} \underbrace{g^{\psi\psi}}_{=a^2+\frac{1}{\rho^2}} \underbrace{\left(\underbrace{\frac{\partial g_{\rho\psi}}{\partial \varsigma}}_{=0} + \underbrace{\frac{\partial g_{\psi\varsigma}}{\partial \rho}}_{=2a\rho} - \underbrace{\frac{\partial g_{\varsigma\rho}}{\partial \psi}}_{=0} \right)}_{=a^3\rho+\frac{a\rho}{\rho^2}}$$

$$+ \quad \frac{1}{2} \underbrace{g^{\psi\varsigma}}_{=-a} \underbrace{\left(\underbrace{\frac{\partial g_{\rho\varsigma}}{\partial \varsigma}}_{=0} + \underbrace{\frac{\partial g_{\varsigma\varsigma}}{\partial \rho}}_{=2a^2\rho} - \underbrace{\frac{\partial g_{\varsigma\rho}}{\partial \varsigma}}_{=0} \right)}_{=-a^3\rho} = \frac{a}{\rho}$$

$$\Gamma^{\psi}{}_{\psi\varsigma} = \frac{1}{2}\underbrace{g^{\psi\rho}}_{=0}\left(\underbrace{\frac{\partial g_{\psi\rho}}{\partial \varsigma}}_{=0}+\underbrace{\frac{\partial g_{\rho\varsigma}}{\partial \psi}}_{=0}-\underbrace{\frac{\partial g_{\varsigma\psi}}{\partial \rho}}_{2a\rho}\right)}_{=0$$

$$+\;\underbrace{\frac{1}{2}\underbrace{g^{\psi\psi}}_{=a^2+\frac{1}{\rho^2}}\left(\underbrace{\frac{\partial g_{\psi\psi}}{\partial \varsigma}}_{=0}+\underbrace{\frac{\partial g_{\psi\varsigma}}{\partial \psi}}_{=0}-\underbrace{\frac{\partial g_{\varsigma\psi}}{\partial \psi}}_{=0}\right)}_{=0}$$

$$+\;\underbrace{\frac{1}{2}\underbrace{g^{\psi\varsigma}}_{=-a}\left(\underbrace{\frac{\partial g_{\psi\varsigma}}{\partial \varsigma}}_{=0}+\underbrace{\frac{\partial g_{\varsigma\varsigma}}{\partial \psi}}_{=0}-\underbrace{\frac{\partial g_{\varsigma\psi}}{\partial \varsigma}}_{=0}\right)}_{=0}=0$$

$$\Gamma^{\psi}{}_{\psi\psi} = \underbrace{\frac{1}{2}\underbrace{g^{\psi\rho}}_{=0}\left(\underbrace{\frac{\partial g_{\psi\rho}}{\partial \psi}}_{=0}+\underbrace{\frac{\partial g_{\rho\psi}}{\partial \psi}}_{=0}-\underbrace{\frac{\partial g_{\psi\psi}}{\partial \rho}}_{=2\rho}\right)}_{=0}$$

$$+\;\underbrace{\frac{1}{2}\underbrace{g^{\psi\psi}}_{=a^2+\frac{1}{\rho^2}}\left(\underbrace{\frac{\partial g_{\psi\psi}}{\partial \psi}}_{=0}+\underbrace{\frac{\partial g_{\psi\psi}}{\partial \psi}}_{=0}-\underbrace{\frac{\partial g_{\psi\psi}}{\partial \psi}}_{=0}\right)}_{=0}$$

$$+\;\underbrace{\frac{1}{2}\underbrace{g^{\psi\varsigma}}_{=-a}\left(\underbrace{\frac{\partial g_{\psi\varsigma}}{\partial \psi}}_{=0}+\underbrace{\frac{\partial g_{\varsigma\psi}}{\partial \psi}}_{=0}-\underbrace{\frac{\partial g_{\psi\psi}}{\partial \varsigma}}_{=0}\right)}_{=0}=0$$

$$\Gamma^{\psi}{}_{\varsigma\varsigma} \quad = \quad \frac{1}{2}\underbrace{g^{\psi\rho}}_{=0}\left(\underbrace{\frac{\partial g_{\varsigma\rho}}{\partial \varsigma}}_{=0}+\underbrace{\frac{\partial g_{\rho\varsigma}}{\partial \varsigma}}_{=0}-\underbrace{\frac{\partial g_{\varsigma\varsigma}}{\partial \rho}}_{=2a^2\rho}\right)}_{=0}$$

$$+ \quad \frac{1}{2}\underbrace{g^{\psi\psi}}_{=a^2+\frac{1}{\rho^2}}\left(\underbrace{\frac{\partial g_{\varsigma\psi}}{\partial \varsigma}}_{=0}+\underbrace{\frac{\partial g_{\psi\varsigma}}{\partial \varsigma}}_{=0}-\underbrace{\frac{\partial g_{\varsigma\varsigma}}{\partial \psi}}_{=0}\right)}_{=0}$$

$$+ \quad \frac{1}{2}\underbrace{g^{\psi\varsigma}}_{=-a}\left(\underbrace{\frac{\partial g_{\varsigma\varsigma}}{\partial \varsigma}}_{=0}+\underbrace{\frac{\partial g_{\varsigma\varsigma}}{\partial \varsigma}}_{=0}-\underbrace{\frac{\partial g_{\varsigma\varsigma}}{\partial \varsigma}}_{=0}\right)}_{=0}=0$$

$$\Gamma^{\varsigma}{}_{\rho\rho} \quad = \quad \frac{1}{2}\underbrace{g^{\varsigma\rho}}_{=0}\left(\underbrace{\frac{\partial g_{\rho\rho}}{\partial \rho}}_{=0}+\underbrace{\frac{\partial g_{\rho\rho}}{\partial \rho}}_{=0}-\underbrace{\frac{\partial g_{\rho\rho}}{\partial \rho}}_{=0}\right)}_{=0}$$

$$+ \quad \frac{1}{2}\underbrace{g^{\varsigma\psi}}_{=-a}\left(\underbrace{\frac{\partial g_{\rho\psi}}{\partial \rho}}_{=0}+\underbrace{\frac{\partial g_{\psi\rho}}{\partial \rho}}_{=0}-\underbrace{\frac{\partial g_{\rho\rho}}{\partial \psi}}_{=0}\right)}_{=0}$$

$$+ \quad \frac{1}{2}\underbrace{g^{\varsigma\varsigma}}_{=1}\left(\underbrace{\frac{\partial g_{\rho\varsigma}}{\partial \rho}}_{=0}+\underbrace{\frac{\partial g_{\varsigma\rho}}{\partial \rho}}_{=0}-\underbrace{\frac{\partial g_{\rho\rho}}{\partial \varsigma}}_{=0}\right)}_{=0}=0$$

$$\Gamma^{\varsigma}{}_{\psi\rho} \;=\; \underbrace{\frac{1}{2}\underbrace{g^{\varsigma\rho}}_{=0}\left(\underbrace{\frac{\partial g_{\psi\rho}}{\partial\rho}}_{=0}+\underbrace{\frac{\partial g_{\rho\rho}}{\partial\psi}}_{=0}-\underbrace{\frac{\partial g_{\rho\psi}}{\partial\rho}}_{0}\right)}_{=0}$$

$$+\;\underbrace{\frac{1}{2}\underbrace{g^{\varsigma\psi}}_{=-a}\left(\underbrace{\frac{\partial g_{\psi\psi}}{\partial\rho}}_{=2\rho}+\underbrace{\frac{\partial g_{\psi\rho}}{\partial\psi}}_{=0}-\underbrace{\frac{\partial g_{\rho\psi}}{\partial\psi}}_{=0}\right)}_{=\,-a\rho}$$

$$+\;\underbrace{\frac{1}{2}\underbrace{g^{\varsigma\varsigma}}_{=1}\left(\underbrace{\frac{\partial g_{\psi\varsigma}}{\partial\rho}}_{=2a\rho}+\underbrace{\frac{\partial g_{\varsigma\rho}}{\partial\psi}}_{=0}-\underbrace{\frac{\partial g_{\rho\psi}}{\partial\varsigma}}_{=0}\right)}_{=a\rho}=0$$

$$\Gamma^{\varsigma}{}_{\psi\psi} \;=\; \underbrace{\frac{1}{2}\underbrace{g^{\varsigma\rho}}_{=0}\left(\underbrace{\frac{\partial g_{\psi\rho}}{\partial\psi}}_{=0}+\underbrace{\frac{\partial g_{\rho\psi}}{\partial\psi}}_{=0}-\underbrace{\frac{\partial g_{\psi\psi}}{\partial\rho}}_{=2\rho}\right)}_{=0}$$

$$+\;\underbrace{\frac{1}{2}\underbrace{g^{\varsigma\psi}}_{=-a}\left(\underbrace{\frac{\partial g_{\psi\psi}}{\partial\psi}}_{=0}+\underbrace{\frac{\partial g_{\psi\psi}}{\partial\psi}}_{=0}-\underbrace{\frac{\partial g_{\psi\psi}}{\partial\psi}}_{=0}\right)}_{=0}$$

$$+\;\underbrace{\frac{1}{2}\underbrace{g^{\varsigma\varsigma}}_{=1}\left(\underbrace{\frac{\partial g_{\psi\varsigma}}{\partial\psi}}_{=0}+\underbrace{\frac{\partial g_{\varsigma\psi}}{\partial\psi}}_{=0}-\underbrace{\frac{\partial g_{\psi\psi}}{\partial\varsigma}}_{=0}\right)}_{=0}=0$$

$$\Gamma^\varsigma{}_{\varsigma\rho} = \frac{1}{2}\underbrace{g^{\varsigma\rho}}_{=0}\left(\underbrace{\frac{\partial g_{\varsigma\rho}}{\partial \rho}}_{=0} + \underbrace{\frac{\partial g_{\rho\rho}}{\partial \varsigma}}_{=0} - \underbrace{\frac{\partial g_{\rho\varsigma}}{\partial \rho}}_{=0}\right)}_{=0}$$

$$+ \;\underbrace{\frac{1}{2}\underbrace{g^{\varsigma\psi}}_{=-a}\left(\underbrace{\frac{\partial g_{\varsigma\psi}}{\partial \rho}}_{=2a\rho} + \underbrace{\frac{\partial g_{\psi\rho}}{\partial \varsigma}}_{=0} - \underbrace{\frac{\partial g_{\rho\varsigma}}{\partial \psi}}_{=0}\right)}_{=-a^2\rho}$$

$$+ \;\underbrace{\frac{1}{2}\underbrace{g^{\varsigma\varsigma}}_{=1}\left(\underbrace{\frac{\partial g_{\varsigma\varsigma}}{\partial \rho}}_{=2a^2\rho} + \underbrace{\frac{\partial g_{\varsigma\rho}}{\partial \varsigma}}_{=0} - \underbrace{\frac{\partial g_{\rho\varsigma}}{\partial \varsigma}}_{=0}\right)}_{=a^2\rho} = 0$$

$$\Gamma^\varsigma{}_{\varsigma\psi} = \underbrace{\frac{1}{2}\underbrace{g^{\varsigma\rho}}_{=0}\left(\underbrace{\frac{\partial g_{\varsigma\rho}}{\partial \psi}}_{=0} + \underbrace{\frac{\partial g_{\rho\psi}}{\partial \varsigma}}_{=0} - \underbrace{\frac{\partial g_{\psi\varsigma}}{\partial \rho}}_{=2a\rho}\right)}_{=0}$$

$$+ \;\underbrace{\frac{1}{2}\underbrace{g^{\varsigma\psi}}_{=-a}\left(\underbrace{\frac{\partial g_{\varsigma\psi}}{\partial \psi}}_{=0} + \underbrace{\frac{\partial g_{\psi\psi}}{\partial \varsigma}}_{=0} - \underbrace{\frac{\partial g_{\psi\varsigma}}{\partial \psi}}_{=0}\right)}_{=0}$$

$$+ \;\underbrace{\frac{1}{2}\underbrace{g^{\varsigma\varsigma}}_{=1}\left(\underbrace{\frac{\partial g_{\varsigma\varsigma}}{\partial \psi}}_{=0} + \underbrace{\frac{\partial g_{\varsigma\psi}}{\partial \varsigma}}_{=0} - \underbrace{\frac{\partial g_{\psi\varsigma}}{\partial \varsigma}}_{=0}\right)}_{=0} = 0$$

$$\Gamma^{\varsigma}{}_{\varsigma\varsigma} = \underbrace{\frac{1}{2}\underbrace{g^{\varsigma\rho}}_{=0}\left(\underbrace{\frac{\partial g_{\varsigma\rho}}{\partial\varsigma}}_{=0}+\underbrace{\frac{\partial g_{\rho\varsigma}}{\partial\varsigma}}_{=0}-\underbrace{\frac{\partial g_{\varsigma\varsigma}}{\partial\rho}}_{=2a^2\rho}\right)}_{=0}$$

$$+\;\underbrace{\frac{1}{2}\underbrace{g^{\varsigma\psi}}_{=-a}\left(\underbrace{\frac{\partial g_{\varsigma\psi}}{\partial\varsigma}}_{=0}+\underbrace{\frac{\partial g_{\psi\varsigma}}{\partial\varsigma}}_{=0}-\underbrace{\frac{\partial g_{\varsigma\varsigma}}{\partial\varsigma}}_{=0}\right)}_{=0}$$

$$+\;\underbrace{\frac{1}{2}\underbrace{g^{\varsigma\varsigma}}_{=1}\left(\underbrace{\frac{\partial g_{\varsigma\varsigma}}{\partial\varsigma}}_{=0}+\underbrace{\frac{\partial g_{\varsigma\varsigma}}{\partial\varsigma}}_{=0}-\underbrace{\frac{\partial g_{\varsigma\varsigma}}{\partial\varsigma}}_{=0}\right)}_{=0}=0$$

3. ($I.5.013$):

You cannot use ($I.5.013$) as the metric tensor is not a diagonal matrix!

∎

($I.5.E.2$) Calculate the Christoffel symbols of the 2nd kind for spherical coordinates using ($I.5.013$).

Solution:

Spherical coordinates are defined by

$$x^1(r,\theta,\phi) = r\sin\theta\cos\phi$$

$$x^2(r,\theta,\phi) = r\sin\theta\sin\phi$$

$$x^3(r,\theta,\phi) = r\cos\theta$$

with $r \geq 0,\ 0 \leq \theta \leq \pi,\ 0 \leq \phi \leq 2\pi$

Building the Jacobian matrix

$$
J(r, \theta, \phi) \quad = \quad \begin{bmatrix} \cos(\phi)\sin(\theta) & r\cos(\theta)\cos(\phi) & -r\sin(\theta)\sin(\phi) \\ \sin(\theta)\sin(\phi) & r\cos(\theta)\sin(\phi) & r\cos(\phi)\sin(\theta) \\ \cos(\theta) & -r\sin(\theta) & 0 \end{bmatrix}
$$

enables us to build up the metric tensor as

$$
[g_{ij}] = J^T J = \begin{bmatrix} 1 & 0 & 0 \\ 0 & r^2 & 0 \\ 0 & 0 & r^2\sin^2(\theta) \end{bmatrix}
$$

Obviously, $[g_{ij}]$ is a diagonal matrix, so we can employ $(\,I.5.013\,)$.

$$
\Gamma^{\theta}{}_{\theta r} \quad = \quad \frac{\partial}{\partial r}\left(\frac{1}{2}\ln |g_{\theta\theta}|\right)
$$

$$
= \quad \frac{\partial}{\partial r}\left(\frac{1}{2}\ln |r^2|\right)
$$

$$
= \quad \frac{1}{2}\frac{2r}{r^2} = \frac{1}{r}
$$

$$
\Gamma^{\phi}{}_{\phi r} \quad = \quad \frac{\partial}{\partial r}\left(\frac{1}{2}\ln |g_{\phi\phi}|\right)
$$

$$
= \quad \frac{\partial}{\partial r}\left(\frac{1}{2}\ln (r^2 \sin^2(\theta))\right)
$$

$$
= \quad \frac{r\sin^2(\theta)}{r^2 \sin^2(\theta)} = \frac{1}{r}
$$

$$
\Gamma^{\phi}{}_{\phi \theta} \quad = \quad \frac{\partial}{\partial \theta}\left(\frac{1}{2}\ln |g_{rr}|\right)
$$

$$
= \quad \frac{\partial}{\partial \theta}\left(\frac{1}{2}\ln (r^2 \sin^2(\theta))\right)
$$

$$= \frac{1}{2} \frac{r^2\, 2\sin(\theta) * \cos(\theta)}{r^2\,\sin^2(\theta)} = \frac{\cos(\theta)}{\sin(\theta)} = \cot(\theta)$$

$$\Gamma^r{}_{\theta\theta} = -\frac{1}{2\,g_{rr}} \frac{\partial}{\partial r}\, g_{\theta\theta}$$

$$= -\frac{1}{2(1)} \frac{\partial}{\partial r}\, r^2 = -r$$

$$\Gamma^r{}_{\phi\phi} = -\frac{1}{2\,g_{rr}} \frac{\partial}{\partial r}\, g_{\phi\phi}$$

$$= -\frac{1}{2(1)} \frac{\partial}{\partial r}\left(r^2\sin^2(\theta)\right) = -r\sin^2(\theta)$$

$$\Gamma^\theta{}_{\phi\phi} = -\frac{1}{2\,g_{\theta\theta}} \frac{\partial}{\partial \theta}\, g_{\phi\phi}$$

$$= -\frac{1}{2(r)^2} \frac{\partial}{\partial \theta}\left(r^2\sin^2(\theta)\right) = -\sin(\theta)\cos(\theta)$$

We note that $\dfrac{\partial \ln(x)}{\partial x} = \dfrac{1}{x}$ and that we have to use the product rule as also the chain rule for derivations.

∎

($I.5.E.3$) Given the squared line element $ds^2 = dx^2 + e^{2x}\,dy^2$, compute the Christoffel symbols of the 2nd kind.

Solution:

The squared line element induces the metric tensor

$$g_{ij} = \begin{bmatrix} 1 & 0 \\ 0 & e^{2x} \end{bmatrix}, \text{ so } g^{ij} = \begin{bmatrix} 1 & 0 \\ 0 & e^{-2x} \end{bmatrix}$$

We calculate the derivatives:

$$\frac{\partial}{\partial x}[g_{ij}] = \begin{bmatrix} 0 & 0 \\ 0 & 2\,e^{2x} \end{bmatrix},$$

$$\frac{\partial}{\partial y}[g_{ij}] = \begin{bmatrix} 0 & 0 \\ 0 & 0 \end{bmatrix}$$

We use ($I.5.008$):

$$\Gamma^m_{ij} = \frac{1}{2} g^{ml} \left(\frac{\partial g_{il}}{\partial x^j} + \frac{\partial g_{lj}}{\partial x^i} - \frac{\partial g_{ji}}{\partial x^l} \right)$$

Expanding the ESN to our need, we get:

$$\Gamma^m_{ij} = \frac{1}{2} g^{mx} \left(\frac{\partial g_{ix}}{\partial x^j} + \frac{\partial g_{xj}}{\partial x^i} - \frac{\partial g_{ji}}{\partial x} \right) + \frac{1}{2} g^{my} \left(\frac{\partial g_{iy}}{\partial x^j} + \frac{\partial g_{yj}}{\partial x^i} - \frac{\partial g_{ji}}{\partial y} \right)$$

Thus, we get:

$$\Gamma^x_{xx} = \frac{1}{2} g^{xx} \left(\underbrace{\frac{\partial g_{xx}}{\partial x}}_{=0} + \underbrace{\frac{\partial g_{xx}}{\partial x}}_{=0} - \underbrace{\frac{\partial g_{xx}}{\partial x}}_{=0} \right) + \frac{1}{2} \underbrace{g^{xy}}_{=0} \left(\frac{\partial g_{xy}}{\partial x} + \frac{\partial g_{yx}}{\partial x} - \frac{\partial g_{xx}}{\partial y} \right) = 0$$

$$\Gamma^x_{xy} = \frac{1}{2} g^{xx} \left(\underbrace{\frac{\partial g_{xx}}{\partial y}}_{=0} + \underbrace{\frac{\partial g_{xy}}{\partial x}}_{=0} - \underbrace{\frac{\partial g_{yx}}{\partial x}}_{=0} \right) + \frac{1}{2} \underbrace{g^{xy}}_{=0} \left(\frac{\partial g_{xy}}{\partial x^j} + \frac{\partial g_{yy}}{\partial x} - \frac{\partial g_{yx}}{\partial y} \right) = 0$$

$$\Gamma^x_{yy} = \frac{1}{2} \underbrace{g^{xx}}_{=1} \left(\underbrace{\frac{\partial g_{yx}}{\partial y}}_{=0} + \underbrace{\frac{\partial g_{xy}}{\partial y}}_{=0} - \underbrace{\frac{\partial g_{yy}}{\partial x}}_{=2\,e^{2x}} \right) + \frac{1}{2} \underbrace{g^{xy}}_{=0} \left(\frac{\partial g_{yy}}{\partial y} + \frac{\partial g_{yy}}{\partial y} - \frac{\partial g_{yy}}{\partial y} \right) = e^{2x}$$

$$\Gamma^y_{yy} = \frac{1}{2} \underbrace{g^{yx}}_{=0} \left(\frac{\partial g_{yx}}{\partial y} + \frac{\partial g_{xy}}{\partial y} - \frac{\partial g_{yy}}{\partial x} \right) + \frac{1}{2} \underbrace{g^{yy}}_{=e^{-2x}} \left(\underbrace{\frac{\partial g_{yy}}{\partial y}}_{=0} + \underbrace{\frac{\partial g_{yy}}{\partial y}}_{=0} - \underbrace{\frac{\partial g_{yy}}{\partial y}}_{=0} \right) = 0$$

$$\Gamma^y_{xy} = \frac{1}{2} \underbrace{g^{yx}}_{=0} \left(\frac{\partial g_{xx}}{\partial y} + \frac{\partial g_{xy}}{\partial x} - \frac{\partial g_{yx}}{\partial x} \right) + \frac{1}{2} \underbrace{g^{yy}}_{=e^{-2x}} \left(\underbrace{\frac{\partial g_{xy}}{\partial x^j}}_{=0} + \underbrace{\frac{\partial g_{yy}}{\partial x}}_{=2\,e^{2x}} - \underbrace{\frac{\partial g_{yx}}{\partial y}}_{=0} \right)$$

$$= e^{-2x}\,e^{2x} = 1$$

$$\Gamma^{y}_{xx} = \frac{1}{2}\underbrace{g^{yx}}_{=0}\left(\frac{\partial g_{xx}}{\partial x} + \frac{\partial g_{xx}}{\partial x} - \frac{\partial g_{xx}}{\partial x}\right) + \frac{1}{2}g^{yy}\left(\underbrace{\frac{\partial g_{xy}}{\partial x}}_{=0} + \underbrace{\frac{\partial g_{yx}}{\partial x}}_{=0} - \underbrace{\frac{\partial g_{xx}}{\partial y}}_{=0}\right) = 0$$

∎

$(\boldsymbol{I.6.E.1})$ Let $\left[g_{ij}\right] = \begin{bmatrix} 0 & 1 \\ 1 & 1 \end{bmatrix}$ be the metric tensor for a 2-dimensional vectorspace over $\mathbb{R}$. Let $T_{ij} = \begin{bmatrix} 0 & 2 \\ 0 & 0 \end{bmatrix}$ be a tensor.

Determine $T^{i}{}_{j}$, $T_{j}{}^{i}$ and T^{ij}

Solution:

We need the inverse metric tensor, which we calculate as:

$$[g^{ij}] = \begin{bmatrix} -1 & 1 \\ 1 & 0 \end{bmatrix}$$

We get:

$$T^{i}{}_{j} \underset{(I.6.016)}{=} \underbrace{\begin{bmatrix} -1 & 1 \\ 1 & 0 \end{bmatrix}}_{g^{-1}} \underbrace{\begin{bmatrix} 0 & 2 \\ 0 & 0 \end{bmatrix}}_{T_{ij}} = \begin{bmatrix} 0 & -2 \\ 2 & 0 \end{bmatrix}$$

$$T_{j}{}^{i} \underset{(I.6.018)}{=} \underbrace{\begin{bmatrix} 0 & 2 \\ 0 & 0 \end{bmatrix}}_{T_{ij}} \underbrace{\begin{bmatrix} -1 & 1 \\ 1 & 0 \end{bmatrix}}_{g^{-1}} = \begin{bmatrix} 2 & 0 \\ 0 & 0 \end{bmatrix}$$

$$T^{ij} \underset{(I.6.020)}{=} \underbrace{\begin{bmatrix} -1 & 1 \\ 1 & 0 \end{bmatrix}}_{g^{-1}} \underbrace{\begin{bmatrix} 0 & 2 \\ 0 & 0 \end{bmatrix}}_{T_{ij}} \underbrace{\begin{bmatrix} -1 & 1 \\ 1 & 0 \end{bmatrix}}_{g^{-1}} = \begin{bmatrix} -2 & 0 \\ 2 & 0 \end{bmatrix}$$

∎

$(I.6.E.2)$ Let $[g_{ij}] = \begin{bmatrix} 2 & -1 \\ -1 & 2 \end{bmatrix}$ be a metric tensor and T^i_{*jkl} a 4-rank-tensor. Let us assume that the only non-zero element of T^i_{*jkl} is

$$T^1_{*212} = \pi$$

Calculate T_{1212} and T_{2212}

Solution:

We lower the upper index i of T^i_{*jkl}:

$$T_{rjkl} = g_{ri} T^i_{*jkl}$$

We set $r = 1, j = 2, k = 1 \text{ and } l = 2$:

$$T_{1212} = g_{1i} T^i_{*212} = \underbrace{g_{11}}_{=2} \underbrace{T^1_{*212}}_{=\pi} + g_{22} \underbrace{T^2_{*212}}_{=0} = 2\pi$$

Now, we set $r = 2, j = 2, k = 1 \text{ and } l = 2$:

$$T_{2212} = g_{2i} T^i_{*212} = \underbrace{g_{21}}_{=-1} \underbrace{T^1_{*212}}_{=\pi} + g_{22} \underbrace{T^2_{*212}}_{=0} = -\pi$$

∎

$(I.6.E.3)$ Let A_{ij} be a symmetric tensor and B^{ij} be an anti-symmetric tensor. Show that $A_{ij} B^{ij} = 0$

Solution:

A_{ij} being symmetric means $A_{ij} = A_{ji}$.

B^{ij} being anti-symmetric means: $B^{ij} = -B^{ji}$.

Furthermore, $B^{ii} = -B^{ii}$, so $B^{ii} = 0$

We calculate:

$$A_{ij} B^{ij} = A_{11}B^{11} + A_{12}B^{12} + A_{13}B^{13}$$

$$+A_{21}B^{21} + A_{22}B^{22} + A_{23}B^{23}$$

$$+A_{31}B^{31} + A_{32}B^{32} + A_{33}B^{33}$$

$$= \quad A_{11}B^{11} - A_{21}B^{21} - A_{31}B^{31}$$

$$+A_{21}B^{21} + A_{22}B^{22} - A_{32}B^{32} \qquad \text{\textit{A} symmetric, \textit{B} anti-symmetric}$$

$$+A_{31}B^{31} + A_{32}B^{32} + A_{33}B^{33}$$

$$= \quad A_{11}\underbrace{B^{11}}_{=0} + A_{22}\underbrace{B^{22}}_{=0} + +A_{33}\underbrace{B^{33}}_{=0} \qquad \text{terms cancel out}$$

$$= \quad 0 \qquad\qquad as\ B^{ii} = 0$$

∎

($\textbf{\textit{I}}.6.\textbf{\textit{E}}.4$) Determine if the following tensor expressions are legitimate and explain your answer.

(a) $T^{k}_{\ i} + S_i$

(b) $T^{k}_{k} - S^{n}_{m}$

(c) $T^{ijm}_{\quad ijk} - T^{lnm}_{\quad lnk}$

Solution:

(a) illegitimate: the tensors do not have the same structure.
(b) Illegitimate: T^{k}_{k} is the trace of the tensor T. The rank of the resulting object depends on the original tensor. If the original tensor is of rank 2 (a matrix), then T^{k}_{k} would represent the sum of the diagonal elements, resulting in a scalar (rank 0 tensor). If the original tensor is of higher rank, the resulting object would still be a tensor but with reduced rank.
(c) Legitimate: both tensors have the same structure.

$(I.6.E.5)$ Write down the transformation laws for (a) covariant and contravariant vectors and (b) for covariant, contravariant, and mixed rank-2 tensors between different coordinate systems.

Solution:

(a) Covariant: $\bar{T}_i = \dfrac{\partial x^j}{\partial \bar{x}^i}\, T_j$, contravariant: $\bar{T}^i = \dfrac{\partial \bar{x}^i}{\partial x^j}\, T^j$

(b) Covariant: $\bar{T}_{ij} = \dfrac{\partial x^k}{\partial \bar{x}^i}\dfrac{\partial x^m}{\partial \bar{x}^j}\, T_{km}$

contravariant: $\bar{T}^{ij} = \dfrac{\partial \bar{x}^i}{\partial x^k}\dfrac{\partial \bar{x}^j}{\partial x^m}\, T^{km}$

mixed: $T_i{}^j = \dfrac{\partial x^k}{\partial \bar{x}^i}\dfrac{\partial \bar{x}^j}{\partial x^m}\, T_m{}^k$

∎

$(I.6.E.6)$ Let $T_{\alpha\beta}$ be a rank-2 tensor and $g_{\mu\nu}$ a metric tensor.

Calculate (1) $T^\alpha{}_\beta$ and (2) $T^{\alpha\beta}$.

Suppose $T_{\alpha\beta}$ has a symmetric part $S_{\alpha\beta}$ and an antisymmetric part $A_{\alpha\beta}$. Write down the raised versions of (3) $S^{\alpha\beta}$ and (4) $A^{\alpha\beta}$.

Solution:

We apply $(I.6.015)$:

(1) $T^\alpha{}_\beta = g^{\alpha\mu}\, T_{\mu\beta}$

Now, we apply again $(I.6.015)$:

(2) $T^{\alpha\beta} = g^{\beta\nu}\, T^\alpha{}_\nu = g^{\beta\nu} g^{\alpha\mu}\, T_{\mu\nu}$

Notice that we had to rename the indices!

We apply (2) and we get:

(3) $S^{\alpha\beta} = g^{\beta\nu}g^{\alpha\mu} S_{\mu\nu}$

(4) $A^{\alpha\beta} = g^{\beta\nu}g^{\alpha\mu} A_{\mu\nu}$

∎

($\boldsymbol{I.6.E.7}$) Let the metric tensor be given by:

$$[g_{ij}] = \begin{bmatrix} 1 & 0 \\ 0 & 4 \end{bmatrix}$$

And a tensor T_{ij} given by:

$$[T_{ij}] = \begin{bmatrix} 2 & 1 \\ 3 & 5 \end{bmatrix}$$

Calculate $T^i{}_j$ and T^{ij}.

Solution:

We note that the inverse metric tensor is:

$$[g^{ij}] = \begin{bmatrix} 1 & 0 \\ 0 & 1/4 \end{bmatrix}$$

Now we apply ($I.6.018$):

$$[T^i{}_j] = [g^{ij}][T_{ij}] = \begin{bmatrix} 1 & 0 \\ 0 & 1/4 \end{bmatrix}\begin{bmatrix} 2 & 1 \\ 3 & 5 \end{bmatrix} = \begin{bmatrix} 2 & 1 \\ 3/4 & 5/4 \end{bmatrix}$$

Apply ($I.6.020$):

$$[T^{ij}] = [g^{ij}][T_{ij}][g^{ij}] = \begin{bmatrix} 1 & 0 \\ 0 & 1/4 \end{bmatrix}\begin{bmatrix} 2 & 1 \\ 3 & 5 \end{bmatrix}\begin{bmatrix} 1 & 0 \\ 0 & 1/4 \end{bmatrix} = \begin{bmatrix} 2 & 1 \\ 3/4 & 5/4 \end{bmatrix}\begin{bmatrix} 1 & 0 \\ 0 & 1/4 \end{bmatrix}$$

$$= \begin{bmatrix} 2 & 1/4 \\ 3/4 & 5/16 \end{bmatrix}$$

∎

($\boldsymbol{I.6.E.8}$) Let g_{ij} be the metric tensor for spherical coordinates and let $A_{\mu\nu}$ be a tensor with the following non-zero components:

$$A_{rr} = r^2$$

$$A_{r\theta} = \theta$$

$$A_{\theta\theta} = sin(\theta)$$

$$A_{\theta\phi} = \phi$$

$$A_{\phi\phi} = 1$$

Determine $A^{\mu}{}_{\nu}$.

Solution:

The metric tensor for spherical coordinates is given by:

$$[g_{ij}] = \begin{bmatrix} 1 & 0 & 0 \\ 0 & r^2 & 0 \\ 0 & 0 & r^2 \, sin^2(\theta) \end{bmatrix}$$

The inverse metric tensor is:

$$[g^{ij}] = \begin{bmatrix} 1 & 0 & 0 \\ 0 & \dfrac{1}{r^2} & 0 \\ 0 & 0 & \dfrac{1}{r^2 \, sin^2(\theta)} \end{bmatrix}$$

Renaming indices in $(I.\,6.015)$ we get the formula:

$$A^{\mu}{}_{\nu} = g^{\mu\alpha} A_{\alpha\nu} = g^{\mu r} A_{r\nu} + g^{\mu\theta} A_{\theta\nu} + g^{\mu\phi} A_{\phi\nu}$$

Replacing now the values for μ and ν gives:

$$A^{r}{}_{r} = g^{r\alpha} A_{\alpha r} = \underbrace{g^{rr}}_{=1} \underbrace{A_{rr}}_{=r^2} + \underbrace{g^{r\theta}}_{=0} A_{\theta r} + \underbrace{g^{r\phi}}_{=0} A_{\phi r} = r^2$$

$$A^{r}{}_{\theta} = g^{r\alpha} A_{\alpha\theta} = \underbrace{g^{rr}}_{=1} \underbrace{A_{r\theta}}_{=\theta} + \underbrace{g^{r\theta}}_{=0} A_{\theta\theta} + \underbrace{g^{r\phi}}_{=0} A_{\phi\theta} = \theta$$

$$A^{r}{}_{\phi} = g^{r\alpha} A_{\alpha\phi} = g^{rr} \underbrace{A_{r\phi}}_{=0} + \underbrace{g^{r\theta}}_{=0} A_{\theta\phi} + \underbrace{g^{r\phi}}_{=0} A_{\phi\phi} = 0$$

$$A^{\theta}{}_{\nu} = g^{\theta\alpha} A_{\alpha\nu} = \underbrace{g^{\theta r}}_{=0} A_{rr} + \underbrace{g^{\theta\theta}}_{=\frac{1}{r^2}} \underbrace{A_{\theta r}}_{=0} + \underbrace{g^{\theta\phi}}_{=0} \underbrace{A_{\phi r}}_{=0} = 0$$

$$A^{\theta}{}_{\theta} = g^{\theta\alpha} A_{\alpha\theta} = \underbrace{g^{\theta r}}_{=0} A_{r\theta} + \underbrace{g^{\theta\theta}}_{=\frac{1}{r^2}} \underbrace{A_{\theta\theta}}_{=sin(\theta)} + \underbrace{g^{\theta\phi}}_{=0} A_{\phi\theta} = \frac{sin(\theta)}{r^2}$$

$$A^{\theta}{}_{\phi} = g^{\theta\alpha} A_{\alpha\phi} = \underbrace{g^{\theta r}}_{=0} A_{r\phi} + \underbrace{g^{\theta\theta}}_{=\frac{1}{r^2}} \underbrace{A_{\theta\phi}}_{\phi} + \underbrace{g^{\theta\phi}}_{=0} A_{\phi\phi} = \frac{\phi}{r^2}$$

$$A^{\phi}{}_{r} = g^{\phi\alpha} A_{\alpha r} = \underbrace{g^{\phi r}}_{=0} A_{rr} + \underbrace{g^{\phi\theta}}_{=0} A_{\theta r} + \underbrace{g^{\phi\phi}}_{=0} A_{\phi r} = 0$$

$$A^{\phi}{}_{\theta} = g^{\phi\alpha} A_{\alpha\theta} = \underbrace{g^{\phi\theta}}_{=0} A_{r\theta} + \underbrace{g^{\phi\theta}}_{=0} A_{\theta\theta} + \underbrace{g^{\phi\phi}}_{=0} A_{\phi\theta} = 0$$

$$A^{\phi}{}_{\phi} = g^{\phi\alpha} A_{\alpha\phi} = \underbrace{g^{\phi r}}_{=0} A_{r\phi} + \underbrace{g^{\phi\theta}}_{=0} A_{\theta\phi} + \underbrace{g^{\phi\phi}}_{=\frac{1}{r^2 sin^2(\theta)}} \underbrace{A_{\phi\phi}}_{=1} = \frac{1}{r^2 sin^2(\theta)}$$

∎

$(I.6.E.9)$ Let (x^i) be a coordinate system defined by

$x = x^1(u, v) = u + v$

$y = x^2(u, v) = u^2 - v^2$

Let $T_{\alpha\beta}$ be a tensor in the (x, y)-base.

(1) *Calculate the metric tensor* $g_{\mu\nu}$ in the (u, v)-base.

(2) Express $T_{\alpha\beta}$ in the (u, v)-base.

Solution:

For building up the metric tensor, we need to calculate the partial derivatives:

$$\frac{\partial x^1}{\partial u} = 1$$

266

$$\frac{\partial x^1}{\partial v} = 1$$

$$\frac{\partial x^2}{\partial u} = 2u$$

$$\frac{\partial x^2}{\partial v} = -2v$$

The Jacobian matrix is therefore:

$$\mathcal{J}_{(u,v)} = \begin{bmatrix} 1 & 1 \\ 2u & -2v \end{bmatrix}$$

Next step is to calculate the metric tensor with the Jacobi matrix:

$$[g_{\mu\nu}] \underset{(I.4.008)}{=} \mathcal{J}^T{}_{(u,v)}\, \mathcal{J}_{(u,v)} = \begin{bmatrix} 1 & 2u \\ 1 & -2v \end{bmatrix}\begin{bmatrix} 1 & 1 \\ 2u & -2v \end{bmatrix} = \begin{bmatrix} 1 + 4u^2 & 1 - 4uv \\ 1 - 4uv & 1 + 4v^2 \end{bmatrix}$$

Using $(I.6.006)$, we can write:

$$\bar{T}_{\alpha\beta} = \frac{\partial x^i}{\partial u^\alpha}\frac{\partial x^j}{\partial u^\beta}\, T_{ij}$$

where $(u, v) = (u^1, u^2)$

∎

$(I.8.2.E.1)$ Given the vector field $V = \begin{pmatrix} V_\sigma \\ V_\tau \\ V_z \end{pmatrix} = \begin{pmatrix} \frac{1}{\sigma^2} \\ 0 \\ 0 \end{pmatrix}$, calculate the divergence in parabolic 3D coordinates given by the set of functions

$$x^1(\sigma, \tau, z) = \sigma\tau\,\cos(\phi)$$
$$x^2(\sigma, \tau, z) = \sigma\tau\,\sin(\phi)$$
$$x^3(\sigma, \tau, z) = \frac{1}{2}(\tau^2 - \sigma^2)$$

Solution:

Referring to section $(I.A.2.12)$ we get for the divergence the formula:

$$div\, V(V_\sigma, V_\tau, V_\phi) = \frac{1}{\sigma\tau(\sigma^2+\tau^2)}\left(\frac{\partial}{\partial\sigma}\left(\sigma\tau\sqrt{\sigma^2+\tau^2}\, V_\sigma\right) + \frac{\partial}{\partial\tau}\left(\sigma\tau\sqrt{\sigma^2+\tau^2}\, V_\tau\right)\right.$$

$$\left. + \frac{\partial}{\partial\phi}\left((\sigma^2+\tau^2)\, V_\phi\right)\right)$$

$$= \frac{1}{\sigma\tau(\sigma^2+\tau^2)}\left(\tau\frac{\partial}{\partial\sigma}\left(\sigma\sqrt{\sigma^2+\tau^2}\, V_\sigma\right) + \sigma\frac{\partial}{\partial\tau}\left(\tau\sqrt{\sigma^2+\tau^2}\, \underbrace{V_\tau}_{=0}\right)\right)$$

$$+ \underbrace{\frac{1}{\sigma\tau}\frac{\partial V_\phi}{\partial\phi}}_{=0} = \frac{1}{\sigma\tau(\sigma^2+\tau^2)}\left(\tau\frac{\partial}{\partial\sigma}\left(\sigma\sqrt{\sigma^2+\tau^2}\,\underbrace{\frac{1}{\sigma^2}}_{=V_\sigma}\right)\right)$$

$$= \frac{1}{\sigma(\sigma^2+\tau^2)}\left(\frac{\partial}{\partial\sigma}\left(\sqrt{\sigma^2+\tau^2}\,\frac{1}{\sigma}\right)\right) = \frac{1}{\sigma(\sigma^2+\tau^2)}\left(\frac{\sigma}{\sqrt{\sigma^2+\tau^2}}\right)$$

$$= \frac{1}{\sigma(\sigma^2+\tau^2)^{\frac{3}{2}}}$$

∎

$(I.8.2.E.2)$ Given the following coordinate system:

$$u(x,y,z) = x^2 + y^2$$

$$v(x,y,z) = xy$$

$$w(x,y,z) = z$$

Calculate curl of the following vector field:

$$V = (V_1, V_2, V_3) = u\hat{e}_u + v^2\hat{e}_v + w\hat{e}_w$$

Solution:

We want to use $(I.8.008)$
$curl\, V(V_1, V_2, V_3)$

$$= \frac{1}{\sqrt{det(g)}}\left[e_1\left(\frac{\partial V_3}{\partial x^2} - \frac{\partial V_2}{\partial x^3}\right) + e_2\left(\frac{\partial V_1}{\partial x^3} - \frac{\partial V_3}{\partial x^1}\right)\right.$$

$$\left. + e_3\left(\frac{\partial V_2}{\partial x^1} - \frac{\partial V_1}{\partial x^2}\right)\right]$$

Therefore, we need to calculate the covariant metric tensor $[g_{ij}]$, using $(I.\,4.008)$.

We calculate the Jacobi-matrix:

$$J = \begin{bmatrix} 2x & 2y & 0 \\ y & x & 0 \\ 0 & 0 & 1 \end{bmatrix}$$

We get:

$$[g_{ij}] = J^T J = \begin{bmatrix} 4x^2 + y^2 & 5xy & 0 \\ 5xy & x^2 + 4y^2 & 0 \\ 0 & 0 & 1 \end{bmatrix}$$

$$det([g_{ij}]) = 4(x^2 - y^2)^2$$

The next step is to express $4(x^2 - y^2)^2$ in terms of the coordinate functions u, v and w. It will be sufficient to do this for the inner term $(x^2 - y^2)$.

We use the identity:

$$(x^2 - y^2) = (x + y)(x - y)$$

Furthermore, we know that:

$$(x^2 + y^2) = x^2 + 2xy + y^2 = \underbrace{x^2 + y^2}_{=u} + 2\underbrace{xy}_{=v} = u + 2v,$$

So $(x + y) = \pm \sqrt{u + 2v}$

Now, let us turn to $(x - y)$:

We know by definition of the coordinate functions that:

$$(x^2 + y^2) = u$$

It follows that:

$$(x - y)^2 = x^2 - 2xy + y^2 = u - 2v, \text{ so:}$$

$$(x - y) = \pm\sqrt{u - 2v}$$

Putting all together, we find that:

$$x^2 - y^2 = (x + y)(x - y) = \pm\left(\sqrt{u + 2v}\,\sqrt{u - 2v}\right) = \pm\sqrt{u^2 - 4v^2}$$

It follows that:

$$det\left(\left[g_{ij}\right]\right) = 4(u^2 - 4v^2)$$

Now, we can calculate curl by substituting the corresponding values:

$$curl\ V(V_1, V_2, V_3) = \frac{1}{4(u^2 - 4v^2)}\left[\hat{e}_u\left(\underbrace{\frac{\partial}{\partial v}(w)}_{=0} - \underbrace{\frac{\partial}{\partial w}(v^2)}_{=0}\right)\right.$$

$$\left. + \hat{e}_v\left(\underbrace{\frac{\partial}{\partial w}(u)}_{=0} - \underbrace{\frac{\partial}{\partial u}(w)}_{=0}\right) + \hat{e}_w\left(\underbrace{\frac{\partial}{\partial u}(v^2)}_{=0} - \underbrace{\frac{\partial}{\partial v}(u)}_{=0}\right)\right] = 0$$

As all partial derivatives are zero, *curl* is zero: in this cases it was not necessary to express the determinant of the metric tensor in terms of (u, v, w): we did it nevertheless because in other situations, you need to do the replacement.

■

$(I.8.3.E.1)$ Show that the following vector field is irrotational:

$$V(r, \theta, \phi) = r^2 sin^2(\theta)\left(\,3sin(\theta)cos(\phi)\hat{e}_r + 3cos(\theta)cos(\phi)\hat{e}_\theta - sin(\phi)\hat{e}_\phi\right)$$

Solution:

We have to show that $curl\ V(r,\theta,\phi)$ is zero. We calculate in spherical coordinates using the scale factors:

$$h_r = 1$$

$$h_\theta = r$$

$$h_\phi = r\sin(\theta)$$

We rewrite $V(r,\theta,\phi)$:

$$V(r,\theta,\phi) = \begin{pmatrix} V_r \\ V_\theta \\ V_\phi \end{pmatrix} = \begin{pmatrix} 3\,r^2 sin^2(\theta)\,sin(\theta)cos(\phi) \\ 3\,r^2 sin^2(\theta)\,cos(\theta)cos \\ -r^2 sin^2(\theta)\,sin(\phi) \end{pmatrix}$$

Using $(I.8.009)$, we obtain:

$$(1)\ curl\ V\left(V_r, V_\theta, V_\phi\right) = \begin{pmatrix} \dfrac{1}{r\,sin\,(\theta)}\left(\dfrac{\partial}{\partial\theta}\left(sin(\theta)\,V_\phi\right) - \dfrac{\partial V_\theta}{\partial\phi}\right) \\[2ex] \dfrac{1}{r\,sin\,(\theta)}\left(\dfrac{\partial}{\partial\phi}V_r - sin\,(\theta)\,(\dfrac{\partial}{\partial r}\,r\,V_\phi)\right) \\[2ex] \dfrac{1}{r}\left(\dfrac{\partial}{\partial r}(r\,V_\theta) - \dfrac{\partial}{\partial\theta}V_r\right) \end{pmatrix}$$

(see also $(I.A.1.3)$). Now, we calculate the partial differentials:

$$\frac{\partial}{\partial\theta}\left(sin(\theta)\,V_\phi\right) = -3\,r^2\,cos(\theta)\,sin(\phi)\,sin^2(\theta)$$

$$\frac{\partial V_\theta}{\partial\phi} = -3\,r^2\,cos(\theta)\,sin(\phi)\,sin^2(\theta)$$

$$\frac{\partial}{\partial\phi}V_r = -3\,r^2\,sin(\phi)\,sin^3(\theta)$$

$$\frac{\partial}{\partial r}\,r\,V_\phi == -3\,r^2\,sin(\phi)\,sin^2(\theta)$$

$$\frac{\partial}{\partial r}(r\,V_\theta) = 9\,r^2\,cos(\theta)\,cos(\phi)\,sin^2(\theta)$$

$$\frac{\partial}{\partial \theta}V_r = 9\,r^2\,cos(\theta)\,cos(\phi)\,sin^2(\theta)$$

Putting these results into (1), abbreviating $c = cos, s = sin$, we get:

$$curl\,V\left(V_r, V_\theta, V_\phi\right)$$

$$= \begin{pmatrix} \dfrac{1}{r\,s(\theta)}\left(-3\,r^2\,c(\theta)\,s(\phi)\,s^2(\theta) - \left(-3\,r^2\,c(\theta)\,s(\phi)\,s^2(\theta)\right)\right) \\[2mm] \dfrac{1}{r\,s(\theta)}\left(-3\,r^2\,s(\phi)\,s^3(\theta) - s\,(\theta)\,(-3\,r^2\,s(\phi)\,s^2(\theta))\right) \\[2mm] \dfrac{1}{r}\left(9\,r^2\,c(\theta)\,c(\phi)\,s^2(\theta) - 9\,r^2\,c(\theta)\,c(\phi)\,s^2(\theta)\right) \end{pmatrix} = \begin{pmatrix} 0 \\ 0 \\ 0 \end{pmatrix}$$

∎

$(I.8.4.E.1)$ Let $\Theta(r,\theta,h) = r\,cos(\theta) + h^2$.

Calculate $\nabla^2\Theta(r,\theta,h)$ in cylindrical coordinates.

Solution:

The Laplacian in cylindrical coordinates (see $(I.A.1.2)$) is given by:

$$(1)\;\; \nabla^2\Theta(r,\theta,h) = \frac{1}{r}\left(\frac{\partial}{\partial r}\left(r\,\frac{\partial\Theta}{\partial r}\right)\right) + \frac{1}{r^2}\frac{\partial^2\Theta}{\partial\theta^2} + \frac{\partial^2\Theta}{\partial h^2}$$

Using $(I.8.013)$ with the scale factors

$$h_r = h_h = 1$$

$$h_\theta = r$$

We calculate the partial differentials:

$$\frac{\partial\Theta}{\partial r} = cos(\theta)$$

$$\frac{\partial}{\partial r}\left(r\,\frac{\partial\Theta}{\partial r}\right) = \frac{\partial}{\partial r}\left(r\,cos(\theta)\right) = cos(\theta)$$

$$\frac{\partial^2\Theta}{\partial\theta^2} = -r\,cos(\theta)$$

$$\frac{\partial^2\Theta}{\partial h^2} = 2$$

Inserting these results in (1):

$$\nabla^2\Theta(\,r,\theta,h\,) = \frac{1}{r}\big(cos(\theta)\big) - \frac{1}{r^2}r\,cos(\theta) + 2 = 2$$

■

$(\mathit{I}.9.6.1)$ Sphere inside of a cylinder

Let $A = \{(x,y,z) \in \mathbb{R}^3 : x^2 + y^2 + z^2 \leq 1\}$,

$$B = \{(x,y,z) \in \mathbb{R}^3 : x^2 + y^2 \leq \frac{1}{2}\}$$

Calculate the volume of $V_{A\cap B} = A \cap B$.

First of all, we note that:

$$(1)\; A \cap B = \left\{(x,y,z) \in \mathbb{R}^3 : x^2 + y^2 + z^2 \leq 1,\; x^2 + y^2 \leq \frac{1}{2}\right\}.$$

A defines a solid sphere, but what about B? If B were a subset of $\mathbb{R}^2$, then B would describe a circle. But $\subseteq \mathbb{R}^3$, so although nothing is said about z, z takes any values and so B defines a cylinder.

In the following picture, we see a solid sphere with radius 1, which is cut through by a cylinder with radius 0.5. The intersection of A with B corresponds to the volume of the sphere inside of the cylinder.

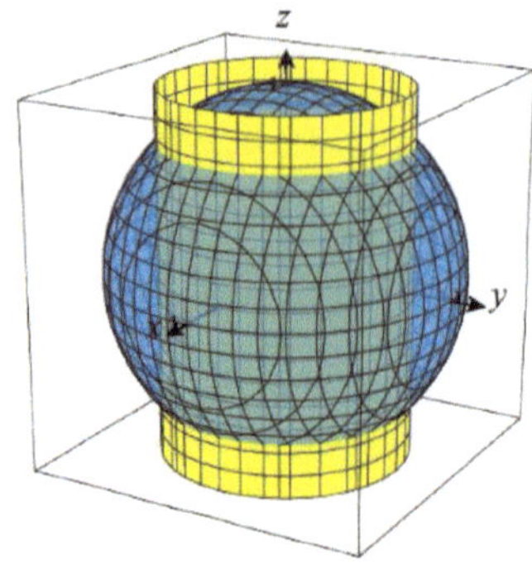

https://c3d.libretexts.org/CalcPlot3D/CalcPlot3D-Help/section-regions.html

We will use cylindrical coordinates $(I.A.1.2)$, so

$$x = r \cos (\varphi)$$

$$y = r \sin (\varphi)$$

$$z = z$$

We will note the transformation with

$$g(r, \varphi, z) = (r \cos(\varphi), r \sin(\varphi), z).$$

Using cylindrical coordinates, we can retrieve boundaries of r, φ and z just by exchanging x, y and z by their transformation. We get:

$g(C) = A \cap B = \left\{ (r, \varphi, z) \in \mathbb{R}^3 : r^2 + z^2 \leq 1, r^2 \leq \frac{1}{2} \right\}$, so we can conclude:

(1) $0 \leq r \leq \sqrt{\frac{1}{2}}$

(2) $|z| \leq \sqrt{1 - r^2}$

(3) $\varphi \in [0, 2\pi]$

The last equation is given by the circumstance that it does not appear in the transformed set of $A \cap B$. It is also seen by the picture that we have to make a full 2π-travel around the cylinder.

274

$$
\begin{aligned}
V_{A\cap B} \quad &= \quad \int\limits_{A\cap B} 1\,dx\,dy\,dz \\[2em]
&= \quad \int\limits_{C} \left| \underbrace{\det \mathcal{J}_g}_{=r} \right| d\varphi\,dz\,dr \\[2em]
&= \quad \int\limits_{r=0}^{\sqrt{\frac{1}{2}}} \int\limits_{z=-\sqrt{1-r^2}}^{\sqrt{1-r^2}} \int\limits_{\varphi=0}^{2\pi} r\,d\varphi\,dz\,dr \\[2em]
&= \quad 2\pi \int\limits_{r=0}^{\sqrt{\frac{1}{2}}} \int\limits_{z=-\sqrt{1-r^2}}^{\sqrt{1-r^2}} r\,dz\,dr \\[2em]
&= \quad 2\pi \int\limits_{0}^{\sqrt{\frac{1}{2}}} 2r\sqrt{1-r^2}\,dr \\[2em]
&= \quad 2\pi \left(-\frac{2}{3}(1-r^2)^{\frac{3}{2}} \right)\Bigg|_{0}^{\sqrt{\frac{1}{2}}} \\[2em]
&= \quad 2\pi \left(-\frac{2}{3}\left(1-\frac{1}{2}\right)^{\frac{3}{2}} \right) \\[2em]
&= \quad \frac{4\pi}{3} \sqrt{\frac{1}{2}}^{\,3}
\end{aligned}
$$

$(\boldsymbol{I}.9.6.2)$ Polar transformation

Calculate the integral

$$\iint_A 3x + 4(y)^2 \; dx \, dy$$

with $A = \{(x,y) \in \mathbb{R}^2, y \geq 0, \; 1 \leq (x)^2 + (y)^2 <= 4\}$

Solution:

The integration area can be visualized by:

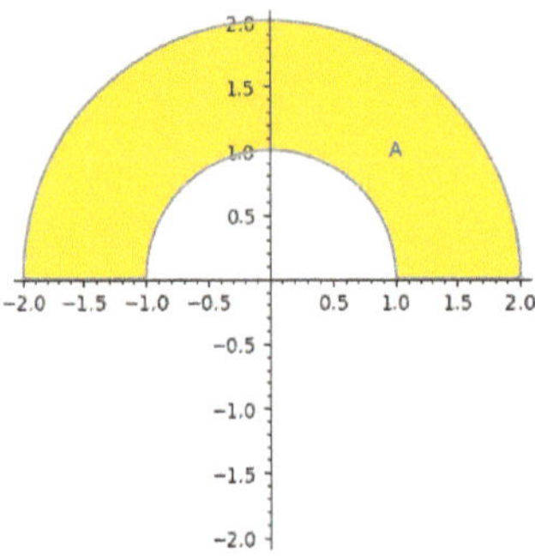

As we are dealing with circles, we transform into polar coordinates. The boundaries are:

$r \in [1,2]$ and $\varphi \in [0,\pi]$

We have:

$$
\begin{aligned}
V_A \;\; &= \;\; \iint_A 3x + 4(y)^2 \; dx \, dy \\[2mm]
&= \;\; \int_{\varphi=0}^{\pi} \int_{r=1}^{2} \left(3(r\cos(\varphi)) + 4(r^2 \sin^2(\varphi)) \right) \, r \, dr \, d\varphi \\[2mm]
&= \;\; \int_{\varphi=0}^{\pi} \int_{r=1}^{2} 3\,r^2 \cos(\varphi) + 4\,r^3 \sin^2(\varphi) \; dr \, d\varphi
\end{aligned}
$$

$$= \int_{\varphi=0}^{\pi} \left. r^3 \cos(\varphi) + (r^4 \sin^2(\varphi)) \right|_{1}^{2} d\varphi$$

$$= \int_{\varphi=0}^{\pi} 7 \cos(\varphi) + 15 \sin^2(\varphi) \, d\varphi$$

$$= 7 \int_{\varphi=0}^{\pi} \cos(\varphi) \, d\varphi + 15 \int_{\varphi=0}^{\pi} \sin^2(\varphi) \, d\varphi$$

$$= 7 \sin(\varphi)|_0^{\pi} \; \left(+15 \left({}^1\!/_2 \varphi - {}^1\!/_4 \sin(2\varphi) \right) \right)\Big|_0^{\pi}$$

$$= {}^{15}\!/_2 \pi$$

∎

$(I.9.6.3)$ Cylindrical transformation

Let $B = \{(x, y, z) \in \mathbb{R}^3, \; (x)^2 + (y)^2 \leq 1, \; 0 \leq z \leq 1 \}$.

Calculate

$$I = \int_B y \sqrt{(x)^2 + (y)^2} + z \, dx \, dy \, dz$$

Solution:

B is appearently 3-dimensional and all parameters x, y and z appear in the integral. The condition $(x)^2 + (y)^2 \leq 1$ shows, that we are dealing, once more, with a circle. As z is part of the integral, we should choose cylindrical coordinates $(I.A.1.2)$:

$x = r \cos(\varphi)$

$y = r \sin(\varphi)$

$z = z$

For the new boundaries, we find:

$r \in [0,1]$

$$\varphi \in \left[0, \frac{\pi}{2}\right]$$

$$z \in [0, 1]$$

$$
\begin{aligned}
I \;&=\; \int_B y \sqrt{(x)^2 + (y)^2} + z \; dx \, dy \, dz \\[2mm]
&=\; \int_{r=0}^{1} \int_{\varphi=0}^{\frac{\pi}{2}} \int_{z=0}^{1} \left(r \sin(\varphi) \sqrt{(r \cos(\varphi))^2 + (r \sin(\varphi))^2} + z \right) r \; dz \, d\varphi \, dr \\[2mm]
&=\; \int_{r=0}^{1} \int_{\varphi=0}^{\frac{\pi}{2}} \int_{z=0}^{1} \left(r^2 \sin(\varphi) + z \right) r \; dr \, d\varphi \, dz \\[2mm]
&=\; \int_{r=0}^{1} \int_{\varphi=0}^{\frac{\pi}{2}} \left(r^3 \sin(\varphi) + \frac{1}{2} z^2 r \Big|_{r=0}^{1} \right) d\varphi \, dr \\[2mm]
&=\; \int_{r=0}^{1} \int_{\varphi=0}^{\frac{\pi}{2}} \left(r^3 \sin(\varphi) + \frac{1}{2} r \right) d\varphi \, dr \\[2mm]
&=\; \int_{r=0}^{1} -r^3 \cos(\varphi) + \frac{1}{2} \varphi r \Big|_{\varphi=0}^{\frac{\pi}{2}} dr \\[2mm]
&=\; \int_{r=0}^{1} r^3 + \frac{1}{4} \pi r \; dr \\[2mm]
&=\; \frac{1}{4} r^3 + \frac{1}{8} \pi r^2 \Big|_{r=0}^{1} \\[2mm]
&=\; \frac{1}{4} + \frac{\pi}{8}
\end{aligned}
$$

∎

($I.9.6.4$) Cardioid

Let $B = \{(x, y) \in \mathbb{R}^2, (x^2 + y^2)^2 - 2x(x^2 + y^2) - y^2 = 0\}$

And let $f(x, y) = 1$.

Calculate the area of f above the region B.

Solution:

Let us have a view on B and f:

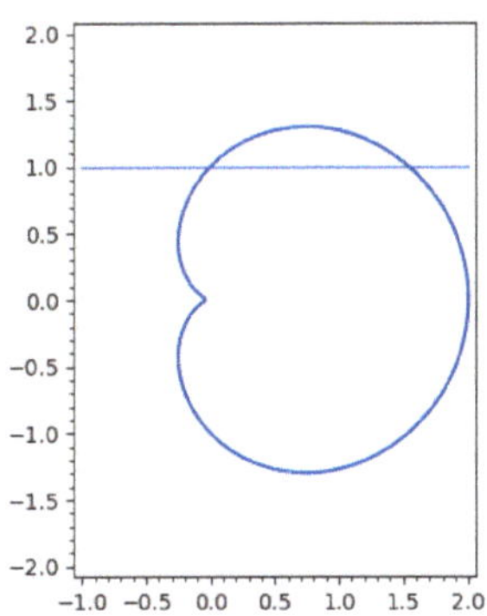

As we are dealing with a two-dimensional area, we will use polar coordinates. Therefore, we need to get information on r and φ. We use the equation in B and use the transformation

$x = r\,\cos(\varphi)$

$y = r\,\sin(\varphi)$

So, we have:

$(x^2 + y^2)^2 - 2x(x^2 + y^2) - y^2 = 0 \iff$

$((r\,\cos(\varphi))^2 + (r\,\sin(\varphi))^2)^2 - 2(r\,\cos(\varphi))((r\,\cos(\varphi))^2 + (r\,\sin(\varphi))^2) - (r\,\sin(\varphi))^2 = 0 \iff$

$$\left(r^2 \underbrace{\left(\cos^2(\varphi) + \sin^2(\varphi) \right)}_{=1} \right)^2 - 2r\cos(\varphi)\left(r^2 \underbrace{\left(\cos^2(\varphi) + \sin^2(\varphi) \right)}_{=1} \right)$$

$$- r^2\sin^2(\varphi) = 0 \iff$$

$$r^4 - 2r^3\cos(\varphi) - r^2\sin^2(\varphi) = 0 \iff$$

$$r^2 - 2r\cos(\varphi) - \sin^2(\varphi) = 0$$

This quadratic equation has the solutions

$$r = \cos(\varphi) \pm \sqrt{\cos^2(\varphi) + \sin^2(\varphi)} = \cos(\varphi) \pm 1$$

As $r > 0$ we have to choose $r = \cos(\varphi) + 1$ as the upper limit.

There is no implication on φ, so we take $\varphi \in [0, 2\pi]$

$$
\begin{aligned}
\iint_B 1 \, dx \, dy \;&=\; \int_{\varphi=0}^{2\pi} \int_{r=0}^{\cos(\varphi)+1} r \, dr \, d\varphi \\[2ex]
&=\; \int_0^{2\pi} \frac{1}{2} r^2 \Big|_0^{\cos(\varphi)+1} \, d\varphi \\[2ex]
&=\; \frac{1}{2} \int_0^{2\pi} (\cos(\varphi) + 1)^2 \, d\varphi \\[2ex]
&=\; \frac{1}{2} \int_0^{2\pi} \cos^2(\varphi) + 2\cos(\varphi) + 1 \, d\varphi \\[2ex]
&=\; \frac{1}{2} \left(\int_0^{2\pi} \cos^2(\varphi) \, d\varphi + \int_0^{2\pi} 2\cos(\varphi) \, d\varphi + \int_0^{2\pi} 1 \, d\varphi \right) \\[2ex]
&=\; \frac{1}{2} \left[\frac{1}{2}\left(\cos(\varphi)\sin(\varphi) + \varphi \right)\Big|_0^{2\pi} + 2\sin(\varphi)\big|_0^{2\pi} + \varphi\big|_0^{2\pi} \right] \\[2ex]
&=\; \frac{3}{2}\pi
\end{aligned}
$$

Notice, that $\sin(2\pi) = 0$ and $sin(0) = 0$, so all terms containing sin vanish.

■

($\textbf{\textit{I}.9.6.5}$) Part of a sphere

Calculate the volume of the section of the sphere $x^2 + y^2 + z^2 \leq R^2$ and the half-space $z \leq R/2$

Solution:

Let us have a view on what we need to calculate:

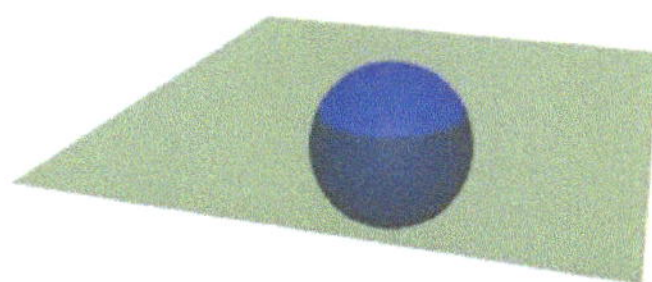

We need to calculate the sphere below the plane.

As we have to deal with a sphere, it might seem that spherical coordinates are convenient to calculate the volume. Nevertheless, the fact that a quarter of the sphere is "removed" is difficult to manage. We will try to turn to cylindrical coordinates.

It is obvious that φ runs from 0 to 2π. In cylindrical coordinates we have $r^2 = x^2 + y^2$, so we can conclude $r^2 + z^2 \leq R^2$, hence r is running from 0 to $\sqrt{R^2 - z^2}$.

So, we get:

$$V \quad = \quad \iiint_B dx\, dy\, dz$$

$$= \quad \int_0^{2\pi} \int_{-R}^{\frac{R}{2}} \int_0^{\sqrt{R^2 - z^2}} r\, dr\, dz\, d\varphi$$

$$= \quad 2\pi \int_{-R}^{\frac{R}{2}} \int_0^{\sqrt{R^2 - z^2}} r \, dr \, dz$$

$$= \quad 2\pi \int_{-R}^{\frac{R}{2}} \left[\frac{1}{2} r^2 \right]_0^{\sqrt{R^2 - z^2}} dz$$

$$= \quad \pi \int_{-R}^{\frac{R}{2}} R^2 - z^2 \, dz$$

$$= \quad \frac{9}{6} \pi R^2$$

∎

I.A List of coordinate systems

In this chapter, we describe several coordinate systems that the reader may encounter. We will give the function sets for Cartesian coordinates, the metric, the derivatives of the metric tensor and the same for the inverse metric tensor, the Jacobi matrix and its inverse, we calculate the non-zero Christoffel symbols of the 2nd kind, we calculate divergence, curl and the Laplacian.

Some remarks:

- For the divergence, curl and Laplacian we give only the non-expanded version. This means that we do not calculate the derivatives as far as possible. Two reasons for this: on the one hand, the expanded version becomes unreadable, on the other hand there might be good chances that some of the scale factors cancel out when working with concrete scalar- or vector fields.
- As far as possible, we give a representation of the Jacobi-matrix and the metric tensor as a matrix. Where this is not possible, we write down the matrix elements in components. For diagonal matrices, we use $diag \begin{cases} g[1,1] \\ g[2,2]. \\ g[3,3] \end{cases}$
- For shortening we will use $[0]$ for a matrix that contains only zeros.
- All values have been calculated in Sage Math. For every coordinate system, there will be a link to a Jupiter Notebook that contains all calculations for the coordinate system. <u>For a description on how and where to download the Jupiter notebooks see ($I.\,1.4$)</u>.If you use the Jupiter notebooks, you will encounter differences between what Sage Math calculates and what you read in this paper. Remember, please, that Sage Math has some difficulties in simplifying (trigonometric) expressions. Where necessary, we simplified expressions by giving a link to Wolfram Simplification App.
- We use $\mathcal{J}$ for the Jacobi matrix, $[g_{ij}]$ for the metric tensors.
- Especially for the metric tensor $[g_{ij}]$ we will use $[g^{ij}]$ for the inverse matrix. So, be aware of the position of the indices!
- As the Christoffel symbols of the 2nd kind are commutative in both lower indices, we only write down one of the symbols.

- Wherever necessary, we will use the following abbreviations:
  - $c = cos$
  - $s = sin$
  - $ch = cosh$
  - $sh = sinh$

I.A.1 List of mostly used coordinate systems

In this section, you find those coordinate systems that are widely used in physics.

I.A.1.1 Polar coordinates

References:

[Neutsch]: p. 1139

Coordinate definition:

$$Cartesian\ functions \begin{cases} x^1(r,\theta) = r\cos\theta \\ x^2(r,\theta) = r\sin\theta \end{cases}$$

$$r >= 0, \qquad 0 \le \theta \le 2\pi$$

Orthogonal   *yes*

Jacobian and metric:

$$J_{(r,\theta)} = \begin{bmatrix} \cos(\theta) & -r\,\sin(\theta) \\ \sin(\theta) & r\,\cos(\theta) \end{bmatrix}$$

$$J^{-1}_{(r,\theta)} = \begin{bmatrix} \cos(\theta) & \sin(\theta) \\ \dfrac{-\sin(\theta)}{r} & \dfrac{\cos(\theta)}{r} \end{bmatrix}$$

$$\det\big(J_{(r,\theta)}\big) = r$$

Metric tensor (covariant) and derivatives:

$$[g_{ij}] \quad = \begin{bmatrix} 1 & 0 \\ 0 & r^2 \end{bmatrix}$$

$$\frac{\partial}{\partial r}[g_{ij}] \quad = \begin{bmatrix} 0 & 0 \\ 0 & 2r \end{bmatrix}$$

$$\frac{\partial}{\partial \theta}[g_{ij}] \quad = [0]$$

Inverse metric tensor (contravariant) and derivatives:

$$[g^{ij}] \quad = \begin{bmatrix} 1 & 0 \\ 0 & 1/r^2 \end{bmatrix}$$

$$\frac{\partial}{\partial r}[g^{ij}] \quad = \begin{bmatrix} 1 & 0 \\ 0 & -2/r^3 \end{bmatrix}$$

$$\frac{\partial}{\partial \theta}[g^{ij}] \quad = [0]$$

Scale factors and Christoffel symbols:

$$Scale\ factors: \quad \begin{cases} h_r = 1 \\ h_\theta = r \end{cases}$$

Non-zero Christoffel symbols of the 2nd kind:

$$\Gamma^{\theta}_{\theta r} = 1/r$$

$$\Gamma^{r}_{\theta\theta} = -r$$

Unit Basis Vectors:

$$\hat{e}_r = \begin{pmatrix} \cos(\theta) \\ \sin(\theta) \end{pmatrix} \quad \hat{e}_\theta = \begin{pmatrix} -\sin(\theta) \\ \cos(\theta) \end{pmatrix}$$

Differential operators:

$$div\,(V_r, V_\theta) = \frac{1}{r}\frac{\partial}{\partial r}\,(\,r\,V_r\,) + \frac{1}{r}\frac{\partial}{\partial \theta}\,(\,V_\theta\,)$$

$$\Delta\,\Theta(\,r,\theta\,) = \frac{1}{r}\frac{\partial\Theta}{\partial r}\left(r\frac{\partial\Theta}{\partial r}\right) + \frac{1}{r^2}\frac{\partial^2\Theta}{\partial^2\theta}$$

I.A.1.2 Cylindrical coordinates

References:

[Neutsch]: p. 1174

Coordinate definition:

$$Cartesian\ functions\ \begin{cases} x^1(r,\theta,h) = r\cos(\theta) \\ x^2(r,\theta,h) = r\sin(\theta) \\ x^3(r,\theta,h) = h \end{cases}$$

$$r >= 0,\ 0 \le \theta \le 2\pi$$

Orthogonal   yes

Jacobian and metric:

$$J_{(r,\theta,h)} \quad = \quad \begin{bmatrix} \cos(\theta) & -r\sin(\theta) & 0 \\ \sin(\theta) & r\cos(\theta) & 0 \\ 0 & 0 & 1 \end{bmatrix}$$

$$J_{(r,\theta,h)}^{-1} \quad = \quad \begin{bmatrix} \cos(\theta) & \sin(\theta) & 0 \\ \dfrac{-\sin(\theta)}{r} & \dfrac{\cos(\theta)}{r} & 0 \\ 0 & 0 & 1 \end{bmatrix}$$

$$\det(J_{(r,\theta,h)}) \quad = \quad r$$

Metric tensor (covariant) and derivatives:

$$[g_{ij}] \quad = \quad \begin{bmatrix} 1 & 0 & 0 \\ 0 & r^2 & 0 \\ 0 & 0 & 1 \end{bmatrix}$$

$$\frac{\partial}{\partial r}[g_{ij}] = \begin{bmatrix} 0 & 0 & 0 \\ 0 & 2r & 0 \\ 0 & 0 & 0 \end{bmatrix}$$

$$\frac{\partial}{\partial \theta}[g_{ij}] = [0]$$

$$\frac{\partial}{\partial h}[g_{ij}] = [0]$$

Inverse metric tensor (contravariant) and derivatives:

$$[g^{ij}] = \begin{bmatrix} 1 & 0 & 0 \\ 0 & 1/r^2 & 0 \\ 0 & 0 & 1 \end{bmatrix}$$

$$\frac{\partial}{\partial r}[g^{ij}] = \begin{bmatrix} 1 & 0 & 0 \\ 0 & -2/r^3 & 0 \\ 0 & 0 & 0 \end{bmatrix}$$

$$\frac{\partial}{\partial \theta}[g^{ij}] = [0]$$

$$\frac{\partial}{\partial h}[g^{ij}] = [0]$$

Scale factors and Christoffel symbols:

$$Scale\ factors: \begin{cases} h_r = h_h = 1 \\ h_\theta = r \end{cases}$$

Non-zero Christoffel symbols of the $\mathbf{2^{nd}}$ kind:

$$\Gamma^r_{\theta\theta} = -r$$

$$\Gamma^r_{\phi\phi} = -r\,sin^2(\theta)$$

$$\Gamma^\theta_{r\theta} = \,^1\!/_r$$

$$\Gamma^\theta_{\phi\phi} = -cos(\theta)\,sin(\theta)$$

$$\Gamma^\phi_{r\phi} = \,^1\!/_r$$

$$\Gamma^\phi_{\theta\phi} = cos(\theta)/\,sin(\theta)$$

Unit Basis Vectors:

$$\hat{e}_r = \begin{pmatrix} cos(\phi)sin(\theta) \\ sin(\theta)sin(\phi) \\ cos(\theta) \end{pmatrix} \quad \hat{e}_\theta = \begin{pmatrix} cos(\phi)\,cos(\theta) \\ cos(\theta)sin(\phi) \\ -sin(\theta) \end{pmatrix}$$

$$\hat{e}_\phi = \begin{pmatrix} -sin(\phi) \\ cos(\phi) \\ 0 \end{pmatrix}$$

Differential operators:

$$div\ V\left(V_r, V_\theta, V_\phi\right) = \frac{1}{r^2}\frac{\partial}{\partial r}(r^2\,V_r) + \frac{1}{r\,sin(\theta)}\frac{\partial}{\partial\theta}(sin(\theta)\,V_\theta) + \frac{1}{r\,sin(\theta)}\frac{\partial V_\phi}{\partial\phi}$$

$$curl\ V\left(V_r, V_\theta, V_\phi\right)$$

$$= \frac{1}{r\,sin(\theta)}\left(\frac{\partial}{\partial\theta}\left(sin(\theta)\,V_\phi\right) - \frac{\partial V_\theta}{\partial\phi}\right)\hat{e}_r$$

$$+ \frac{1}{r\,sin\,(\theta)}\left(\frac{\partial}{\partial\phi}V_r - sin(\theta)\left(\frac{\partial}{\partial r}\,r\,V_\phi\right)\right)\hat{e}_\theta$$

$$+ \frac{1}{r}\left(\frac{\partial}{\partial r}\left(r\,V_\theta\right) - \frac{\partial}{\partial\theta}V_r\right)\hat{e}_\phi$$

$$\Delta\,\Theta\,(\,r, \theta, \phi\,) = \frac{1}{r^2}\frac{\partial}{\partial r}\left(r^2\frac{\partial\Theta}{\partial r}\right) + \frac{1}{r^2 sin^2(\theta)}\frac{\partial}{\partial\theta}\left(sin(\theta)\frac{\partial\Theta}{\partial\theta}\right) + \frac{1}{r^2 sin^2(\theta)}\frac{\partial^2\Theta}{\partial\phi^2}$$

I.A.1.3 Spherical coordinates

References:

https://mathworld.wolfram.com/SphericalCoordinates.html

Coordinate definition:

$$Cartesian\ functions \begin{cases} x^1(r,\theta,\phi) = r\,sin(\theta)\,cos(\phi) \\ x^2(r,\theta,\phi) = r\,sin(\theta)\,sin(\phi) \\ x^3(r,\theta,\phi) = r\,cos(\theta) \end{cases}$$

$$r \geq 0, 0 \leq \theta \leq \pi, 0 \leq \phi \leq 2\pi$$

Orthogonal   yes

Jacobian and metric:

$$\mathcal{J}_{(r,\theta,\phi)} = \begin{bmatrix} cos(\phi)\,sin(\theta) & r\,cos(\theta)\,cos(\phi) & -r\,sin(\phi) \\ sin(\theta)\,sin(\phi) & r\,cos(\theta)\,sin(\phi) & r\,cos(\phi)\,sin(\theta) \\ cos(\theta) & -r\,sin(\theta) & 0 \end{bmatrix}$$

$$\mathcal{J}_{(r,\theta,\phi)}^{-1} = \begin{bmatrix} cos(\phi)\,sin(\theta) & sin(\theta)\,sin(\phi) & cos(\theta) \\ \dfrac{cos(\theta)\,cos(\phi)}{r} & \dfrac{cos(\theta)\,sin(\phi)}{r} & \dfrac{-sin(\theta)}{r} \\ -\dfrac{sin(\phi)}{r\,sin(\theta)} & \dfrac{cos(\phi)}{r\,sin(\theta)} & 0 \end{bmatrix}$$

$$det\left(\mathcal{J}_{(r,\theta,\phi)}\right) = r^2\,sin(\theta)$$

Metric tensor (covariant) and derivatives:

$$[g_{ij}] = \begin{bmatrix} 1 & 0 & 0 \\ 0 & r^2 & 0 \\ 0 & 0 & r^2 \sin^2(\theta) \end{bmatrix}$$

$$\frac{\partial}{\partial r}[g_{ij}] = \begin{bmatrix} 1 & 0 & 0 \\ 0 & 2\,r & 0 \\ 0 & 0 & 2\,r\,\sin^2(\theta) \end{bmatrix}$$

$$\frac{\partial}{\partial \theta}[g_{ij}] = \begin{bmatrix} 0 & 0 & 0 \\ 0 & 0 & 0 \\ 0 & 0 & r^2 \sin(2\,\theta) \end{bmatrix}$$

$$\frac{\partial}{\partial \phi}[g_{ij}] = [0]$$

Inverse metric tensor (contravariant) and derivatives:

$$[g^{ij}] = \begin{bmatrix} 1 & 0 & 0 \\ 0 & \dfrac{1}{r^2} & 0 \\ 0 & 0 & \dfrac{1}{r^2 \sin^2(\theta)} \end{bmatrix}$$

$$\frac{\partial}{\partial r}[g^{ij}] = \begin{bmatrix} 1 & 0 & 0 \\ 0 & -\dfrac{2}{r^3} & 0 \\ 0 & 0 & -\dfrac{2}{r^3 \sin^2(\theta)} \end{bmatrix}$$

$$\frac{\partial}{\partial \theta}[g^{ij}] = \begin{bmatrix} 1 & 0 & 0 \\ 0 & -\dfrac{2}{r^3} & 0 \\ 0 & 0 & -\dfrac{2 \cos(\theta)}{r^2 \sin^3(\theta)} \end{bmatrix}$$

$$\frac{\partial}{\partial \phi}[g^{ij}] \quad = \quad [0]$$

Scale factors and Christoffel symbols:

$$Scale\ factors:\ \begin{cases} h_r = 1 \\ h_\theta = r \\ h_\phi = r\ sin(\theta) \end{cases}$$

Non-zero Christoffel symbols of the 2nd kind:

$$\Gamma^r_{\theta\theta} = -r$$

$$\Gamma^r_{\phi\phi} = -r\ sin^2(\theta)$$

$$\Gamma^\theta_{r\theta} = {}^1\!/_r$$

$$\Gamma^\theta_{\phi\phi} = -cos(\theta)\ sin(\theta)$$

$$\Gamma^\phi_{r\phi} = {}^1\!/_r$$

$$\Gamma^\phi_{\theta\phi} = cos(\theta)/\ sin(\theta)$$

Unit Basis Vectors:

$$\hat{e}_r = \begin{pmatrix} cos(\phi)\ sin(\theta) \\ sin(\theta)\ sin(\phi) \\ cos(\theta) \end{pmatrix} \quad \hat{e}_\theta = \begin{pmatrix} cos(\phi)\ cos(\theta) \\ cos(\theta)\ sin(\phi) \\ -sin(\theta) \end{pmatrix} \quad \hat{e}_\phi = \begin{pmatrix} -sin(\phi) \\ cos(\phi) \\ 0 \end{pmatrix}$$

Differential operators:

$$div\ V\left(V_r, V_\theta, V_\phi\right) = \frac{1}{r^2}\frac{\partial}{\partial r}(r^2\ V_r) + \frac{1}{r\ sin(\theta)}\frac{\partial}{\partial \theta}(sin(\theta)\ V_\theta\ ) + \frac{1}{r\ sin(\theta)}\frac{\partial V_\phi}{\partial \phi}$$

$$curl\ V\left(V_r, V_\theta, V_\phi\right)$$

$$= \frac{1}{r\ sin(\theta)}\left(\frac{\partial}{\partial \theta}\left(sin(\theta)\ V_\phi\right) - \frac{\partial V_\theta}{\partial \phi}\right)\hat{e}_r$$

$$+ \frac{1}{r\ sin\ (\theta)}\left(\frac{\partial}{\partial \phi}V_r - sin(\theta)\ (\frac{\partial}{\partial r}\ r\ V_\phi\right)\hat{e}_\theta$$

$$+ \frac{1}{r}\left(\frac{\partial}{\partial r}(r\ V_\theta) - \frac{\partial}{\partial \theta}V_r\right)\hat{e}_\phi$$

$$\Delta\ \Theta(\,r, \theta, \phi\,) = \frac{1}{r^2}\frac{\partial}{\partial r}\left(r^2\frac{\partial \Theta}{\partial r}\right) + \frac{1}{r^2 sin^2(\theta)}\frac{\partial}{\partial \theta}\left(sin(\theta)\frac{\partial \Theta}{\partial \theta}\right) + \frac{1}{r^2 sin^2(\theta)}\frac{\partial^2 \Theta}{\partial \phi^2}$$

I.A.2 List of further used coordinate systems

In this section we present further coordinate systems which you may encounter.

I.A.2.1 Astroid coordinates

References:

[Werner]: p. 296

Coordinate definition:

$$Cartesian\ functions\ \begin{cases} x^1(\,u,\varphi,z\,) = \ u\,cos^3(\varphi) \\ x^2(\,u,\varphi,z\,) = \ u\,sin^3(\varphi) \\ x^3(\,u,\varphi,z\,) = \ z \end{cases}$$

Orthogonal   yes

Jacobian and metric:

$$\mathcal{J}_{(\,u,\varphi,z\,)} = \begin{bmatrix} cos^3(\varphi) & -3u\,sin(\varphi)\,cos^2(\varphi) & 0 \\ sin^3(\varphi) & 3u\,sin^2(\varphi)\,cos(\varphi) & 0 \\ 0 & 0 & 1 \end{bmatrix}$$

$$J_{(u,\varphi,z)}^{-1} = \begin{bmatrix} \dfrac{1}{\cos(\varphi)} & \dfrac{1}{\sin(\varphi)} & 0 \\[2ex] -\dfrac{\sin(\varphi)}{3\,u\,\cos^2(\varphi)} & \dfrac{\cos(\varphi)}{3\,u\,\sin^2(\varphi)} & 0 \\[2ex] 0 & 0 & 1 \end{bmatrix}$$

$$\det\left(J_{(u,\varphi,z)}\right) = 3\,u\,\left(\cos^2(\varphi) - \cos^4(\varphi)\right)$$

Metric tensor (covariant) and derivatives:

$$\left[g_{ij}\right] = \begin{bmatrix} s^6 + c^6 & 3\,u\,s\,c\,(s^2 - c^2) & 0 \\[1ex] 3\,u\,s\,c\,(s^2 - c^2) & 9\,u^2 s^2 c^2 & 0 \\[1ex] 0 & 0 & 1 \end{bmatrix}$$

$$\frac{\partial}{\partial u}\left[g_{ij}\right] = \begin{bmatrix} 0 & -3\,s\,(2c^3 - c) & 0 \\[1ex] -3\,s\,(2c^3 - c) & 18\,u\,c^2\,s^2 & 0 \\[1ex] 0 & 0 & 0 \end{bmatrix}$$

$$\frac{\partial}{\partial \varphi}\left[g_{ij}\right] = \begin{bmatrix} -6\,s\,c\,(2\,c^2 - 1) & -24\,u\,s^2(s^2 + 1) - 3u & 0 \\[1ex] -24\,u\,s^2(s^2 + 1) - 3u & 18\,s\,(\,2\,u^2\,c\,(c^2 - 1)) & 0 \\[1ex] 0 & 0 & 0 \end{bmatrix}$$

$$\frac{\partial}{\partial z}\left[g_{ij}\right] = [0]$$

Inverse metric tensor (contravariant) and derivatives:

$$\left[g^{ij}\right] = \begin{bmatrix} \dfrac{1}{c^2 s^2} & \dfrac{2\,c^2 - 1}{3\,u\,s^3 c^3} & 0 \\[2ex] \dfrac{2\,c^2 - 1}{3\,u\,s^3 c^3} & \dfrac{1 - 3\,c^2 s^2}{9u^2 c^4 s^4} & 0 \\[2ex] 0 & 0 & 1 \end{bmatrix}$$

$$\frac{\partial}{\partial u}[g^{ij}] \quad = \quad \begin{bmatrix} 0 & \dfrac{2\,s^2 - 1}{3\,u^2\,s^3 c^3} & 0 \\[3mm] \dfrac{2\,s^2 - 1}{3\,u^2\,s^3 c^3} & -\dfrac{2\,(33\,c^2 s^2 - 1)}{9\,u^3 c^4 s^4} & 0 \\[3mm] 0 & 0 & 0 \end{bmatrix}$$

$$\frac{\partial}{\partial \varphi}[g^{ij}] \quad = \quad \begin{bmatrix} \dfrac{2\,(2 s^2 - 1)}{s^3 c^3} & \dfrac{8\,c^2 s^2 + 3}{3\,u\,s^4 c^4} & 0 \\[3mm] \dfrac{8\,c^2 s^2 + 3}{3\,u\,s^4 c^4} & \dfrac{2(6\,s^6 - 9\,s^4 + 7\,s^2 - 2)}{9\,u^2 c^5 s^5} & 0 \\[3mm] 0 & 0 & 0 \end{bmatrix}$$

$$\frac{\partial}{\partial z}[g^{ij}] \quad = \quad [0]$$

Scale factors and Christoffel symbols:

Scale factors: $\quad \begin{cases} h_u = \sqrt{\cos^6(\varphi) + \sin^6(\varphi)} \\[2mm] h_\varphi = 3u\,\sqrt{\sin^2(\varphi) - \sin^4(\varphi)} \\[2mm] h_z = 1 \end{cases}$

Non-zero Christoffel symbols of the 2nd kind and derivatives:

$$\Gamma^u_{\varphi\varphi} = 3u$$

$$\Gamma^\varphi_{u\varphi} = \frac{1}{u}$$

$$\Gamma^\varphi_{\varphi\varphi} = \frac{2\,(2\cos^2(\varphi) - 1)}{\cos(\varphi)\,\sin(\varphi)}$$

Unit Basis Vectors:

$$\hat{e}_u = \begin{pmatrix} \dfrac{cos^3(\varphi)}{\sqrt{cos^6(\varphi) + sin^6(\varphi)}} \\[3ex] \dfrac{sin^3(\varphi)}{\sqrt{cos^6(\varphi) + sin^6(\varphi)}} \\[3ex] 0 \end{pmatrix} \qquad \hat{e}_\varphi = \begin{pmatrix} -\dfrac{cos^2(\varphi)\,sin(\varphi)}{\sqrt{sin^2(\varphi) - sin^4(\varphi)}} \\[3ex] \dfrac{cos(\varphi)\,sin^2(\varphi)}{\sqrt{sin^2(\varphi) - sin^4(\varphi)}} \\[3ex] 0 \end{pmatrix}$$

$$\hat{e}_z = \begin{pmatrix} 0 \\ 0 \\ 1 \end{pmatrix}$$

Differential operators:

$$div\ V(V^u, V^\varphi, V^z) =$$

$$= \frac{1}{3u\,|sin^2(\varphi) - 1|\,sin^2(\varphi)}\left(\frac{\partial}{\partial u}(3u\,|sin^2(\varphi)\right.$$

$$\left. - 1|\,sin^2(\varphi)\,V^u)\right)$$

$$+ \frac{1}{3u\,|sin^2(\varphi) - 1|\,sin^2(\varphi)}\left(\frac{\partial}{\partial \varphi}(3u\,|sin^2(\varphi)\right.$$

$$\left. - 1|\,sin^2(\varphi)\,V^\varphi)\right)$$

$$+ \frac{1}{3u\,|sin^2(\varphi) - 1|\,sin^2(\varphi)}\left(\frac{\partial}{\partial z}(3u\,|sin^2(\varphi)\right.$$

$$\left. - 1|\,sin^2(\varphi)\,V^z)\right)$$

$$= \frac{1}{u}\left(\frac{\partial}{\partial u}(u\,V^u)\right)$$

$$+ \frac{1}{|sin^2(\varphi) - 1|\,sin^2(\varphi)}\left(\frac{\partial}{\partial \varphi}(\,|sin^2(\varphi) - 1|\,sin^2(\varphi)\,V^\varphi)\right)$$

$$+ \left(\frac{\partial V^z}{\partial z}\right)$$

Note that the divergence is given for contravariant vector fields.

$$curl\ V(V^u, V^\varphi, V^z)$$

$$= \frac{1}{3\,u|sin^2(\varphi) - 1|\,sin^2(\varphi)}\left[e_u\left(\frac{\partial V_z}{\partial \varphi} - \frac{\partial V_\varphi}{\partial z}\right)\right.$$

$$\left. + e_\varphi\left(\frac{\partial V_u}{\partial z} - \frac{\partial V_z}{\partial u}\right) + e_z\left(\frac{\partial V_\varphi}{\partial u} - \frac{\partial V_u}{\partial \varphi}\right)\right]$$

Note that curl is given for contravariant vector fields.

$$\Delta\,\Theta(u,\varphi,z) = \frac{1}{3u\,|s^2(\varphi)-1|\,s^2(\varphi)}\frac{\partial}{\partial u}\left(3u\,|s^2(\varphi)-1|\,s^2(\varphi)\,\frac{1}{c^2(\varphi)s^2(\varphi)}\,\frac{\partial\Theta}{\partial u}\right.$$

$$\left.+\,3u\,|s^2(\varphi)-1|\,s^2(\varphi)\,\frac{2\,c^2(\varphi)-1}{3\,u\,s^3(\varphi)\,c^3(\varphi)}\,\frac{\partial\Theta}{\partial\varphi}\right)$$

$$+\,\frac{1}{3u\,|s^2(\varphi)-1|\,s^2(\varphi)}\frac{\partial}{\partial\varphi}\left(3u\,|s^2(\varphi)\right.$$

$$-\,1|\,s^2(\varphi)\,\frac{2\,c^2(\varphi)-1}{3\,u\,s^3(\varphi)\,c^3(\varphi)}\,\frac{\partial\Theta}{\partial\varphi}$$

$$\left.+\,3u\,|s^2(\varphi)-1|\,s^2(\varphi)\,\frac{1-3\,c^2(\varphi)\,s^2(\varphi)}{9u^2c^4(\varphi)\,s^4(\varphi)}\,\frac{\partial\Theta}{\partial\varphi}\right)$$

$$+\,\frac{1}{3u\,|s^2(\varphi)-1|\,s^2(\varphi)}\left(\frac{\partial}{\partial z}3u\,|s^2(\varphi)-1|\,s^2(\varphi)\,\frac{\partial\Theta}{\partial z}\right)$$

$$=\,\frac{1}{u}\frac{\partial}{\partial u}\left(u\,\frac{1}{c(\varphi)^2 s(\varphi)^2}\,\frac{\partial\Theta}{\partial u}+u\,\frac{2\,c^2(\varphi)-1}{3\,u\,s^3(\varphi)\,cos^3(\varphi)}\,\frac{\partial\Theta}{\partial\varphi}\right)$$

$$+\,\frac{1}{|s^2(\varphi)-1|\,s^2(\varphi)}\frac{\partial}{\partial\varphi}\left(\,|s^2(\varphi)\right.$$

$$-\,1|\,s^2(\varphi)\,\frac{2\,c^2(\varphi)-1}{3\,u\,s^3(\varphi)\,c^3(\varphi)}\,\frac{\partial\Theta}{\partial u}$$

$$\left.+\,|s^2(\varphi)-1|\,s^2(\varphi)\,\frac{1-3\,c^2(\varphi)\,s^2(\varphi)}{9u^2c^4(\varphi)\,s^4(\varphi)}\,\frac{\partial\Theta}{\partial\varphi}\right)+\frac{\partial^2\Theta}{\partial z^2}$$

I.A.2.2 Bispherical coordinates

References:

[Neutsch], p. 1238

Coordinate definition:

$$Cartesian\ functions \begin{cases} x^1(\tau,\sigma,\varphi) = a\,\dfrac{sin(\sigma)}{cosh(\tau)-cos(\sigma)}cos(\varphi) \\[2mm] x^2(\tau,\sigma,\varphi) = a\,\dfrac{sin(\sigma)}{cosh(\tau)-cos(\sigma)}sin(\varphi) \\[2mm] x^3(\tau,\sigma,\varphi) = a\,\dfrac{sinh(\tau)}{cosh(\tau)-cos(\sigma)} \end{cases}$$

$$\tau \in (-\infty,+\infty),\ \ \sigma \in [0,\pi], \varphi \in [0,2\pi], a > 0, fixed$$

Orthogonal   yes

Jacobian and metric:

$$
J_{(\tau,\sigma,\varphi)} = \begin{cases}
[1,1]: -\dfrac{a\cos(\varphi)\sin(\sigma)\sinh(\tau)}{(\cos(\sigma)-\cosh(\tau))^2} \\[2mm]
[1,2]: \dfrac{a\cos(\varphi)(\cos(\sigma)\cosh(\tau)-1)}{(\cos(\sigma)-\cosh(\tau))^2} \\[2mm]
[1,3]: \dfrac{a\sin(\sigma)\sin(\varphi)}{\cos(\sigma)-\cosh(\tau)} \\[2mm]
[2,1]: -\dfrac{a\sin(\sigma)\sin(\varphi)\sinh(\tau)}{(\cos(\sigma)-\cosh(\tau))^2} \\[2mm]
[2,2]: \dfrac{a\sin(\varphi)(\cos(\sigma)\cosh(\tau)-1)}{(\cos(\sigma)-\cosh(\tau))^2} \\[2mm]
[2,3]: -\dfrac{a\cos(\varphi)\sin(\sigma)}{\cos(\sigma)-\cosh(\tau)} \\[2mm]
[3,1]: -\dfrac{a(\cos(\sigma)\cosh(\tau)-1)}{(\cos(\sigma)-\cosh(\tau))^2} \\[2mm]
[3,2]: -\dfrac{a\sin(\sigma)\sinh(\tau)}{(\cos(\sigma)-\cosh(\tau))^2} \\[2mm]
0
\end{cases}
$$

$$
J^{-1}{}_{(\tau,\sigma,\varphi)} = \begin{cases}
[1,1]: -\dfrac{\cos(\varphi)\sin(\sigma)\sinh(\tau)}{a} \\[2mm]
[1,2]: -\dfrac{\sin(\sigma)\sin(\varphi)\sinh(\tau)}{a} \\[2mm]
[1,3]: -\dfrac{\cos(\sigma)\cosh(\tau)-1}{a} \\[2mm]
[2,1]: \dfrac{(\cos(\sigma)\cosh(\tau)-1)\cos(\varphi)}{a} \\[2mm]
[2,2]: \dfrac{(\cos(\sigma)\cosh(\tau)-1)\sin(\varphi)}{a} \\[2mm]
[2,3]: -\dfrac{\sin(\sigma)\sinh(\tau)}{a} \\[2mm]
[3,1]: \dfrac{(\cos(\sigma)-\cosh(\tau))\sin(\varphi)}{a\sin(\sigma)} \\[2mm]
[3,2]: -\dfrac{(\cos(\sigma)-\cosh(\tau))\cos(\varphi)}{a\sin(\sigma)} \\[2mm]
0
\end{cases}
$$

$$det\left(\mathcal{J}_{(\tau,\sigma,\varphi)}\right) = \frac{a^3 \, sin(\sigma)}{(cos(\sigma) \, - \, cosh(\tau))^3}$$

Metric tensor (covariant) and derivatives:

$$\left[g_{ij}\right] = diag \left\{ \begin{array}{c} \dfrac{a^2}{(cos(\sigma) \, - \, cosh(\tau))^2} \\[2mm] \dfrac{a^2}{(cos(\sigma) \, - \, cosh(\tau))^2} \\[2mm] \dfrac{a^2 sin^2(\sigma)}{(cos(\sigma) \, - \, cosh(\tau))^2} \end{array} \right.$$

$$\frac{\partial}{\partial\tau}\left[g_{ij}\right] = diag \left\{ \begin{array}{c} \dfrac{2\,a^2\,sinh(\tau)}{(cos(\sigma) - cosh(\tau))^3} \\[2mm] \dfrac{2\,a^2\,sinh(\tau)}{(cos(\sigma) - cosh(\tau))^3} \\[2mm] \dfrac{2\,a^2\,sin^2(\sigma)\,sinh(\tau)}{(cos(\sigma) - cosh(\tau))^3} \end{array} \right.$$

$$\frac{\partial}{\partial\sigma}\left[g_{ij}\right] = diag \left\{ \begin{array}{c} \dfrac{2\,a^2\,sin(\sigma)}{(cos(\sigma) - cosh(\tau))^3} \\[2mm] \dfrac{2\,a^2\,sin(\sigma)}{(cos(\sigma) - cosh(\tau))^3} \\[2mm] -\dfrac{2\,a^2\,sin(\sigma)(cos(\sigma)\,cosh(\tau) - 1)}{(cos(\sigma) - cosh(\tau))^3} \end{array} \right.$$

$$\frac{\partial}{\partial\varphi}\left[g_{ij}\right] = [0]$$

Inverse metric tensor (contravariant) and derivatives:

$$[g^{ij}] = diag \left\{ \begin{array}{c} \dfrac{(cos(\sigma) - cosh(\tau))^2}{a^2} \\[2mm] \dfrac{(cos(\sigma) - cosh(\tau))^2}{a^2} \\[2mm] \dfrac{(cos(\sigma) - cosh(\tau))^2}{a^2 \, sin^2(\sigma)} \end{array} \right.$$

$$\frac{\partial}{\partial \tau}[g^{ij}] = diag \left\{ \begin{array}{c} -\dfrac{2\left(cos(\sigma) - cosh(\tau)\right) sinh(\tau)}{a^2} \\[2mm] -\dfrac{2\left(cos(\sigma) - cosh(\tau)\right) sinh(\tau)}{a^2} \\[2mm] -\dfrac{2\left(cos(\sigma) - cosh(\tau)\right) sinh(\tau)}{a^2 \, sin^2(\sigma)} \end{array} \right.$$

$$\frac{\partial}{\partial \sigma}[g^{ij}] = diag \left\{ \begin{array}{c} -\dfrac{2\left(cos(\sigma) - cosh(\tau)\right) sin(\sigma)}{a^2} \\[2mm] -\dfrac{2\left(cos(\sigma) - cosh(\tau)\right) sin(\sigma)}{a^2} \\[2mm] \dfrac{2\left(cos(\sigma) cosh(\tau) - 1\right)(cos(\sigma) - cosh(\tau))}{a^2 \, sin^3(\sigma)} \end{array} \right.$$

$$\frac{\partial}{\partial \varphi}[g^{ij}] = [0]$$

Scale factors and Christoffel symbols:

$$Scale\ factors: \left\{ \begin{array}{l} h_\sigma = h_\tau = \dfrac{a}{cosh(\tau) - cos(\sigma)} \\[3mm] h_\varphi = \dfrac{a\ sin\ (\sigma)}{cosh(\tau) - cos(\sigma)} \end{array} \right.$$

Non-zero Christoffel symbols of the 2nd kind and derivatives:

$$\Gamma^{\tau}_{\tau\tau} = \frac{sinh(\tau)}{cos(\sigma) - cosh(\tau)}$$

$$\Gamma^{\tau}_{\tau\sigma} = \frac{sin(\sigma)}{cos(\sigma) - cosh(\tau)}$$

$$\Gamma^{\tau}_{\sigma\sigma} = -\frac{sinh(\tau)}{cos(\sigma) - cosh(\tau)}$$

$$\Gamma^{\tau}_{\varphi\varphi} = -\frac{sin^2(\sigma)\ sinh(\tau)}{cos(\sigma) - cosh(\tau)}$$

$$\Gamma^{\sigma}_{\tau\tau} = -\frac{sin(\sigma)}{cos(\sigma) - cosh(\tau)}$$

$$\Gamma^{\sigma}_{\tau\sigma} = \frac{sinh(\tau)}{cos(\sigma) - cosh(\tau)}$$

$$\Gamma^{\sigma}_{\sigma\sigma} = \frac{sin(\sigma)}{cos(\sigma) - cosh(\tau)}$$

$$\Gamma^{\sigma}_{\varphi\varphi} = \frac{(cos(\sigma)\ cosh(\tau) - 1)\ sin(\sigma)}{cos(\sigma) - cosh(\tau)}$$

$$\Gamma^{\varphi}_{\tau\varphi} = \frac{sinh(\tau)}{cos(\sigma) - cosh(\tau)}$$

$$\Gamma^{\varphi}_{\sigma\varphi} = -\frac{cos(\sigma)\ cosh(\tau) - 1}{sin(\sigma)(\ cos(\sigma) - cosh(\tau))}$$

Unit Basis Vectors:

$$\hat{e}_\tau = \begin{pmatrix} \dfrac{\cos(\varphi)\sin(\sigma)\sinh(\tau)}{\cos(\sigma) - \cosh(\tau)} \\[2ex] \dfrac{\sin(\sigma)\sin(\varphi)\sinh(\tau)}{\cos(\sigma) - \cosh(\tau)} \\[2ex] \dfrac{\cos(\sigma)\cosh(\tau) - 1}{\cos(\sigma) - \cosh(\tau)} \end{pmatrix}$$

$$\hat{e}_\sigma = \begin{pmatrix} -\dfrac{(\cos(\sigma)\cosh(\tau) - 1)\cos(\varphi)}{\cos(\sigma) - \cosh(\tau)} \\[2ex] -\dfrac{(\cos(\sigma)\cosh(\tau) - 1)\sin(\varphi)}{\cos(\sigma) - \cosh(\tau)} \\[2ex] \dfrac{\sin(\sigma)\sinh(\tau)}{\cos(\sigma) - \cosh(\tau)} \end{pmatrix}$$

$$\hat{e}_\varphi = \begin{pmatrix} -\sin(\varphi) \\ \cos(\varphi) \\ 0 \end{pmatrix}$$

Differential operators:

$$\begin{aligned}
\operatorname{div} V(V_\tau, V_\sigma, V_\varphi) = \frac{(\cosh(\tau) - \cos(\sigma))^3}{a\,\sin(\sigma)} & \left(\frac{\partial}{\partial\sigma}\left(\frac{\sin(\sigma)}{(\cos(\sigma) - \cosh(\tau))^2}\, V_\sigma \right) \right. \\[2ex]
& + \frac{\partial}{\partial\tau}\left(\frac{\sin(\sigma)}{(\cos(\sigma) - \cosh(\tau))^2}\, V_\tau \right) \\[2ex]
& \left. + \frac{\partial}{\partial\varphi}\left(\frac{1}{(\cos(\sigma) - \cosh(\tau))^2}\, V_\varphi \right) \right)
\end{aligned}$$

$$curl\, V\left(V_\tau, V_\sigma, V_\varphi\right) = \frac{\left(cos(\sigma) - cosh(\tau)\right)^2}{a\, sin(\sigma)} \left(\frac{\partial}{\partial\sigma}\left(\frac{V_\varphi\, sin(\sigma)}{cosh(\tau) - cos(\sigma)}\right)\right.$$

$$\left. - \frac{\partial}{\partial\varphi}\left(\frac{V_\sigma}{cosh(\tau) - cos(\sigma)}\right)\right) \hat{e}_\tau$$

$$+ \frac{\left(cos(\sigma) - cosh(\tau)\right)^2}{a\, sin(\sigma)} \left(\frac{\partial}{\partial\tau}\left(\frac{V_\varphi\, sin(\sigma)}{cosh(\tau) - cos(\sigma)}\right)\right.$$

$$\left. - \frac{\partial}{\partial\varphi}\left(\frac{V_\tau}{cosh(\tau) - cos(\sigma)}\right)\right) \hat{e}_\sigma$$

$$+ \frac{\left(cos(\sigma) - cosh(\tau)\right)^2}{a} \left(\frac{\partial}{\partial\sigma}\left(\frac{V_\tau}{cosh(\tau) - cos(\sigma)}\right)\right.$$

$$\left. - \frac{\partial}{\partial\tau}\left(\frac{V_\sigma}{cosh(\tau) - cos(\sigma)}\right)\right) \hat{e}_\varphi$$

$$\Delta\, \Theta(u, \varphi, z) = \frac{\left(cos(\sigma) - cosh(\tau)\right)^3}{a^2\, sin(\sigma)} \left(sin(\sigma)\frac{\partial}{\partial\tau}\left(\frac{1}{cosh(\tau) - cos(\sigma)}\frac{\partial\Theta}{\partial\tau}\right)\right.$$

$$\left. + \frac{\partial}{\partial\sigma}\left(\frac{sin(\sigma)}{cosh(\tau) - cos(\sigma)}\frac{\partial\Theta}{\partial\sigma}\right)\right) + \frac{\left(cos(\sigma) - cosh(\tau)\right)^2}{a^2\, sin^2(\sigma)}\frac{\partial^2\Theta}{\partial\varphi^2}$$

I.A.2.3 Bipolar Cylindrical Coordinates

References:

[Neutsch]: p. 1234

Coordinate definition:

$$Cartesian\ functions \begin{cases} x^1(u,v,z) = \dfrac{a\ sinh(v)}{cosh(v) - cos(u)} \\[2mm] x^2(u,v,z) = \dfrac{a\ sin(u)}{cosh(v) - cos(u)} \\[2mm] x^3(u,v,z) = z \end{cases}$$

$$\text{with } u \in [0, 2\pi), v, z \in (-\infty, \infty)$$

$$\tau \in (-\infty, +\infty),\ \ \sigma \in [0, \pi], \varphi \in [0, 2\pi], a > 0, fixed$$

Orthogonal   yes

308

Jacobian and metric:

$$\mathcal{J}_{(u,v,z)} = \begin{bmatrix} -\dfrac{a\,sin(u)\,sinh(v)}{(cosh(v)-cos(u))^2} & -\dfrac{a\,(cos(u)\,cosh(v)-1)}{(cosh(v)-cos(u))^2} & 0 \\[2ex] \dfrac{a\,(cos(u)\,cosh(v)-1)}{(cosh(v)-cos(u))^2} & -\dfrac{a\,sin(u)\,sinh(v)}{(cosh(v)-cos(u))^2} & 0 \\[2ex] 0 & 0 & 1 \end{bmatrix}$$

$$\mathcal{J}^{-1}{}_{(u,v,z)} = \begin{bmatrix} -\dfrac{sin(u)\ sinh(v)}{a} & \dfrac{cos(u)\ cosh(v)-1}{a} & 0 \\[2ex] -\dfrac{cos(u)\ cosh(v)-1}{a} & -\dfrac{sin(u)\ sinh(v)}{a} & 0 \\[2ex] 0 & 0 & 1 \end{bmatrix}$$

$$det\,(\mathcal{J}_{(u,v,z)}) = \dfrac{a^2}{\left(cosh(v)-cos(u)\right)^2}$$

Metric tensor (covariant) and derivatives:

$$\left[g_{ij}\right] = \begin{bmatrix} \dfrac{a^2}{\left(cosh(v)-cos(u)\right)^2} & 0 & 0 \\[3ex] 0 & \dfrac{a^2}{\left(cosh(v)-cos(u)\right)^2} & 0 \\[3ex] 0 & 0 & 1 \end{bmatrix}$$

$$\dfrac{\partial}{\partial u}\left[g_{ij}\right] = \begin{bmatrix} \dfrac{2\,a^2\,sin(u)}{\left(cos(u)-cosh(v)\right)^3} & 0 & 0 \\[3ex] 0 & \dfrac{2\,a^2\,sin(u)}{\left(cos(u)-cosh(v)\right)^3} & 0 \\[3ex] 0 & 0 & 0 \end{bmatrix}$$

$$\dfrac{\partial}{\partial v}\left[g_{ij}\right] = \begin{bmatrix} \dfrac{2\,a^2\,sinh(v)}{\left(cos(u)-cosh(v)\right)^3} & 0 & 0 \\[3ex] 0 & \dfrac{2\,a^2\,sinh(v)}{\left(cos(u)-cosh(v)\right)^3} & 0 \\[3ex] 0 & 0 & 0 \end{bmatrix}$$

$$\frac{\partial}{\partial z}[g_{ij}] \quad = \quad [0]$$

Inverse metric tensor (contravariant) and derivatives:

$$[g^{ij}] \quad = \quad \begin{bmatrix} \dfrac{\big(\cosh(v) - \cos(u)\big)^2}{a^2} & 0 & 0 \\[2em] 0 & \dfrac{\big(\cosh(v) - \cos(u)\big)^2}{a^2} & 0 \\[2em] 0 & 0 & 1 \end{bmatrix}$$

$$\frac{\partial}{\partial u}[g^{ij}] \quad = \quad \begin{bmatrix} \dfrac{2\big(\cosh(v) - \cos(u)\big)\sin(u)}{a^2} & 0 & 0 \\[2em] 0 & \dfrac{2\big(\cosh(v) - \cos(u)\big)\sin(u)}{a^2} & 0 \\[2em] 0 & 0 & 0 \end{bmatrix}$$

$$\frac{\partial}{\partial v}[g^{ij}] \quad = \quad \begin{bmatrix} \dfrac{2\big(\cosh(v) - \cos(u)\big)\sinh(v)}{a^2} & 0 & 0 \\[2em] 0 & \dfrac{2\big(\cosh(v) - \cos(u)\big)\sinh(v)}{a^2} & 0 \\[2em] 0 & 0 & 0 \end{bmatrix}$$

$$\frac{\partial}{\partial z}[g^{ij}] \quad = \quad [0]$$

Scale factors and Christoffel symbols:

$$\text{Scale factors:} \quad \begin{cases} h_u = h_v = \dfrac{a}{\cosh(v) - \cos(u)} \\[1.2em] h_z = 1 \end{cases}$$

Non-zero Christoffel symbols of the 2nd kind and derivatives:

$$\Gamma^u_{uu} = \frac{sin(u)}{cos(u) - cosh(v)}$$

$$\Gamma^u_{uv} = \frac{sinh(v)}{cos(u) - cosh(v)}$$

$$\Gamma^u_{vv} = -\frac{sin(u)}{cos(u) - cosh(v)}$$

$$\Gamma^v_{uu} = -\frac{sinh(v)}{cos(u) - cosh(v)}$$

$$\Gamma^v_{uv} = \frac{sin(u)}{cos(u) - cosh(v)}$$

$$\Gamma^v_{vv} = \frac{sinh(v)}{cos(u) - cosh(v)}$$

$$\Gamma^u_{uu} = \frac{sin(u)}{cos(u) - cosh(v)}$$

$$\Gamma^u_{uv} = \frac{sinh(v)}{cos(u) - cosh(v)}$$

$$\Gamma^u_{vv} = -\frac{sin(u)}{cos(u) - cosh(v)}$$

$$\Gamma^v_{uu} = -\frac{sinh(v)}{cos(u) - cosh(v)}$$

Unit Basis Vectors:

$$\hat{e}_u = \begin{pmatrix} \dfrac{sin(u)\,sinh(v)}{cos(u) - cosh(v)} \\[2ex] -\dfrac{cos(u)\,cosh(v) - 1}{cos(u) - cosh(v)} \\[2ex] 0 \end{pmatrix} \qquad \hat{e}_v = \begin{pmatrix} \dfrac{cos(u)\,cosh(v) - 1}{cos(u) - cosh(v)} \\[2ex] \dfrac{sin(u)\,sinh(v)}{cos(u) - cosh(v)} \\[2ex] 0 \end{pmatrix}$$

$$\hat{e}_z = \begin{pmatrix} 0 \\ 0 \\ 1 \end{pmatrix}$$

Differential operators:

$$div\, V\,(V_u, V_v, V_z) = \frac{\left(cosh(v) - cos(u)\right)^2}{a} \left[\frac{\partial}{\partial u} \left(\frac{1}{cosh(v) - cos(u)} \, V_u \right) \right.$$
$$\left. + \frac{\partial}{\partial v} \left(\frac{1}{cosh(v) - cos(u)} \, V_v \right) \right] + \frac{\partial V_z}{\partial z}$$

$$curl\, V\,(V_u, V_v, V_z) = \left(\frac{cosh(v) - cos(u)}{a} \frac{\partial V_z}{\partial v} - \frac{\partial V_v}{\partial z} \right) \hat{e}_u$$
$$+ \frac{\left(cos(u) - cosh(v)\right)}{a} \left(\frac{\partial V_z}{\partial u} \right.$$
$$\left. - \frac{\partial}{\partial z} \left(\frac{a}{cosh(v) - cos(u)} \, V_u \right) \right) \hat{e}_v$$
$$+ \frac{\left(cos(u) - cosh(v)\right)^2}{a} \left(\frac{\partial}{\partial u} \left(\frac{1}{cosh(v) - cos(u)} \, V_v \right) \right.$$
$$\left. - \frac{\partial}{\partial v} \left(\frac{1}{cosh(v) - cos(u)} \, V_u \right) \right) \hat{e}_z$$

$$\Delta\, \Theta(\,u, v, z\,) = \frac{\left(cosh(v) - cos(u)\right)^2}{a^2} \left(\frac{\partial^2 \Theta}{\partial u^2} + \frac{\partial^2 \Theta}{\partial v^2} \right) + \frac{\partial^2 \Theta}{\partial z^2}$$

I.A.2.4 Cardioid Coordinates

References:

https://mathworld.wolfram.com/CardioidCoordinates.html

Coordinate definition:

$$Cartesian\ functions \begin{cases} x^1(\mu, \nu, \psi) = \dfrac{\mu\nu}{(\mu^2 + \nu^2)^2}\, cos(\psi) \\[2mm] x^2(\mu, \nu, \psi) = \dfrac{\mu\nu}{(\mu^2 + \nu^2)^2}\, sin\,(\psi) \\[2mm] x^3(\mu, \nu, \psi) = \dfrac{1}{2}\, \dfrac{\mu^2 - \nu^2}{(\mu^2 + \nu^2)^2} \end{cases}$$

$$\mu, \nu\ \geq 0, \psi \in [\,0, 2\pi\,)$$

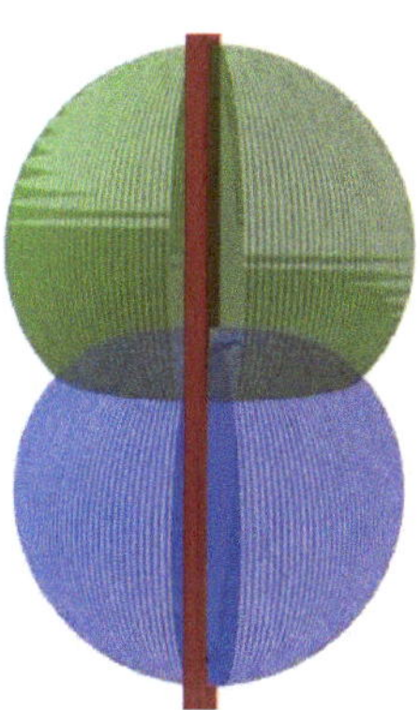

Orthogonal   yes

Jacobian and metric:

$$
\mathcal{J}_{(\mu,v,\psi)} =
\begin{bmatrix}
\dfrac{(v^2 - 3\,\mu^2)\,v\,\cos(\psi)}{(\mu^2 + v^2)^3} & \dfrac{(\mu^2 - 3\,v^2)\,\mu\,\cos(\psi)}{(\mu^2 + v^2)^3} & -\dfrac{\mu v\,\sin(\psi)}{(\mu^2 + v^2)^2} \\[3mm]
\dfrac{(v^2 - 3\,\mu^2)\,v\,\sin(\psi)}{(\mu^2 + v^2)^3} & \dfrac{(\mu^2 - 3\,v^2)\,\mu\,\sin(\psi)}{(\mu^2 + v^2)^3} & \dfrac{\mu v\,\cos(\psi)}{(\mu^2 + v^2)^2} \\[3mm]
\dfrac{(3\,v^2 - \mu^2)\,\mu}{(\mu^2 + v^2)^3} & \dfrac{(v^2 - 3\mu^2)\,v}{(\mu^2 + v^2)^3} & 0
\end{bmatrix}
$$

$$
\mathcal{J}^{-1}{}_{(\mu,v,\psi)} =
\begin{bmatrix}
(v^2 - 3\mu^2)\,v\,\cos(\psi) & (v^2 - 3\mu^2)\,v\,\sin(\psi) & (3v^2 - \mu^2)\,\mu \\[2mm]
(\mu^2 - 3v^2)\,\mu\,\cos(\psi) & (\mu^2 - 3v^2)\,\mu\,\sin(\psi) & (v^2 - 3\mu^2)\,v \\[2mm]
-\dfrac{(\mu^2 + v^2)^2\,\sin(\psi)}{\mu v} & \dfrac{(\mu^2 + v^2)^2\,\cos(\psi)}{\mu v} & 0
\end{bmatrix}
$$

$$
\det\!\left(\mathcal{J}_{(\mu,v,\psi)}\right) = -\frac{\mu v}{(\mu^2 + v^2)^5}
$$

Metric tensor (covariant) and derivatives:

$$
[g_{ij}] =
\begin{bmatrix}
\dfrac{1}{(\mu^2 + v^2)^3} & 0 & 0 \\[3mm]
0 & \dfrac{1}{(\mu^2 + v^2)^3} & 0 \\[3mm]
0 & 0 & \dfrac{\mu^2\,v^2}{(\mu^2 + v^2)^4}
\end{bmatrix}
$$

$$
\frac{\partial}{\partial \mu}[g_{ij}] =
\begin{bmatrix}
-\dfrac{6\,\mu}{(\mu^2 + v^2)^4} & 0 & 0 \\[3mm]
0 & -\dfrac{6\,\mu}{(\mu^2 + v^2)^4} & 0 \\[3mm]
0 & 0 & -\dfrac{2\,(3\mu^2 - v^2)\,\mu\,v^2}{(\mu^2 + v^2)^5}
\end{bmatrix}
$$

$$\frac{\partial}{\partial v}[g_{ij}] \quad = \quad \begin{bmatrix} -\dfrac{6\,v}{(\mu^2 + v^2)^4} & 0 & 0 \\[2ex] 0 & -\dfrac{6\,v}{(\mu^2 + v^2)^4} & 0 \\[2ex] 0 & 0 & -\dfrac{2\,(\mu^2 - 3v^2)\,\mu^2\,v}{(\mu^2 + v^2)^5} \end{bmatrix}$$

$$\frac{\partial}{\partial \psi}[g_{ij}] \quad = \quad [0]$$

Inverse metric tensor (contravariant) and derivatives:

$$[g^{ij}] \quad = \quad \begin{bmatrix} (\mu^2 + v^2)^3 & 0 & 0 \\[2ex] 0 & (\mu^2 + v^2)^3 & 0 \\[2ex] 0 & 0 & \dfrac{(\mu^2 + v^2)^4}{\mu^2 v^2} \end{bmatrix}$$

$$\frac{\partial}{\partial \mu}[g^{ij}] \quad = \quad \begin{bmatrix} 6(\mu^2 + v^2)^2\mu & 0 & 0 \\[2ex] 0 & 6(\mu^2 + v^2)^2\mu & 0 \\[2ex] 0 & 0 & \dfrac{2\,(3\,\mu^2 - v^2)\,(\mu^2 + v^2)^3}{\mu^3\,v^2} \end{bmatrix}$$

$$\frac{\partial}{\partial v}[g^{ij}] \quad = \quad \begin{bmatrix} 6(\mu^2 + v^2)^2 v & 0 & 0 \\[2ex] 0 & 6(\mu^2 + v^2)^2 v & 0 \\[2ex] 0 & 0 & -\dfrac{2\,(\mu^2 - 3\,v^2)\,(\mu^2 + v^2)^3}{\mu^2\,v^3} \end{bmatrix}$$

$$\frac{\partial}{\partial \psi}[g^{ij}] \quad = \quad [0]$$

Scale factors and Christoffel symbols:

Scale factors:
$$\begin{cases} h_\mu = h_v = \dfrac{1}{\sqrt{(\mu^2+v^2)^3}} \\[2ex] h_\psi = \dfrac{\mu v}{(\mu^2+v^2)^2} \end{cases}$$

Non-zero Christoffel symbols of the 2nd kind and derivatives:

$$\Gamma^\mu_{\mu\mu} = -\frac{3\mu}{\mu^2+v^2}$$

$$\Gamma^\mu_{\mu v} = -\frac{3v}{\mu^2+v^2}$$

$$\Gamma^\mu_{vv} = \frac{3\mu}{\mu^2+v^2}$$

$$\Gamma^\mu_{\psi\psi} = \frac{(3\mu^3\,v^2 - \mu\,v^4)}{(\mu^2+v^2)^2}$$

$$\Gamma^v_{vv} = -\frac{3v}{\mu^2+v^2}$$

$$\Gamma^v_{\mu\mu} = \frac{3v}{\mu^2+v^2}$$

$$\Gamma^v_{v\mu} = -\frac{3\mu}{\mu^2+v^2}$$

$$\Gamma^v_{\psi\psi} = -\frac{\mu^2 v\,(\mu^2 - 3\,v^2)}{(\mu^2+v^2)^2}$$

$$\Gamma^\psi_{\psi\mu} = -\frac{(3\mu^2 - v^2)}{\mu^3 + \mu v^2}$$

$$\Gamma^\psi_{\psi v} = \frac{\mu^2 - 3\,v^2}{\mu^2 v + v^3}$$

Unit Basis Vectors:

$$\hat{e}_\mu = \begin{pmatrix} \dfrac{(v^2 - 3\,\mu^2)v\cos(\psi)}{\sqrt{(\mu^2 + v^2)^3}} \\[2ex] \dfrac{(v^2 - 3\,\mu^2)v\sin(\psi)}{\sqrt{(\mu^2 + v^2)^3}} \\[2ex] \dfrac{(\mu^2 - 3v^2)\mu}{\sqrt{(\mu^2 + v^2)^3}} \end{pmatrix} \qquad \hat{e}_v = \begin{pmatrix} \dfrac{(\mu^2 - 3v^2)\mu\cos(\psi)}{\sqrt{(\mu^2 + v^2)^3}} \\[2ex] \dfrac{(\mu^2 - 3v^2)\mu\sin(\psi)}{\sqrt{(\mu^2 + v^2)^3}} \\[2ex] \dfrac{(v^2 - 3\mu^2)\,v}{\sqrt{(\mu^2 + v^2)^3}} \end{pmatrix}$$

$$\hat{e}_\psi = \begin{pmatrix} -\sin(\psi) \\ \cos(\psi) \\ 0 \end{pmatrix}$$

Differential operators:

$$\mathrm{div}\, V\left(V_\mu, V_v, V_\psi\right) = \left(\frac{(\mu^2 + v^2)^5}{\mu v}\right)\left(\frac{\partial}{\partial\mu}\left(\frac{\mu v}{\sqrt{(\mu^2 + v^2)^7}}\,V_\mu\right)\right.$$

$$\left. + \frac{\partial}{\partial v}\left(\frac{\mu v}{\sqrt{(\mu^2 + v^2)^7}}\,V_v\right) + \frac{\partial}{\partial\psi}\left(\frac{1}{(\mu^2 + v^2)^3}\,V_\psi\right)\right)$$

$$curl\, V\left(V_\mu, V_\nu, V_\psi\right) = \left(\frac{\sqrt{(\mu^2 + \upsilon^2)^7}}{\mu\nu}\right)\left(\frac{\partial}{\partial\nu}\left(\frac{\mu\nu}{(\mu^2 + \nu^2)^2}\, V_\psi\right)\right.$$

$$\left.- \frac{\partial}{\partial\psi}\left(\frac{1}{\sqrt{(\mu^2 + \upsilon^2)^3}}\, V_\nu\right)\right)\hat{e}_\mu$$

$$+ \left(\frac{\sqrt{(\mu^2 + \upsilon^2)^7}}{\mu\nu}\right)\left(\frac{\partial}{\partial\mu}\left(\frac{\mu\nu}{(\mu^2 + \nu^2)^2}\, V_\psi\right)\right.$$

$$\left.- \frac{\partial}{\partial\psi}\left(\frac{1}{\sqrt{(\mu^2 + \upsilon^2)^3}}\, V_\mu\right)\right)\hat{e}_\nu$$

$$+ (\mu^2 + \upsilon^2)^3 \left(\frac{\partial}{\partial\mu}\left(\frac{1}{\sqrt{(\mu^2 + \upsilon^2)^3}}\, V_\nu\right)\right.$$

$$\left.- \frac{\partial}{\partial\nu}\left(\frac{1}{\sqrt{(\mu^2 + \upsilon^2)^3}}\, V_\mu\right)\right)\hat{e}_\psi$$

$$\Delta\,\Theta == \frac{(\mu^2 + v^2)^5}{\mu\nu}\left(\frac{\partial}{\partial\mu}\left(\frac{\mu\nu}{(\mu^2 + v^2)^2}\frac{\partial\Theta}{\partial\mu}\right) + \frac{\partial}{\partial\nu}\left(\frac{\mu\nu}{(\mu^2 + v^2)^2}\frac{\partial\Theta}{\partial\nu}\right)\right.$$

$$\left.+ \frac{1}{(\mu^2 + v^2)\,\mu\nu}\frac{\partial^2\Theta}{\partial\psi^2}\right)$$

I.A.2.5 Conical coordinates

References:

https://mathworld.wolfram.com/ConicalCoordinates.html

Coordinate definition:

$$Cartesian\ functions \begin{cases} x^1(r,\mu,v) = \dfrac{r\mu v}{bc} \\[2em] x^2(r,\mu,v) = \dfrac{r}{b}\sqrt{\dfrac{(\mu^2 - b^2)(v^2 - b^2)}{b^2 - c^2}} \\[2em] x^3(r,\mu,v) = \dfrac{r}{c}\sqrt{\dfrac{(\mu^2 - c^2)(v^2 - c^2)}{c^2 - b^2}} \end{cases}$$

$$r,\mu,v \in \mathbb{R}, b,c \in \mathbb{R}, fixed, v^2 < c^2 < \mu^2 < b^2$$

Orthogonal   yes

Jacobian and metric:

$$
\mathcal{J}_{(r,\mu,\nu)} = \begin{cases}
[1,1]: \dfrac{\mu\nu}{bc} \\[2ex]
[1,2]: \dfrac{r\nu}{bc} \\[2ex]
[1,3]: \dfrac{r\mu}{bc} \\[2ex]
[2,1]: \dfrac{1}{b}\sqrt{\dfrac{(b^2-\mu^2)(b^2-\nu^2)}{b^2-c^2}} \\[3ex]
[2,2]: -\dfrac{(b^2-\nu^2)\,r\mu}{b\,(b^2-c^2)\sqrt{\dfrac{(b^2-\mu^2)(b^2-\nu^2)}{b^2-c^2}}} \\[3ex]
[2,3]: -\dfrac{(b^2-\mu^2)\,r\nu}{b\,(b^2-c^2)\sqrt{\dfrac{(b^2-\mu^2)(b^2-\nu^2)}{b^2-c^2}}} \\[3ex]
[3,1]: \dfrac{1}{c}\sqrt{\dfrac{(c^2-\mu^2)(c^2-\nu^2)}{c^2-b^2}} \\[3ex]
[3,2]: \dfrac{(c^2-\nu^2)\,r\mu}{c\,(b^2-c^2)\sqrt{\dfrac{(c^2-\mu^2)(c^2-\nu^2)}{b^2-c^2}}} \\[3ex]
[3,3]: \dfrac{(c^2-\mu^2)\,r\,\nu}{c\,(b^2-c^2)\sqrt{\dfrac{(c^2-\mu^2)(c^2-\nu^2)}{b^2-c^2}}}
\end{cases}
$$

$$J^{-1}{}_{(r,\mu,v)} = \begin{cases}
[1,1]: \dfrac{\mu v}{bc} \\[2em]
[1,2]: \dfrac{1}{b}\sqrt{\dfrac{(b^2-\mu^2)(b^2-v^2)}{b^2-c^2}} \\[2em]
[1,3]: \dfrac{1}{c}\sqrt{\dfrac{(c^2-\mu^2)(c^2-v^2)}{c^2-b^2}} \\[2em]
[2,1]: \dfrac{v\,(b^2-\mu^2)(c^2-\mu^2)}{bcr\,(\mu^2-v^2)} \\[2em]
[2,2]: \dfrac{\mu\,(c^2-\mu^2)\sqrt{\dfrac{(b^2-\mu^2)(b^2-v^2)}{b^2-c^2}}}{br\,(\mu^2-v^2)} \\[2em]
[2,3]: \dfrac{\mu\,\sqrt{\dfrac{(c^2-\mu^2)(c^2-v^2)}{c^2-b^2}}\,(b^2-\mu^2)}{cr\,(\mu^2-v^2)} \\[2em]
[3,1]: \dfrac{\mu\,(b^2-v^2)(c^2-v^2)}{bcr\,(\mu^2-v^2)} \\[2em]
[3,2]: -\dfrac{v\,(c^2-\mu^2)\sqrt{\dfrac{(b^2-\mu^2)(b^2-v^2)}{b^2-c^2}}}{br\,(\mu^2-v^2)} \\[2em]
[3,3]: -\dfrac{v\,(b^2-v^2)\sqrt{\dfrac{(c^2-\mu^2)(c^2-v^2)}{c^2-b^2}}}{cr\,(\mu^2-v^2)}
\end{cases}$$

$$\det\!\left(\mathcal{J}_{(r,\mu,v)}\right) = \frac{r^2\,(\mu^2-v^2)}{\sqrt{(\mu^2-b^2)(b^2-v^2)(c^2-\mu^2)(c^2-v^2)}}$$

Metric tensor (covariant) and derivatives:

$$[g_{ij}] = \begin{bmatrix} 1 & 0 & 0 \\ 0 & \dfrac{r^2(\mu^2 - v^2)}{(\mu^2 - b^2)(c^2 - \mu^2)} & 0 \\ 0 & 0 & \dfrac{r^2(\mu^2 - v^2)}{(b^2 - v^2)(c^2 - v^2)} \end{bmatrix}$$

$$\frac{\partial}{\partial r}[g_{ij}] = \begin{bmatrix} 0 & 0 & 0 \\ 0 & \dfrac{2r(\mu^2 - v^2)}{(\mu^2 - b^2)(c^2 - \mu^2)} & 0 \\ 0 & 0 & \dfrac{2r(\mu^2 - v^2)}{(b^2 - v^2)(c^2 - v^2)} \end{bmatrix}$$

$$\frac{\partial}{\partial \mu}[g_{ij}] = diag \left\{ 0, \; \frac{2\,r^2\mu\left(1 - (\mu^2 - v^2)\right)}{(\mu^2 - b^2)(c^2 - \mu^2)} - \frac{2\,(\mu^2 - v^2)r^2\mu}{(b^2 - \mu^2)(c^2 - \mu^2)^2}, \; \frac{2r^2\mu}{(b^2 - v^2)(c^2 - v^2)} \right\}$$

$$\frac{\partial}{\partial v}[g_{ij}] = diag \left\{ 0, \; \frac{2r^2v}{(b^2 - \mu^2)(c^2 - \mu^2)}, \; -\frac{2\,r^2v\left(1 + (\mu^2 - v^2)\right)}{(b^2 - v^2)(c^2 - v^2)} + \frac{2\,(\mu^2 - v^2)r^2v}{(b^2 - v^2)(c^2 - v^2)^2} \right\}$$

Inverse metric tensor (contravariant) and derivatives:

$$[g^{ij}] = \begin{bmatrix} 1 & 0 & 0 \\ 0 & -\dfrac{(b^2 - \mu^2)(c^2 - \mu^2)}{r^2(\mu^2 - v^2)} & 0 \\ 0 & 0 & \dfrac{(b^2 - v^2)(c^2 - v^2)}{r^2(\mu^2 - v^2)} \end{bmatrix}$$

$$\frac{\partial}{\partial r}[g^{ij}] = \begin{bmatrix} 0 & 0 & 0 \\ 0 & \dfrac{2(b^2 - \mu^2)(c^2 - \mu^2)}{r^3(\mu^2 - v^2)} & 0 \\ 0 & 0 & -\dfrac{2(b^2 - v^2)(c^2 - v^2)}{r^3(\mu^2 - v^2)} \end{bmatrix}$$

$$\frac{\partial}{\partial \mu}[g^{ij}] \quad = \quad diag \left\{ \begin{array}{l} 0 \\[4pt] \dfrac{2\mu}{r^2}\left(\dfrac{(b^2-\mu^2)(c^2-\mu^2)}{(\mu^2-v^2)^2} + \dfrac{(b^2-\mu^2)+(c^2-\mu^2)}{(\mu^2-v^2)} \right) \\[12pt] -\dfrac{2\mu\,(b^2-v^2)(c^2-v^2)}{r^2(\mu^2-v^2)} \end{array} \right.$$

$$\frac{\partial}{\partial v}[g^{ij}] \quad = \quad diag \left\{ \begin{array}{l} 0 \\[4pt] -\dfrac{2v\,(b^2-\mu^2)(c^2-\mu^2)}{r^2(\mu^2-v^2)} \\[12pt] \dfrac{2v}{r^2}\left(\dfrac{(b^2-v^2)(c^2-v^2)}{(\mu^2-v^2)^2} + \dfrac{(b^2-v^2)+(c^2-v^2)}{(\mu^2-v^2)} \right) \end{array} \right.$$

Scale factors and Christoffel symbols:

$$\textit{Scale factors:} \quad \left\{ \begin{array}{l} h_r = 1 \\[6pt] h_\mu = r\,\sqrt{\dfrac{\mu^2-v^2}{(b^2-\mu^2)(\mu^2-c^2)}} \\[12pt] h_v = r\,\sqrt{\dfrac{\mu^2-v^2}{(b^2-v^2)(c^2-v^2)}} \end{array} \right.$$

Non-zero Christoffel symbols of the 2nd kind and derivatives:

$$\Gamma^{r}_{\mu\mu} \quad = \quad \frac{r\,(\mu^2-v^2)}{(b^2-\mu^2)\,(c^2-\mu^2)}$$

$$\Gamma^{r}_{vv} \quad = \quad -\frac{r\,(\mu^2-v^2)}{(b^2-v^2)\,(c^2-v^2)}$$

$$\Gamma^{\mu}_{r\mu} \quad = \quad \frac{1}{r}$$

$$\Gamma^{\mu}_{\mu\mu} \quad = \quad \frac{1}{2}\left(-\frac{1}{b+\mu} + \frac{1}{b-\mu} + \frac{1}{c-\mu} - \frac{1}{c+\mu} + \frac{1}{\mu-v} + \frac{1}{\mu+v} \right)$$

$$\Gamma^{\mu}_{\mu\nu} = -\frac{\nu}{\mu^2 - \nu^2}$$

$$\Gamma^{\mu}_{\nu\nu} = \frac{\mu\,(b^2 - \mu^2)(c^2 - \mu^2)}{(b^2 - \nu^2)(c^2 - \nu^2)(\mu^2 - \nu^2)}$$

$$\Gamma^{\nu}_{r\nu} = \frac{1}{r}$$

$$\Gamma^{\nu}_{\mu\mu} = \frac{\nu\,(b^2 - \nu^2)(c^2 - \nu^2)}{(b^2 - \mu^2)(c^2 - \mu^2)(\mu^2 - \nu^2)}$$

$$\Gamma^{\nu}_{\mu\nu} = \frac{\mu}{\mu^2 - \nu^2}$$

$$\Gamma^{\nu}_{\nu\nu} = \frac{1}{2}\left(-\frac{1}{b+\nu} + \frac{1}{b-\nu} + \frac{1}{c-\nu} - \frac{1}{c+\nu} + \frac{1}{\nu-\mu} + \frac{1}{\mu+\nu}\right)$$

Unit Basis Vectors:

$$\hat{e}_r = \begin{pmatrix} \dfrac{\mu\nu}{bc} \\[2em] \dfrac{1}{b}\sqrt{\dfrac{(b^2-\mu^2)(b^2-\nu^2)}{b^2-c^2}} \\[2em] \dfrac{1}{c}\sqrt{-\dfrac{(c^2-\mu^2)(c^2-\nu^2)}{b^2-c^2}} \end{pmatrix} \qquad \hat{e}_\mu = \begin{pmatrix} bc\sqrt{\dfrac{\nu^2-\mu^2}{(b^2-\mu^2)(c^2-\mu^2)}}\,\nu \\[2em] \dfrac{\mu\sqrt{\nu^2-b^2}}{b\sqrt{\dfrac{(c^2-b^2)(\nu^2-\mu^2)}{(c^2-\mu^2)}}} \\[2em] \dfrac{\mu\sqrt{(c^2-\nu^2)}}{c\sqrt{\dfrac{(b^2-c^2)(\mu^2-\nu^2)}{(b^2-\mu^2)}}} \end{pmatrix}$$

$$\hat{e}_\nu = \begin{pmatrix} bc\sqrt{\dfrac{\mu^2-\nu^2}{(b^2-\nu^2)(c^2-\nu^2)}}\,\mu \\[2em] -\dfrac{\sqrt{b^2-\mu^2}\,\nu}{b\sqrt{\dfrac{(b^2-c^2)(\mu^2-\nu^2)}{(c^2-\nu^2)}}} \\[2em] \dfrac{\sqrt{(c^2-\mu^2)}\,\nu}{c\sqrt{\dfrac{(c^2-b^2)(\mu^2-\nu^2)}{(b^2-\nu^2)}}} \end{pmatrix}$$

Differential operators:

$$div\,V\left(V_r, V_\mu, V_\nu\right)$$

$$= \frac{\sqrt{(b^2-\mu^2)(b^2-\nu^2)(\mu^2-c^2)(c^2-\nu^2)}}{r^2\,(\mu^2-\nu^2)}\left(\frac{\partial}{\partial r}\left(\frac{r^2\,(\mu^2-\nu^2)}{\sqrt{(b^2-\mu^2)(\mu^2-c^2)(b^2-\nu^2)(c^2-\nu^2)}}\,V_r\right)\right.$$

$$\left. + \frac{\partial}{\partial \mu}\left(r\sqrt{\frac{\mu^2-\nu^2}{(b^2-\nu^2)(c^2-\nu^2)}}\,V_\mu\right) + \frac{\partial}{\partial \nu}\left(r\sqrt{\frac{\mu^2-\nu^2}{(b^2-\mu^2)(\mu^2-c^2)}}\,V_\nu\right)\right)$$

$$= \frac{\partial}{\partial r}\,(V_r) + \frac{\sqrt{(b^2-\mu^2)(\mu^2-c^2)}}{r\,(\mu^2-\nu^2)}\frac{\partial}{\partial \mu}\left(\sqrt{\frac{\mu^2-\nu^2}{(b^2-\nu^2)(c^2-\nu^2)}}\,V_\mu\right)$$

$$+ \frac{\sqrt{(b^2-\nu^2)(c^2-\nu^2)}}{r\,(\mu^2-\nu^2)}\frac{\partial}{\partial \nu}\left(\sqrt{\mu^2-\nu^2}\,V_\nu\right)$$

$$curl\, V\left(V_r, V_\mu, V_v\right) = \left(\frac{\sqrt{(\mu^2 - b^2)(c^2 - \mu^2)}}{r\,(\mu^2 - v^2)} \frac{\partial}{\partial\mu}\left(\sqrt{\mu^2 - v^2}\, V_v\right)\right.$$

$$\left. - \frac{\sqrt{(b^2 - v^2)(c^2 - v^2)}}{r\,(\mu^2 - v^2)} \frac{\partial}{\partial v}\left(\sqrt{\mu^2 - v^2}\, V_\mu\right)\right) \hat{e}_r$$

$$- \left(\frac{\partial V_v}{\partial r} - \frac{\sqrt{(b^2 - v^2)(c^2 - v^2)}}{r\sqrt{\mu^2 - v^2}}\frac{\partial V_r}{\partial v}\right)\hat{e}_\mu$$

$$+ \left(\frac{\partial V_\mu}{\partial r} - \frac{\sqrt{(\mu^2 - b^2)(c^2 - \mu^2)}}{r\sqrt{\mu^2 - v^2}}\frac{\partial V_r}{\partial \mu}\right)\hat{e}_v$$

$$\Delta\,\Theta = \frac{\partial^2 \Theta}{\partial r^2} + \frac{\sqrt{(\mu^2 - b^2)(c^2 - \mu^2)}}{r^2(\mu^2 - v^2)} \frac{\partial}{\partial\mu}\left((\mu^2 - b^2)(c^2 - \mu^2)\frac{\partial\Theta}{\partial\mu}\right)$$

$$+ \frac{\sqrt{(b^2 - v^2)(c^2 - v^2)}}{r^2(\mu^2 - v^2)} \frac{\partial}{\partial v}\left(\sqrt{(b^2 - v^2)(c^2 - v^2)}\frac{\partial\Theta}{\partial v}\right)$$

I.A.2.6 Elliptic cylindrical coordinates

References:

[Neutsch]: p. 1190

Coordinate definition:

$$Cartesian\ functions \begin{cases} x^1(u,v,z) = a\,cosh(u)\,cos(v) \\ x^2(u,v,z) = a\,sinh(u)\,sin(v) \\ x^3(u,v,z) = z \end{cases}$$

$$u \in [\,0, \infty), v \in [0, 2\pi), z \in (-\infty, \infty\,), a \in \mathbb{R}, fixed$$

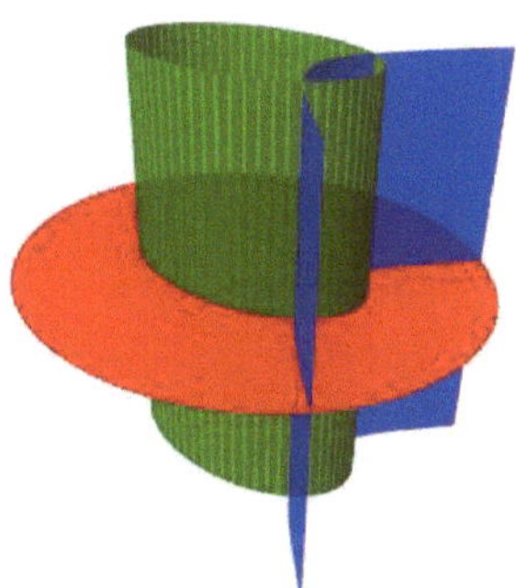

Orthogonal   *yes*

Jacobian and metric:

$$\mathcal{J}_{(u,v,z)} \quad = \quad \begin{bmatrix} a\,cos(v)\,sinh(u) & -a\,cosh(u)\,sin(v) & 0 \\ a\,cosh(u)\,sin(v) & a\,cos(v)\,sinh(u) & 0 \\ 0 & 0 & 1 \end{bmatrix}$$

$$\mathcal{J}^{-1}{}_{(u,v,z)} \quad = \quad \begin{bmatrix} \dfrac{cos(v)\,sinh(u)}{a\,(sinh^2(u) + sin^2(v))} & \dfrac{cosh(u)\,sin(v)}{a\,(sinh^2(u) + sin^2(v))} & 0 \\ \dfrac{cosh(u)\,sin(v)}{a\,(sinh^2(u) + sin^2(v))} & \dfrac{cos(v)\,sinh(u)}{a\,(sinh^2(u) + sin^2(v))} & 0 \\ 0 & 0 & 1 \end{bmatrix}$$

$$\det\big(\mathcal{J}_{(u,v,z)}\big) \quad = \quad a^2\left(\sinh^2(u) + \sin^2(v)\right)$$

Metric tensor (covariant) and derivatives:

$$\left[g_{ij}\right] \quad = \quad \begin{bmatrix} a^2\left(\sinh^2(u) + \sin^2(v)\right) & 0 & 0 \\ 0 & a^2\left(\sinh^2(u) + \sin^2(v)\right) & 0 \\ 0 & 0 & 1 \end{bmatrix}$$

$$\frac{\partial}{\partial u}\left[g_{ij}\right] \quad = \quad \begin{bmatrix} 2a^2\cosh(u)\sinh(u) & 0 & 0 \\ 0 & 2a^2\cosh(u)\sinh(u) & 0 \\ 0 & 0 & 0 \end{bmatrix}$$

$$\frac{\partial}{\partial v}\left[g_{ij}\right] \quad = \quad \begin{bmatrix} 2a^2\cos(v)\sin(v) & 0 & 0 \\ 0 & 2a^2\cos(v)\sin(v) & 0 \\ 0 & 0 & 0 \end{bmatrix}$$

$$\frac{\partial}{\partial z}\left[g_{ij}\right] \quad = \quad [0]$$

Inverse metric tensor (contravariant) and derivatives:

$$\left[g^{ij}\right] \quad = \quad \begin{bmatrix} \dfrac{1}{a^2\left(\sinh^2(u) + \sin^2(v)\right)} & 0 & 0 \\ 0 & \dfrac{1}{a^2\left(\sinh^2(u) + \sin^2(v)\right)} & 0 \\ 0 & 0 & 1 \end{bmatrix}$$

$$\frac{\partial}{\partial u}\left[g^{ij}\right] \quad = \quad \begin{bmatrix} -\dfrac{2\cosh(u)\sinh(u)}{a^2\left(\sinh^2(u) + \sin^2(v)\right)^2} & 0 & 0 \\ 0 & -\dfrac{2\cosh(u)\sinh(u)}{a^2\left(\sinh^2(u) + \sin^2(v)\right)^2} & 0 \\ 0 & 0 & 0 \end{bmatrix}$$

$$\frac{\partial}{\partial v}[g^{ij}] = \begin{bmatrix} -\dfrac{4\sin(2v)}{a^2\big(cos(2v)-cosh(2u)\big)^2} & 0 & 0 \\[2em] 0 & -\dfrac{4\sin(2v)}{a^2\big(cos(2v)-cosh(2u)\big)^2} & 0 \\[2em] 0 & 0 & 0 \end{bmatrix}$$

$$\frac{\partial}{\partial z}[g^{ij}] = [0]$$

Scale factors and Christoffel symbols:

Scale factors: $\begin{cases} h_u = a\,\sqrt{sinh^2(u)+sin^2(v)} \\ h_v = a\,\sqrt{sinh^2(u)+sin^2(v)} \\ h_z = 1 \end{cases}$

Non-zero Christoffel symbols of the 2nd kind and derivatives:

$$\Gamma^u_{uu} = \frac{cosh(u)\,sinh(u)}{sinh^2(u)+sin^2(v))}$$

$$\Gamma^u_{uv} = \frac{cos(v)\,sin(v)}{sinh^2(u)+sin^2(v)}$$

$$\Gamma^u_{vv} = -\frac{cosh(u)\,sinh(u)}{sin^2(v)+sinh^2(u)}$$

$$\Gamma^v_{uu} = -\frac{cos(v)\,sin(v)}{sinh^2(u)+sin^2(v)}$$

$$\Gamma^v_{uv} = \frac{cosh(u)\,sinh(u)}{sinh^2(u)+sin^2(v)}$$

$$\Gamma^v_{vv} = \frac{cos(v)\,sin(v)}{sinh^2(u)+sin^2(v)}$$

Unit Basis Vectors:

$$\hat{e}_u = \begin{pmatrix} \dfrac{\cos(v)\,\sinh(u)}{\sqrt{\sinh^2(u)+\sin^2(v)}} \\[4mm] \dfrac{\cosh(u)\,\sin(v)}{\sqrt{\sinh^2(u)+\sin^2(v)}} \\[2mm] 0 \end{pmatrix} \qquad \hat{e}_v = \begin{pmatrix} -\dfrac{\cosh(u)\,\sin(v)}{\sqrt{\sinh^2(u)+\sin^2(v)}} \\[4mm] \dfrac{\cos(v)\,\sinh(u)}{\sqrt{\sinh^2(u)+\sin^2(v)}} \\[2mm] 0 \end{pmatrix}$$

$$\hat{e}_z = \begin{pmatrix} 0 \\ 0 \\ 1 \end{pmatrix}$$

Differential operators:

$$div\,V(V_u, V_v, V_z) = \frac{1}{a\,(\sinh^2(u)+\sin^2(v))}\left(\frac{\partial}{\partial u}\left(\sqrt{\sinh^2(u)+\sin^2(v)}\;V_u\right)\right.$$

$$\left.+\frac{\partial}{\partial v}\left(\sqrt{\sinh^2(u)+\sin^2(v)}\;V_v\right)\right)+\frac{\partial V_z}{\partial z}$$

$$curl\,V(V_u, V_v, V_z) = \left(\frac{1}{a\,\sqrt{\sinh^2+\sin^2}}\frac{\partial V_z}{\partial v}-\frac{\partial V_v}{\partial z}\right)\hat{e}_u$$

$$-\left(\left(\frac{1}{a\,\sqrt{\sinh^2(u)+\sin^2(v)}}\left(\frac{\partial V_u}{\partial z}-\left(\frac{\partial V_z}{\partial u}\right)\right)\right)\right)\hat{e}_v$$

$$+\frac{1}{a\,(\sinh^2(u)+\sin^2(v))}\left(\frac{\partial}{\partial u}\left(\sqrt{sh^2+s^2}\;V_v\right)\right.$$

$$\left.-\frac{\partial}{\partial v}\left(\sqrt{\sinh^2(u)+\sin^2(v)}\;V_u\right)\right)\hat{e}_z$$

$$\Delta\,\Theta(u,v,z) = \frac{1}{a^2(\sinh^2(u)+\sin^2(v))}\left(\frac{\partial^2\Theta}{\partial u^2}+\frac{\partial^2\Theta}{\partial v^2}\right)+\frac{\partial^2\Theta}{\partial z^2}$$

I.A.2.7 Helical coordinates

References:

[Werner]: p.295

Coordinate definition:

$$Cartesian\ functions\ \begin{cases} x^1(\rho,\psi,\varsigma) = \rho\,cos\,(\,\psi + a\varsigma\,) \\ x^2(\rho,\psi,\varsigma) = \rho\,sin\,(\,\psi + a\varsigma\,) \\ x^3(\rho,\psi,\varsigma) = \varsigma \end{cases}$$

$$\rho \geq 0, a \in \mathbb{R}\ fixed,\ \ \psi \in [0, 2\pi], -\infty\,<\varsigma<\infty$$

Orthogonal   yes

Jacobian and metric:

$$\mathcal{J}_{(\rho,\psi,\varsigma)} = \begin{bmatrix} \cos{(\psi + a\varsigma)} & -\rho \sin{(\psi + a\varsigma)} & -a\rho \sin{(\psi + a\varsigma)} \\ \sin{(\psi + a\varsigma)} & \rho \cos{(\psi + a\varsigma)} & a\rho \cos{(\psi + a\varsigma)} \\ 0 & 0 & 1 \end{bmatrix}$$

$$\mathcal{J}_{(\rho,\psi,\varsigma)}^{-1} = \begin{bmatrix} \cos{(\psi + a\varsigma)} & \sin{(\psi + a\varsigma)} & 0 \\ (-1/\rho) \sin{(\psi + a\varsigma)} & (1/\rho) \cos{(\psi + a\varsigma)} & -a \\ 0 & 0 & 1 \end{bmatrix}$$

$$\det\left(\mathcal{J}_{(\rho,\psi,\varsigma)}\right) = \rho$$

Metric tensor (covariant) and derivatives:

$$[g_{ij}] = \begin{bmatrix} 1 & 0 & 0 \\ 0 & \rho^2 & a\rho^2 \\ 0 & a\rho^2 & 1 + a^2\rho^2 \end{bmatrix}$$

$$\frac{\partial}{\partial \rho}[g_{ij}] = \begin{bmatrix} 1 & 0 & 0 \\ 0 & 2\rho & 2a\rho \\ 0 & 2a\rho & 2a^2\rho \end{bmatrix}$$

$$\frac{\partial}{\partial \psi}[g_{ij}] = [0]$$

$$\frac{\partial}{\partial \varsigma}[g_{ij}] = [0]$$

Inverse metric tensor (contravariant) and derivatives:

$$[g^{ij}] = \begin{bmatrix} 1 & 0 & 0 \\ 0 & \dfrac{a^2\rho^2 + 1}{\rho^2} & -a \\ 0 & -a & 1 \end{bmatrix}$$

$$\frac{\partial}{\partial\rho}[g^{ij}] = \begin{bmatrix} 1 & 0 & 0 \\ 0 & -\dfrac{2}{\rho^3} & 0 \\ 0 & 0 & 1 \end{bmatrix}$$

$$\frac{\partial}{\partial\psi}[g^{ij}] = [0]$$

$$\frac{\partial}{\partial\varsigma}[g^{ij}] = [0]$$

Scale factors and Christoffel symbols:

$$\text{Scale factors:} \quad \begin{cases} h_\rho = 1 \\ h_\psi = \\ h_\varsigma = \sqrt{a^2\,\rho^2 + 1} \end{cases}$$

Non-zero Christoffel symbols of the 2nd kind and derivatives:

$$\Gamma^{\rho}_{\psi\psi} = -\rho$$

$$\Gamma^{\rho}_{\psi\varsigma} = -ap$$

$$\Gamma^{\rho}_{\varsigma\varsigma} = -a^2\rho$$

$$\Gamma^{\psi}_{\psi\rho} = \frac{1}{\rho}$$

$$\Gamma^{\psi}_{\varsigma\rho} = \frac{a}{p}$$

Unit Basis Vectors:

$$\hat{e}_{\rho} = \begin{pmatrix} \cos(a\varsigma + \psi) \\ \sin(a\varsigma + \psi) \\ 0 \end{pmatrix} \quad \hat{e}_{\psi} = \begin{pmatrix} -\sin(a\varsigma + \psi) \\ \cos(a\varsigma + \psi) \\ 0 \end{pmatrix} \quad \hat{e}_{\varsigma} = \begin{pmatrix} \dfrac{-a\rho\,\sin(a\varsigma + \psi)}{\sqrt{(a^2\rho^2 + 1)}} \\[2mm] \dfrac{a\rho\,\cos(a\varsigma + \psi)}{\sqrt{(a^2\rho^2 + 1)}} \\[2mm] \dfrac{1}{\sqrt{(a^2\rho^2 + 1)}} \end{pmatrix}$$

Differential operators:

$$div\, V = \left(V^{\rho}, V^{\psi}, V^{\varsigma}\right) = \frac{V^{\rho}}{\rho} + \frac{\partial}{\partial\rho}V^{\rho} + \frac{\partial}{\partial\varsigma}V^{\varsigma} + \frac{\partial}{\partial\psi}V^{\psi}$$

$$curl\, V\left(V^{\rho}, V^{\psi}, V^{\varsigma}\right)$$
$$= \frac{1}{\rho}\left[\left(\frac{\partial V^{\varsigma}{}_{\varsigma}}{\partial x^{\psi}} - \frac{\partial V^{\psi}}{\partial x^{\varsigma}}\right)e_{\rho} + \left(\frac{\partial V^{\rho}}{\partial x^{\varsigma}} - \frac{\partial V^{\varsigma}}{\partial x^{\rho}}\right)e_{\psi}\right.$$
$$\left. + \left(\frac{\partial V^{\psi}}{\partial x^{\rho}} - \frac{\partial V^{\rho}}{\partial x^{\psi}}\right)e_{\varsigma}\right]$$

$$\Delta\,\Theta(\rho, \psi, \varsigma) = a^2\frac{\partial^2\Theta}{\partial\psi^2} - 2a\frac{\partial^2\Theta}{\partial\psi\partial\varsigma} + \frac{1}{\rho}\frac{\partial\Theta}{\partial\rho} + \frac{1}{\rho^2}\frac{\partial^2\Theta}{\partial\psi^2} + \frac{\partial^2\Theta}{\partial\rho^2} + \frac{\partial^2\Theta}{\partial\varsigma^2}$$

I.A.2.8 Inverse Oblate Spheroidal Coordinates

References:

[Moon/Spencer]: p. 119

Coordinate definition:

$$Cartesian\ functions\ \begin{cases} x^1(\eta,\theta,\psi) = \dfrac{a\ cosh(\eta)\ sin(\theta)\ cos(\psi)}{cosh^2(\eta) - cos^2(\theta)} \\[2mm] x^2(\eta,\theta,\psi) = \dfrac{a\ cosh(\eta)\ sin(\theta)\ sin(\psi)}{cosh^2(\eta) - cos^2(\theta)} \\[2mm] x^3(\eta,\theta,\psi) = \dfrac{a\ sinh(\eta)\ cos(\theta)}{cosh^2(\eta) - cos^2(\theta)} \end{cases}$$

$$\eta \geq 0, \theta \in [0,\pi], \psi \in [0,2\pi), a > 0, fixed$$

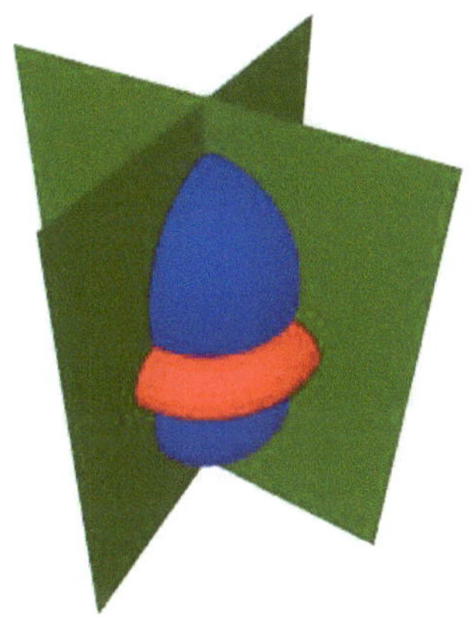

We define:

$$c = cos(\theta), s = sin(\theta), ch = cosh(\eta), sh = sinh(\eta)$$

Orthogonal   yes

Jacobian and metric:

$$
\mathcal{J}_{(\eta,\theta,\psi)} = \begin{cases}
[1,1] & -\dfrac{a\left(\cosh^2(\eta) + \cos^2(\theta)\right)\cos(\psi)\sin(\theta)\sinh(\eta)}{\left(\cos^2(\theta) - \cosh^2(\eta)\right)^2} \\[3ex]
[1,2] & -\dfrac{a\left(\sin^2(\theta) - \sinh^2(\eta)\right)\cos(\theta)\cos(\psi)\cosh(\eta)}{(\cos^2(\theta) - \cosh^2(\eta))^2} \\[3ex]
[1,3] & \dfrac{a\cosh(\eta)\sin(\theta)\sin(\psi)}{\cos^2(\theta) - \cosh^2(\eta)} \\[3ex]
[2,1] & -\dfrac{a\left(\cosh^2(\eta) + \cos^2(\theta)\right)\sin(\psi)\sin(\theta)\sinh(\eta)}{\left(\cos^2(\theta) - \cosh^2(\eta)\right)^2} \\[3ex]
[2,2] & -\dfrac{a\left(\sin^2(\theta) - \sinh^2(\eta)\right)\cos(\theta)\cosh(\eta)\sin(\psi)}{(\cos^2(\theta) - \cosh^2(\eta))^2} \\[3ex]
[2,3] & -\dfrac{a\cos(\psi)\cosh(\eta)\sin(\theta)}{\cos^2(\theta) - \cosh^2(\eta)} \\[3ex]
[3,1] & \dfrac{a\left(\sin^2(\theta) - \sinh^2(\eta)\right)\cos(\theta)\cosh(\eta))}{(\cos^2(\theta) - \cosh^2(\eta))^2} \\[3ex]
[3,2] & -\dfrac{a\left(\cos^2(\theta) + \cosh^2(\eta)\right)\sinh(\eta)\sin(\theta)}{\left(\cos^2(\theta) - \cosh^2(\eta)\right)^2} \\[3ex]
[3,3] & 0
\end{cases}
$$

$$
\mathcal{J}^{-1}{}_{(\eta,\theta,\psi)} = \begin{cases}
[1,1] & -\dfrac{(cos^2(\theta) + cosh^2(\eta))\, cos(\psi)\, sin(\theta)\, sinh(\eta)}{a\,(cosh^2(\eta) - sin^2(\theta))} \\[2ex]
[1,2] & -\dfrac{(cos^2(\theta) + cosh^2(\eta))\, sin(\theta)\, sin(\psi)\, sinh(\eta)}{a\,(cosh^2(\eta) - sin^2(\theta))} \\[2ex]
[1,3] & \dfrac{(sin^2(\theta) - sinh(\eta))\, cos(\theta)\, cosh(\eta)}{a\,(cosh^2(\eta) - sin^2(\theta))} \\[2ex]
[2,1] & -\dfrac{(sin^2(\theta) - sinh^2(\eta))\, cos(\theta)\, cos(\psi)\, cosh(\eta)}{a\,(cosh^2(\eta) - sin^2(\theta))} \\[2ex]
[2,2] & -\dfrac{(sin^2(\theta) - sinh^2(\eta))\, cos(\theta)\, sin(\psi)\, cosh(\eta)}{a\,(cosh^2(\eta) - sin^2(\theta))} \\[2ex]
[2,3] & -\dfrac{(cos^2(\theta) + cosh^2(\eta))\, sin(\theta)\, sinh(\eta)}{a\,(cosh^2(\eta) - sin^2(\theta))} \\[2ex]
[3,1] & \dfrac{\left(\dfrac{cos^2(\theta)}{cosh(\eta)} - cosh(\eta)\right) sin(\psi)}{a\, sin(\theta)} \\[3ex]
[3,2] & \dfrac{\left(\dfrac{sinh^2(\eta)}{sin(\theta)} + sin(\theta)\right) cos(\psi}{a\, cosh(\eta)} \\[3ex]
[3,3] & 0
\end{cases}
$$

$$
det\big(\mathcal{J}_{(\eta,\theta,\psi)}\big) = a^3 ch\, s\, \frac{(c^2 s^2 + ch^2 sh^2)(ch^2 s^2 + c^2 sh^2)}{(c + ch)^5 (c - ch)^5}
$$

Metric tensor (covariant) and derivatives:

$$
[g_{ij}] = diag \begin{cases}
\dfrac{a^2(cosh^2(\eta) - sin^2(\theta))}{(cosh^2(\eta) - cos^2(\theta))^2} \\[3ex]
\dfrac{a^2(cosh^2(\eta) - sin^2(\theta))}{(cosh^2(\eta) - cos^2(\theta))^2} \\[3ex]
\dfrac{a^2\, cosh^2(\eta)\, sin^2(\theta)}{(cosh^2(\eta) - cos^2(\theta))^2}
\end{cases}
$$

$$\frac{\partial}{\partial \eta}\left[g_{ij}\right] = diag \begin{cases} \dfrac{2\,a^2\left(\cosh^2(\eta) - 3\,\sin^2(\theta) + 1\right)\cosh(\eta)\,\sinh(\eta)}{\left(\cos^2(\theta) - \cosh^2(\eta)\right)^3} \\[2ex] \dfrac{2\,a^2\left(\cosh^2(\eta) - 3\,\sin^2(\theta) + 1\right)\cosh(\eta)\,\sinh(\eta)}{\left(\cos^2(\theta) - \cosh^2(\eta)\right)^3} \\[2ex] \dfrac{2\,a^2\left(\cosh^2(\eta) + \cos^2(\theta)\right)\cosh(\eta)\,\sin^2(\theta)\,\sinh(\eta)}{\left(\cos^2(\theta) - \cosh^2(\eta)\right)^3} \end{cases}$$

$$\frac{\partial}{\partial \theta}\left[g_{ij}\right] = diag \begin{cases} \dfrac{2\,a^2\left(\cos^2(\theta) + 3\,\sinh^2(\eta) + 1\right)\cos(\theta)\,\sin(\theta)}{\left(\cos^2(\theta) - \cosh^2(\eta)\right)^3} \\[2ex] \dfrac{2\,a^2\left(\cos^2(\theta) + 3\,\sinh^2(\eta) + 1\right)\cos(\theta)\,\sin(\theta)}{\left(\cos^2(\theta) - \cosh^2(\eta)\right)^3} \\[2ex] -\dfrac{2\,a^2\left(\cos^2(\theta) + \sinh^2(\eta) - 1\right)\cos(\theta)\,\sin(\theta)}{\left(\cos^2(\theta) - \cosh^2(\eta)\right)^3} \end{cases}$$

$$\frac{\partial}{\partial \psi}\left[g_{ij}\right] = [0]$$

Inverse metric tensor (contravariant) and derivatives:

$$[g^{ij}] = diag \begin{cases} \dfrac{\left(\cos^2(\theta) - \cosh^2(\eta)\right)^2}{a^2\left(\cosh^2(\eta) - \sin^2(\theta)\right)} \\[2ex] \dfrac{\left(\cos^2(\theta) - \cosh^2(\eta)\right)^2}{a^2\left(\cosh^2(\eta) - \sin^2(\theta)\right)} \\[2ex] \dfrac{\left(\cos^2(\theta) - \cosh^2(\eta)\right)^2}{a^2\,\cosh^2(\eta)\,\sin^2(\theta)} \end{cases}$$

$$\frac{\partial}{\partial \eta}[g^{ij}] = diag \begin{cases} -\dfrac{2\,(c^2 - ch^2)\,(ch^2 - 3\,s^2 + 1)\,(ch^2 - s^2)\,ch\,sh}{a^2(c^2 + sh^2)^3} \\[2ex] -\dfrac{2\,(c^2 - ch^2)\,(ch^2 - 3\,s^2 + 1)\,(ch^2 - s^2)\,ch\,sh)}{a^2(c^2 + sh^2)^3} \\[2ex] -\dfrac{2\left(\dfrac{c^4}{ch^3} - ch\right)sh}{a^2\,s^2} \end{cases}$$

$$\frac{\partial}{\partial \theta}[g^{ij}] \quad = \quad diag \begin{cases} \dfrac{2\,(c^2 + 3sh^2 + 1)s^2 + sh^2)\,c\,s}{a^2(c^2 + sh^2)^2} \\[3mm] \dfrac{2\,(c^2 + 3sh^2 + 1)s^2 + sh^2)\,c\,s}{a^2(c^2 + sh^2)^2} \\[3mm] \dfrac{2\left(\dfrac{sh^4}{s^3} - s\right)c}{a^2\,ch^2} \end{cases}$$

$$\frac{\partial}{\partial \psi}[g^{ij}] \quad = \quad [0]$$

Scale factors and Christoffel symbols:

$$Scale\ factors: \begin{cases} h_\eta = h_\theta = \dfrac{a\,\sqrt{\cosh^2(\eta) - \sin^2(\theta)}}{\cosh^2(\eta) - \cos^2(\theta)} \\[4mm] h_\psi = \dfrac{a\,\cosh(\eta)\,\sin(\theta)}{\cosh^2(\eta) - \cos^2(\theta)} \end{cases}$$

Non-zero Christoffel symbols of the 2nd kind and derivatives:

$$\Gamma^\eta_{\eta\eta} = \cosh(\eta)\,\sinh^2(\eta)\left(\frac{2}{\cos^2(\theta) - \cosh^2(\eta)} + \frac{1}{\cosh^2(\eta) - \sin^2(\theta)}\right)$$

$$\Gamma^\eta_{\eta\theta} = \cos(\theta)\,\sin^2(\theta)\left(\frac{2}{\cos^2(\theta) - \cosh^2(\eta)} + \frac{1}{\sin^2(\theta) - \cosh^2(\eta)}\right)$$

$$\Gamma^\eta_{\theta\theta} = \cosh(\eta)\,\sinh^2(\eta)\left(\frac{2}{\cosh^2(\eta) - \cos^2(\theta)} + \frac{1}{\sin^2(\theta) - \cosh^2(\eta)}\right)$$

$$\Gamma^\eta_{\psi\psi} = -\frac{(\cos^2(\theta) + \cosh^2(\eta))\,\cosh(\eta)\,\sin(\theta)^2\,\sinh(\eta)}{(\cos^2(\theta) + \sinh^2(\eta))(\cos^2(\theta) - \sinh^2(\eta) - 1)}$$

$$\Gamma^\theta_{\eta\eta} = -\frac{\left(\cos^3(\theta) + (3\,\cosh^3(\eta) - 2)\,\cos(\theta)\right)\sin(\theta)}{-\cos^2(\theta)\,\sin^2(\theta) - \cosh^2(\eta)\,\sinh^2(\eta)}$$

$$\Gamma^\theta_{\eta\theta} = \cos(\theta)\,\sin^2(\theta)\left(\frac{2}{\cos^2(\theta) - \cosh^2(\eta)} + \frac{1}{\sin^2(\theta) - \cosh^2(\eta)}\right)$$

$$\Gamma^\theta_{\theta\theta} = \cos(\theta)\,\sin^2(\theta)\left(\frac{2}{\cos^2(\theta) - \cosh^2(\eta)} + \frac{1}{\sin^2(\theta) - \cosh^2(\eta)}\right)$$

$$\Gamma^{\theta}_{\psi\psi} \;=\; \frac{(cosh^2(\eta)\,cos^3(\theta) + (sinh^4(\eta) - 1)\,cos(\theta))\,sin(\theta)}{-\,cos^2(\theta)\,sin^2(\theta) - cosh^2(\eta)\,sinh^2(\eta)}$$

$$\Gamma^{\psi}_{\eta\psi} \;=\; sinh(\eta) + \frac{2\,sinh^2(\eta)}{cos^2(\theta) - cosh^2(\eta)}$$

$$\Gamma^{\psi}_{\theta\psi} \;=\; cos(\theta) + \frac{2\,cos^2(\theta)}{cos^2(\theta) - cosh^2(\eta)}$$

Unit Basis Vectors:

$$\hat{e}_{\eta} = \begin{pmatrix} \dfrac{(cos^2(\theta)\,sinh(\eta) + sinh^3(\eta) + sinh(\eta))\,cos(\psi)\,sin(\theta)}{(cos^2(\theta) - cosh^2(\eta))\sqrt{cosh^2(\eta) - sin^2(\theta)}} \\[2.5ex] \dfrac{(cos^2(\theta)\,sinh(\eta) + sinh^3(\eta) + sinh(\eta))\,sin(\theta)\,sin(\psi)}{(cos^2(\theta) - cosh^2(\eta))\sqrt{cosh^2(\eta) - sin^2(\theta)}} \\[2.5ex] -\dfrac{cos(\theta)\,cosh(\eta)\,sin^2(\theta) - cos(\theta)\,cosh(\eta)\,sinh^2(\eta)}{(cos^2(\theta) - cosh^2(\eta))\sqrt{cosh^2(\eta) - sin^2(\theta)}} \end{pmatrix}$$

$$\hat{e}_{\theta} = \begin{pmatrix} \dfrac{(cos(\theta)\,cosh(\eta)\,sin^2(\theta) - cos(\theta)\,cosh(\eta)\,sinh^2(\eta))\,cos(\psi)}{(cos^2(\theta) - cosh^2(\eta))\sqrt{cosh^2(\eta) - sin^2(\theta)}} \\[2.5ex] \dfrac{(cos(\theta)\,cosh(\eta)\,sin^2(\theta) - cos(\theta)\,cosh(\eta)\,sinh^2(\eta))\,sin(\psi)}{(cos^2(\theta) - cosh^2(\eta))\sqrt{cosh^2(\eta) - sin^2(\theta)}} \\[2.5ex] \dfrac{(cos^2(\theta)\,sinh(\eta) + sinh^3(\eta) + sinh(\eta))\,sin(\theta)}{(cos^2(\theta) - cosh^2(\eta))\sqrt{cosh^2(\eta) - sin^2(\theta)}} \end{pmatrix}$$

$$\hat{e}_{\psi} = \begin{pmatrix} -sin(\psi) \\ cos(\psi) \\ 0 \end{pmatrix}$$

Differential operators:

$$div\ V\left(V_\eta, V_\theta, V_\psi\right)$$

$$= -\frac{\left(\cosh^2(\eta) - \cos^2(\theta)\right)^3}{a^3\left(\cosh^2(\eta) - \sin^2(\theta)\right)\cosh(\eta)\sin(\theta)}\left(\frac{\partial}{\partial\eta}\left(\frac{a^2\sqrt{\cosh^2(\eta) - \sin^2(\theta)}\,\cosh(\eta)\sin(\theta)}{\left(\cosh^2(\eta) - \cos^2(\theta)\right)^2}V_\eta\right)\right.$$

$$+\frac{\partial}{\partial\theta}\left(\frac{a^2\sqrt{\cosh^2(\eta) - \sin^2(\theta)}\,\cosh(\eta)\sin(\theta)}{\left(\cosh^2(\eta) - \cos^2(\theta)\right)^2}V_\theta\right)$$

$$\left.+\frac{\partial}{\partial\psi}\left(\frac{a^2\left(\cosh^2(\eta) - \sin^2(\theta)\right)}{\left(\cosh^2(\eta) - \cos^2(\theta)\right)^2}V_\psi\right)\right)$$

$$= \frac{\left(\cosh^2(\eta) - \cos^2(\theta)\right)^3}{a\left(\cosh^2(\eta) - \sin^2(\theta)\right)\cosh(\eta)\sin(\theta)}\left(\frac{\partial}{\partial\eta}\left(\frac{\sqrt{\cosh^2(\eta) - \sin^2(\theta)}\,\cosh(\eta)\sin(\theta)}{\left(\cosh^2(\eta) - \cos^2(\theta)\right)^2}V_\eta\right)\right.$$

$$\left.+\frac{\partial}{\partial\theta}\left(\frac{\sqrt{\cosh^2(\eta) - \sin^2(\theta)}\,\cosh(\eta)\sin(\theta)}{\left(\cosh^2(\eta) - \cos^2(\theta)\right)^2}V_\theta\right)\right) - \frac{\left(\cosh^2(\eta) - \cos^2(\theta)\right)}{a\,\cosh(\eta)\sin(\theta)}\frac{\partial V_\psi}{\partial\psi}$$

$$curl\ V\left(V_\eta, V_\theta, V_\psi\right)$$

$$= \frac{(cos^2(\theta) - cosh^2(\eta))^2}{a\sqrt{cosh^2(\eta) - sin^2(\theta)}\ cosh(\eta)\ sin(\theta)} \left(\frac{\partial}{\partial\theta}\left(\frac{cosh(\eta)\ sin(\theta)}{cosh^2(\eta) - cos^2(\theta)} V_\psi \right) \right.$$

$$\left. - \frac{\partial}{\partial\psi}\left(\frac{\sqrt{cos^2(\theta) + cosh^2(\eta) - 1}}{cosh^2(\eta) - cos^2(\theta)} V_\theta \right) \right) \hat{e}_\eta$$

$$+ \frac{(cos^2(\theta) - cosh^2(\eta))^2}{a\sqrt{cosh^2(\eta) - sin^2(\theta)}\ cosh(\eta)\ sin(\theta)} \left(\frac{\partial}{\partial\eta}\left(\frac{cosh(\eta)\ sin(\theta)}{cosh^2(\eta) - cos^2(\theta)} V_\psi \right) \right.$$

$$\left. - \frac{\partial}{\partial\psi}\left(\frac{\sqrt{cosh^2(\eta) - sin^2(\theta)}}{cosh^2(\eta) - cos^2(\theta)} V_\eta \right) \right) \hat{e}_\theta$$

$$+ \frac{(cos^2(\theta) - cosh^2(\eta))^2}{a(cosh^2(\eta) - sin^2(\theta))} \left(\frac{\partial}{\partial\eta}\left(\frac{\sqrt{cos^2(\theta) + cosh^2(\eta) - 1}}{cosh^2(\eta) - cos^2(\theta)} V_\theta \right) \right.$$

$$\left. - \frac{\partial}{\partial\theta}\left(\frac{\sqrt{cos^2(\theta) + cosh^2(\eta) - 1}}{cosh^2(\eta) - cos^2(\theta)} V_\eta \right) \right) \hat{e}_\psi$$

$$\Delta\ \Theta(\eta, \theta, \psi)$$

$$= \frac{(cosh^2(\eta) - cos^2(\theta))^3}{a^2(cosh^2(\eta) - sin^2(\theta))\ cosh(\eta)} \frac{\partial}{\partial\eta}\left(\frac{cosh(\eta)}{cosh^2(\eta) - cos^2(\theta)} \frac{\partial\Theta}{\partial\eta} \right)$$

$$+ \frac{(cosh^2(\eta) - cos^2(\theta))^3}{a^2(cosh^2(\eta) - sin^2(\theta))\ sin(\theta)} \frac{\partial}{\partial\theta}\left(\frac{sin(\theta)}{cosh^2(\eta) - cos^2(\theta)} \frac{\partial\Theta}{\partial\theta} \right)$$

$$+ \frac{(cosh^2(\eta) - cos^2(\theta))^2}{a^2\ cosh^2(\eta)\ sin^2(\theta)} \frac{\partial^2\Theta}{\partial\theta^2}$$

I.A.2.9 Inverse Prolate Spheroidal Coordinates

References:

[Moon/Spencer]: p. 115

Coordinate definition:

$$Cartesian\ functions \begin{cases} x^1(\eta,\theta,\psi) = \dfrac{a\ sinh(\eta)\ sin(\theta)\ cos(\psi)}{cosh^2(\eta) - sin^2(\theta)} \\[2ex] x^2(\eta,\theta,\psi) = \dfrac{a\ sinh(\eta)\ sin(\theta)\ cos(\psi)}{cosh^2(\eta) - sin^2(\theta)} \\[2ex] x^3(\eta,\theta,\psi) = \dfrac{a\ cosh(\eta)\ cos(\theta)}{cosh^2(\eta) - sin^2(\theta)} \end{cases}$$

$$\eta \geq 0, \theta \in [0,\pi], \psi \in [0,2\pi), a > 0, fixed$$

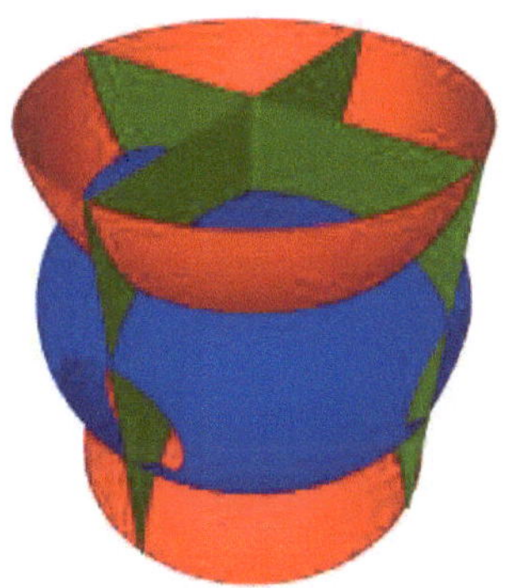

We define:

$$c(x) = cos(x), s(x) = sin(x), ch = cosh(\eta), sh = sinh(\eta)$$

Orthogonal   yes

Jacobian and metric:

$$
\mathcal{J}_{(\eta,\theta,\psi)} = \begin{cases}
[1{,}1] = -\dfrac{a\,(ch^2 + s^2(\theta) - 2)\,c(\psi)\,s(\theta)\,ch}{\left(ch^2 - s^2(\theta)\right)^2} \\[2ex]
[1{,}2] = \dfrac{a\,\left(ch^2 + s^2(\theta)\right)\,c(\psi)\,c(\theta)\,sh}{\left(ch^2 - s^2(\theta)\right)^2} \\[2ex]
[1{,}3] = -\dfrac{a\,sh\,s(\theta)\,s(\psi)}{ch^2 - s^2(\theta)} \\[2ex]
[2{,}1] = -\dfrac{a\,(ch^2 + s^2(\theta) - 2)\,s(\psi)\,s(\theta)\,ch}{\left(ch^2 - s^2(\theta)\right)^2} \\[2ex]
[2{,}2] = \dfrac{a\,\left(ch^2 + s^2(\theta)\right)\,s(\psi)\,c(\theta)\,sh}{\left(ch^2(\eta) - s^2(\theta)\right)^2} \\[2ex]
[2{,}3] = \dfrac{a\,c(\psi)\,sh\,s(\theta)}{ch^2 - s^2(\theta)} \\[2ex]
[3{,}1] = \dfrac{a\,(c^2(\theta) - ch^2 - 1)\,c(\theta)\,sh)}{\left(ch^2 - s^2(\theta)\right)^2} \\[2ex]
[3{,}2] = \dfrac{a\,(c^2(\theta) - ch^2 + 1)\,ch\,s(\theta)}{\left(ch^2 - s^2(\theta)\right)^2} \\[2ex]
[3{,}3] = 0
\end{cases}
$$

$$\mathcal{J}^{-1}{}_{(\eta,\theta,\psi)} \;=\; \begin{cases} [1,1] = \dfrac{(c^2(\theta) - sh^2)\, c(\psi)\, ch\, s(\theta)}{a\,(s^2(\theta) + sh^2)} \\[2mm] [1,2] = \dfrac{(c^2(\theta) - sh^2)\, s(\psi)\, ch\, s(\theta)}{a\,(s^2(\theta) + sh^2)} \\[2mm] [1,3] = \dfrac{(s(\theta) + ch^2)\, c(\theta)\, sh}{a\,(c^2(\theta) - ch^2)} \\[2mm] [2,1] = \dfrac{(s^2(\theta) + ch^2)\, c(\theta)\, c(\psi)\, sh}{a(s^2(\theta) + sh^2)^2} \\[2mm] [2,2] = \dfrac{a\,(ch^2 + s^2(\theta))\, s(\psi)\, c(\theta)\, sh}{a(s^2(\theta) + sh^2)^2} \\[2mm] [2,3] = \dfrac{a\,(ch^2 + s^2(\theta) - 2)\, ch\, s(\theta)}{a\,(c^2(\theta) - ch^2)} \\[2mm] [3,1] = \dfrac{s(\psi)}{a\, sh}\left(\dfrac{ch^2}{s(\theta)} - s(\theta)\right) \\[2mm] [3,2] = \dfrac{c(\psi)}{a\, s(\theta)}\left(\dfrac{c^2(\theta)}{sh(\eta)} + sh\right) \\[2mm] [3,3] = 0 \end{cases}$$

$$det\big(\mathcal{J}_{(\eta,\theta,\psi)}\big) = \dfrac{a^3\big(\cosh^2(\eta) - \cos^2(\theta)\big)\, \sin(\theta)\, \sinh(\eta)}{\big((\sin^3(\theta) - \cosh^3(\eta)\big)^2}$$

Metric tensor (covariant) and derivatives:

$$[g_{ij}] \;=\; diag\begin{cases} \dfrac{a^2(s^2(\theta) + sh^2)}{\big(ch^2(\eta) - s^2(\theta)\big)^2} \\[3mm] \dfrac{a^2(s^2(\theta) + sh^2)}{\big(ch^2 - s^2(\theta)\big)^2} \\[3mm] \dfrac{a^2 s^2(\theta)\, sh^2}{\big(ch^2 - s^2(\theta)\big)^2} \end{cases}$$

$$\frac{\partial}{\partial \eta}[g_{ij}] \quad = \quad diag \left\{ \begin{array}{l} -\dfrac{2a^2(ch^2 + 3s^2(\theta) - 2)\,ch\,sh}{\left(ch^2 - s^2(\theta)\right)^3} \\[3ex] -\dfrac{2a^2(ch^2 + 3s^2(\theta) - 2)\,ch\,sh}{\left(ch^2 - s^2(\theta)\right)^3} \\[3ex] -\dfrac{2\,a^2(c^2(\theta) - sh^2)\,(ch^2 - s^2(\theta))^2\,ch\,s^2(\theta)\,sh}{(s^2(\theta) - sh^2 - 1)^5} \end{array} \right.$$

$$\frac{\partial}{\partial \theta}[g_{ij}] \quad = \quad diag \left\{ \begin{array}{l} -\dfrac{2a^2(c^2(\theta) - 3ch^2 + 1)\,c(\theta)\,s(\theta)}{\left(ch^2 - s^2(\theta)\right)^3} \\[3ex] -\dfrac{2a^2(cos^2(\theta) - 3ch^2 + 1)\,c(\theta)\,s(\theta)}{\left(ch^2 - s^2(\theta)\right)^3} \\[3ex] \dfrac{2\,a^2(c^2(\theta) - sh^2 - 2)\,c(\theta)\,s(\theta)\,sh^2}{(s^2(\theta) - si^2 - 1)^3} \end{array} \right.$$

$$\frac{\partial}{\partial \psi}[g_{ij}] \quad = \quad [0]$$

Inverse metric tensor (contravariant) and derivatives:

$$[g^{ij}] \quad = \quad diag \left\{ \begin{array}{l} \dfrac{\left(ch^2 - s^2(\theta)\right)^2}{a^2(s^2(\theta) + sh^2)} \\[3ex] \dfrac{\left(ch^2 - s^2(\theta)\right)^2}{a^2(s^2(\theta) + sh^2)} \\[3ex] \dfrac{\left(ch^2 - s^2(\theta)\right)^2}{a^2(s^2(\theta) + sh^2)} \end{array} \right.$$

$$\frac{\partial}{\partial \eta}[g^{ij}] \quad = \quad diag \left\{ \begin{array}{l} -\dfrac{2\left(ch^2 - s^2(\theta)\right)(ch^2 - 3\,s^2(\theta) - 2\,sh^2)\,ch\,sh}{a^2((s^2(\theta) + sh^2)^2)} \\[3ex] -\dfrac{2\left(ch^2 - s^2(\theta)\right)(ch^2 - 3\,s^2(\theta) - 2\,sh^2)\,ch\,sh}{a^2((s^2(\theta) + sh^2)^2)} \\[3ex] -\dfrac{2(ch^2 - s^2(\theta) - 2\,sh^2)\left(ch^2 - s^2(\theta)\right)\,ch}{a^2\,s^2(\theta)\,sh^3} \end{array} \right.$$

$$\frac{\partial}{\partial\theta}[g^{ij}] \quad = \quad diag \left\{ \begin{array}{l} -\dfrac{2\,(ch^2 + s^2(\theta) + 2\,sh^2)\,(ch^2 - s^2(\theta))\,c(\theta)\,s(\theta)}{a^2((s^2(\theta) + sh^2)^2)} \\[2ex] -\dfrac{2\,(ch^2 + s^2(\theta) + 2\,sh^2)\,(ch^2 - s^2(\theta))\,c(\theta)\,s(\theta)}{a^2((s^2(\theta) + sh^2)^2)} \\[2ex] -\dfrac{2\left(\dfrac{ch^4}{s^3(\theta)} - s(\theta)\right)c(\theta)}{a^2\,sh^2} \end{array} \right.$$

$$\frac{\partial}{\partial\psi}[g^{ij}] \quad = \quad [0]$$

Scale factors and Christoffel symbols:

$$Scale\ factors: \quad \left\{ \begin{array}{l} h_\eta = h_\theta = \dfrac{a\,\sqrt{sin^2(\theta) + sinh^2(\eta)}}{cosh^2(\eta) - sin^2(\theta)} \\[2ex] h_\psi = \dfrac{a\,sin(\theta)\,sinh(\eta)}{cosh^2(\eta) - sin^2(\theta)} \end{array} \right.$$

Non-zero Christoffel symbols of the 2nd kind and derivatives:

$$\Gamma^\eta_{\eta\eta} \quad = \quad ch\,sh^2\left(\frac{2}{s^2(\theta) - ch^2} + \frac{1}{sh^2 + s^2(\theta)}\right)$$

$$\Gamma^\eta_{\eta\theta} \quad = \quad -\frac{(c^2(\theta) - 3\,ch^2 + 1)\,c(\theta)\,s(\theta)}{(ch^2 + s^2(\theta) - 1)(ch^2 - s^2(\theta))}$$

$$\Gamma^\eta_{\theta\theta} \quad = \quad -\frac{(3\,cos^2(\theta) - ch^2 - 1)\,ch\,sh}{(ch^2 + s^2(\theta) - 1)(ch^2 - s^2(\theta))}$$

$$\Gamma^\eta_{\psi\psi} \quad = \quad \frac{(c^2(\theta) - ch^2 + 1)(c(\theta) + 1)(c(\theta) - 1)\,ch\,sh}{(ch^2 + s^2(\theta) - 1)(ch^2 - s^2(\theta))}$$

$$\Gamma^\theta_{\eta\eta} \quad = \quad \frac{(c^2(\theta) - 3\,ch^2 + 1)\,c(\theta)\,s(\theta)}{(ch^2 + s^2(\theta) - 1)(ch^2 - s^2(\theta))}$$

$$\Gamma^\theta_{\eta\theta} \quad = \quad \frac{(3\,c^2(\theta) - ch^2 - 1)\,ch\,sh}{(ch^2 + s^2(\theta) - 1)(ch^2 - s^2(\theta))}$$

$$\Gamma^{\theta}_{\theta\theta} = -\frac{(c^2(\theta) - 3\,ch^2 + 1)\,c(\theta)\,s(\theta)}{(ch^2 + s^2(\theta) - 1)(ch^2 - s^2(\theta))}$$

$$\Gamma^{\theta}_{\psi\psi} = -\frac{(ch^2 + s^2(\theta))\,c(\theta)\,s(\theta)\,sh^2}{(ch^2 + s^2(\theta) - 1)(ch^2 - s^2(\theta))}$$

$$\Gamma^{\psi}_{\eta\psi} = \frac{(c^2(\theta) - ch^2 + 1)\,ch}{(c^2(\theta) + sh^2)\,sh}$$

$$\Gamma^{\psi}_{\theta\psi} = -\frac{(c^2(\theta) - ch^2 - 1)\,c(\theta)}{s(\theta)(ch^2 - s^2(\theta))}$$

Unit Basis Vectors:

$$\hat{e}_{\eta} = \begin{pmatrix} -\dfrac{(cos^2(\theta) - sinh^2(\eta))\,cos(\psi)\,cosh(\eta)\,sin(\theta)}{\sqrt{sin^2(\theta) + sinh^2(\eta)}\,(cosh^2(\eta) - sin^2(\theta))} \\[4mm] -\dfrac{(cos^2(\theta) - sinh^2(\eta))\,cosh(\eta)\,sin(\theta)\,sin(\psi)}{\sqrt{sin^2(\theta) + sinh^2(\eta)}\,(cosh^2(\eta) - sin^2(\theta))} \\[4mm] \dfrac{(cosh^2(\eta) + sin^2(\theta))\,cos(\theta)\,sinh(\eta)}{\sqrt{sin^2(\theta) + sinh^2(\eta)}\,(cosh^2(\eta) - sin^2(\theta))} \end{pmatrix}$$

$$\hat{e}_{\theta} = \begin{pmatrix} -\dfrac{(cosh^2(\eta) + sin^2(\theta))\,cos(\theta)\,cos(\psi)\,sinh(\eta)}{\sqrt{sin^2(\theta) + sinh^2(\eta)}\,(cosh^2(\eta) - sin^2(\theta))} \\[4mm] -\dfrac{(cosh^2(\eta) + sin^2(\theta))\,cos(\theta)\,sin(\psi)\,sinh(\eta)}{\sqrt{sin^2(\theta) + sinh^2(\eta)}\,(cosh^2(\eta) - sin^2(\theta))} \\[4mm] -\dfrac{(cos^2(\theta) - cosh^2(\eta) + 1)\,cosh(\eta)\,sin(\theta)}{\sqrt{sin^2(\theta) + sinh^2(\eta)}\,(cosh^2(\eta) - sin^2(\theta))} \end{pmatrix}$$

$$\hat{e}_{\psi} = \begin{pmatrix} sin(\psi) \\ cos(\psi) \\ 0 \end{pmatrix}$$

Differential operators:

$$div\ V\left(V_\eta, V_\theta, V_\psi\right)$$

$$= \frac{\left(ch^2 - s^3(\theta)\right)^3}{a\left(s^2(\theta) + sh^2\right)s(\theta)\,sh(\eta)}\left[\frac{\partial}{\partial\eta}\left(\frac{\sqrt{s^2(\theta) + sh^2}\ sh\ s(\theta)}{\left(ch^2 - s^2(\theta)\right)^2}\,V_\eta\right)\right.$$

$$\left. + \frac{\partial}{\partial\theta}\left(\frac{\sqrt{s^2(\theta) + sh^2}\ sh\ s(\theta)}{\left(ch^2 - s^2(\theta)\right)^2}\,V_\theta\right) + \frac{\partial}{\partial\psi}\left(\frac{\left(s^2(\theta) + sh^2\right)}{\left(ch^2 - s^2(\theta)\right)^2}\,V_\psi\right)\right]$$

$$= \frac{\left(ch^2 - s^3(\theta)\right)^3}{a\left(s^2(\theta) + sh^2\right)s(\theta)\,sh}\left[\frac{\partial}{\partial\eta}\left(\frac{\sqrt{s^2(\theta) + sh^2}\ sh\ s(\theta)}{\left(ch^2 - s^2(\theta)\right)^2}\,V_\eta\right)\right.$$

$$\left. + \frac{\partial}{\partial\theta}\left(\frac{\sqrt{s^2(\theta) + sh^2}\ sh\ s(\theta)}{\left(ch^2 - s^2(\theta)\right)^2}\,V_\theta\right)\right] + \frac{\left(ch^2 - s^3(\theta)\right)}{a\,s(\theta)\,sh}\frac{\partial V_\psi}{\partial\psi}$$

$$curl\ V\left(V_\eta, V_\theta, V_\psi\right)$$

$$= \frac{\left(ch^2 - s^2(\theta)\right)^2}{a\sqrt{s^2(\theta) + sh^2}\ sh\ s(\theta)}\left(\frac{\partial}{\partial\theta}\left(\frac{s(\theta)\,sh}{s^2(\theta) - ch^2}\,V_\psi\right)\right.$$

$$\left. - \frac{\partial}{\partial\psi}\left(\frac{a\sqrt{s^2(\theta) + sh^2}}{s^2(\theta) - ch^2}\,V_\theta\right)\right)\hat{e}_\eta$$

$$- \frac{\left(ch^2 - s^2(\theta)\right)^2}{a\sqrt{s^2(\theta) + sh^2}\ sh\ s(\theta)}\left(\frac{\partial}{\partial\eta}\left(\frac{s(\theta)\,sh}{s^2(\theta) - ch^2}\,V_\psi\right)\right.$$

$$\left. - \frac{\partial}{\partial\psi}\left(\frac{\sqrt{s^2(\theta) + sh^2}}{s^2(\theta) - ch^2}\,V_\eta\right)\right)\hat{e}_\theta$$

$$+ \frac{\left(s^2(\theta) - ch^2\right)^2}{a\left(s^2(\theta) + sh^2\right)}\left(\frac{\partial}{\partial\eta}\left(\frac{\sqrt{s^2(\theta) + sh^2}}{s^2(\theta) - ch^2}\,V_\theta\right)\right.$$

$$\left. - \frac{\partial}{\partial\theta}\left(\frac{\sqrt{s^2(\theta) + sh^2}}{s^2(\theta) - ch^2}\,V_\eta\right)\right)\hat{e}_\psi$$

$$\Theta(\eta,\theta,\psi) = \frac{\left(ch^2 - s^2(\theta)\right)^3}{a^2(s^2(\theta) + sh^2)\; s(\theta)\, sh}\left(\frac{\partial}{\partial\eta}\left(\frac{s(\theta)\, sh}{ch^2 - s^2(\theta)}\, \frac{\partial\Theta}{\partial\eta}\right)\right.$$

$$\left. + \frac{\partial}{\partial\theta}\left(\frac{s(\theta)\, sh}{ch^2 - s^2(\theta)}\, \frac{\partial\Theta}{\partial\theta}\right) + \frac{\partial}{\partial\psi}\left(\frac{(s^2(\theta) + sh^2)}{\left(ch^2 - s^2(\theta)\right)\, s(\theta)\, sh}\, \frac{\partial\Theta}{\partial\psi}\right)\right)$$

$$= \frac{\left(ch^2 - s^2(\theta)\right)^3}{a^2(s^2(\theta) + sh^2)\; s(\theta)\, sh}\left(\frac{\partial}{\partial\eta}\left(\frac{s(\theta)\, sh}{ch^2 - s^2(\theta)}\, \frac{\partial\Theta}{\partial\eta}\right)\right.$$

$$\left. + \frac{\partial}{\partial\theta}\left(\frac{s(\theta)\, sh}{ch^2 - s^2(\theta)}\, \frac{\partial\Theta}{\partial\theta}\right)\right) + \frac{\left(ch^2 - s^2(\theta)\right)^2}{a^2\; s^2(\theta)\, sh^2}\, \frac{\partial^2\Theta}{\partial\psi^2}$$

I.A.2.10 Oblate Spheroidal Coordinates

References:

[Moon/Spencer]: p. 31

[Neutsch]: p.1211

Coordinate definition:

$$Cartesian\ functions \begin{cases} x^1(\mu,v,\phi) = a\,cosh(\mu)\,cos(v)\,cos(\phi) \\ x^2(\mu,v,\phi) = a\,cosh(\mu)\,cos(v)\,sin(\phi) \\ x^3(\mu,v,\phi) = a\,sinh(\mu)\,sin(v) \end{cases}$$

$$\mu \in [0,\infty),\, v \in \left[-\tfrac{\pi}{2},\tfrac{\pi}{2}\right],\, \phi \in [-\pi,\pi],\, a > 0,\, fixed$$

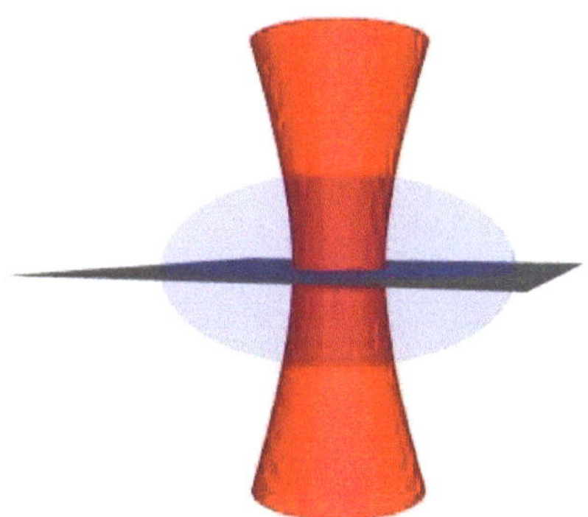

We define: $c(x) = cos(x), s(x) = sin(x), sh = sinh(\mu), ch = cosh(\mu)$

Orthogonal   yes

Jacobian and metric:

$$J(\mu, v, \phi) = \begin{bmatrix} a\,c(v)\,c(\phi)\,sh & -a\,c(\phi)\,ch\,s(v) & -a\,c(v)\,ch\,s(\phi) \\ a\,c(v)\,s(\phi)\,sh & -a\,ch\,s(v)\,s(\phi) & a\,c(v)\,c(\phi)\,ch \\ a\,ch\,s(v) & a\,c(v)\,sh & 0 \end{bmatrix}$$

$$J^{-1}(\mu, v, \phi) = \begin{bmatrix} \dfrac{c(v)\,c(\phi)\,sh}{a\,(s^2(v)+sh^2)} & \dfrac{c(v)\,s(\phi)\,sh}{a\,(s^2(v)+sh^2)} & \dfrac{s(v)\,ch}{a\,(s^2(v)+sh^2)} \\[2mm] -\dfrac{s(v)\,c(\phi)\,ch}{a\,(s^2(v)+sh^2)} & -\dfrac{s(v)\,s(\phi)\,ch}{a\,(s^2(v)+sh^2)} & \dfrac{c(v)\,sh(\mu)}{a\,(s^2(v)+sh^2)} \\[2mm] -\dfrac{s(\phi)}{a\,c(v)\,ch} & \dfrac{c(\phi)}{a\,c(v)\,ch} & 0 \end{bmatrix}$$

$$det\big(J_{(\mu, v, \phi)}\big) = a^3(sin^2(v)+sinh^2(\mu))\,cos(v)\,cosh(\mu)$$

Metric tensor (covariant) and derivatives:

$$[g_{ij}] = \begin{bmatrix} a^2(s^2(v)+sh^2) & 0 & 0 \\ 0 & a^2(s^2(v)+sh^2) & 0 \\ 0 & 0 & a^2c^2(v)\,ch^2 \end{bmatrix}$$

$$\frac{\partial}{\partial \mu}[g_{ij}] = \begin{bmatrix} 2\,a^2ch\,sh & 0 & 0 \\ 0 & 2\,a^2ch\,sh & 0 \\ 0 & 0 & a^2c^2(v)\,ch\,sh \end{bmatrix}$$

$$\frac{\partial}{\partial v}[g_{ij}] = \begin{bmatrix} 2\,a^2c(v)\,s(v) & 0 & 0 \\ 0 & 2\,a^2c(v)\,s(v) & 0 \\ 0 & 0 & -2\,a^2c(v)\,ch^2\,s(v) \end{bmatrix}$$

$$\frac{\partial}{\partial \phi}[g_{ij}] = [0]$$

Inverse metric tensor (contravariant) and derivatives:

$$[g^{ij}] = \begin{bmatrix} \dfrac{1}{a^2(s^2(v)+sh^2)} & 0 & 0 \\[3mm] 0 & \dfrac{1}{a^2(s^2(v)+sh^2)} & 0 \\[3mm] 0 & 0 & \dfrac{1}{a^2\,c^2(v)\,ch^2} \end{bmatrix}$$

$$\frac{\partial}{\partial\mu}[g^{ij}] = \begin{bmatrix} -\dfrac{2\,ch\,sh}{a^2(s^2(v)+sh^2)^2} & 0 & 0 \\[3mm] 0 & -\dfrac{2\,ch(\mu)\,sh}{a^2(s^2(v)+sh^2)^2} & 0 \\[3mm] 0 & 0 & -\dfrac{2\,sh(\mu)}{a^2\,c^2(v)\,ch^3} \end{bmatrix}$$

$$\frac{\partial}{\partial v}[g^{ij}] = \begin{bmatrix} -\dfrac{4\,s(2v)}{a^2(c(2v)-ch\,)^2} & 0 & 0 \\[3mm] 0 & -\dfrac{4\,s(2v)}{a^2(c(2v)-ch\,)^2} & 0 \\[3mm] 0 & 0 & \dfrac{2\,s(v)}{a^2\,c^3(v)\,ch} \end{bmatrix}$$

$$\frac{\partial}{\partial\phi}[g^{ij}] = [0]$$

Scale factors and Christoffel symbols:

$$Scale\ factors: \begin{cases} h_\mu = h_v = a\,\sqrt{sin^2(v)+sinh^2(\mu)} \\ h_\phi = a\,cos(v)\,cosh(\mu) \end{cases}$$

Non-zero Christoffel symbols of the 2nd kind and derivatives:

$$\Gamma^{\mu}_{\mu\mu} = \frac{\cosh(\mu)\,\sinh(\mu)}{\sin^2(v) + \sinh^2(\mu)}$$

$$\Gamma^{\mu}_{\mu v} = \frac{\cos(v)\,\sin(v)}{\sin^2(v) + \sinh^2(\mu)}$$

$$\Gamma^{\mu}_{vv} = -\frac{\cosh(\mu)\,\sinh(\mu)}{\sin^2(v) + \sinh^2(\mu)}$$

$$\Gamma^{\mu}_{\phi\phi} = -\frac{\cos^2(v)\,\cosh(\mu)\,\sinh(\mu)}{\sin^2(v) + \sinh^2(\mu)}$$

$$\Gamma^{v}_{\mu\mu} = -\frac{\cos(v)\,\sin(v)}{\sin^2(v) + \sinh^2(\mu)}$$

$$\Gamma^{v}_{\mu v} = \frac{\cosh(\mu)\,\sinh(\mu)}{\sin^2(v) + \sinh^2(\mu)}$$

$$\Gamma^{v}_{vv} = \frac{\cos(v)\,\sin(v)}{\sin^2(v) + \sinh^2(\mu)}$$

$$\Gamma^{v}_{\phi\phi} = \frac{\cos(v)\,\cosh^2(\mu)\,\sin(v)}{\sin^2(v) + \sinh^2(\mu)}$$

$$\Gamma^{\phi}_{\mu\phi} = \frac{\sinh(\mu)}{\cosh(\mu)}$$

$$\Gamma^{\phi}_{v\phi} = -\frac{\sin(v)}{\cos(v)}$$

Unit Basis Vectors:

$$\hat{e}_\mu = \begin{pmatrix} \dfrac{\cos(v)\,\cos(\phi)\,\sinh(\mu)}{\sqrt{\sin^2(v) + \sinh^2(\mu)}} \\[2mm] \dfrac{\cos(v)\,\sin(\phi)\,\sinh(\mu)}{\sqrt{\sin^2(v) + \sinh^2(\mu)}} \\[2mm] \dfrac{\sin(v)\,\cosh(\mu)}{\sqrt{\sin^2(v) + \sinh^2(\mu)}} \end{pmatrix}$$

$$\hat{e}_v = \begin{pmatrix} -\dfrac{\cos(\phi)\,\sin(v)\,\cosh(\mu)}{\sqrt{\sin^2(v) + \sinh^2(\mu)}} \\[2mm] -\dfrac{\sin(\phi)\,\sin(v)\,\cosh(\mu)}{\sqrt{\sin^2(v) + \sinh^2(\mu)}} \\[2mm] \dfrac{\cos(v)\,\sinh(\mu)}{\sqrt{\sin^2(v) + \sinh^2(\mu)}} \end{pmatrix} \qquad \hat{e}_\phi = \begin{pmatrix} -\sin(\phi) \\ \cos(\phi) \\ 0 \end{pmatrix}$$

Differential operators:

$$div\,V\left(V_\mu, V_v, V_\phi\right) = \frac{1}{a\,(s^2(v) + sh^2)\,c(v)\,ch}\left(\frac{\partial}{\partial\mu}\left(\sqrt{s^2(v) + sh^2}\;c(v)\,ch\,V_\mu\right)\right.$$

$$\left. + \frac{\partial}{\partial v}\left(\sqrt{s^2(v) + sh^2}\,c(v)\,ch\,V_v\right)\right) + \frac{1}{a\;c(v)\,ch(\mu)}\frac{\partial V_\phi}{\partial\phi}$$

$$curl\, V\left(V_\mu, V_v, V_\phi\right) = \frac{1}{a\sqrt{s^2(v)+sh^2}c(v)\,ch}\left(\frac{\partial}{\partial v}\left(c(v)\,ch\,V_\phi\right)\right.$$

$$\left. -\frac{\partial}{\partial\phi}\left(\sqrt{s^2(v)+sh^2}\,V_v\right)\right)\hat{e}_\mu$$

$$-\frac{1}{a\sqrt{s^2(v)+sh^2}\,c(v)\,ch}\left(\frac{\partial}{\partial\mu}\left(c(v)\,ch\,V_\phi\right)\right.$$

$$\left. -\frac{\partial}{\partial\phi}\left(\sqrt{s^2(v)+sh^2}\,V_\mu\right)\right)\hat{e}_v$$

$$+\frac{1}{a\,(s^2(v)+sh^2)}\left(\frac{\partial}{\partial\mu}\left(\sqrt{s^2(v)+sh^2}\,V_v\right)\right.$$

$$\left. -\frac{\partial}{\partial v}\left(\sqrt{s^2(v)+sh^2}\,V_\mu\right)\right)$$

$$\Delta\,\Theta(\mu, v, \phi) = \frac{1}{a^2(s^2(v)+sh^2)\,c(v)\,ch}\left(\frac{\partial}{\partial\mu}\left(c(v)\,ch\,\frac{\partial\Theta}{\partial\mu}\right)\right.$$

$$\left. +\frac{\partial}{\partial v}\left(c(v)\,ch\,\frac{\partial\Theta}{\partial v}\right)\right) + \frac{1}{a^2\,c^2(v)\,ch^2}\frac{\partial^2\Theta}{\partial\phi^2}$$

I.A.2.11 Parabolic 2D Coordinates

References:

[Neutsch]: p. 1142

Coordinate definition:

$$\text{Cartesian functions} \quad \begin{cases} x^1(\sigma, \tau, z) = \sigma\tau \\ x^2(\sigma, \tau, z) = \dfrac{1}{2}(\tau^2 - \sigma^2) \end{cases}$$

$$\tau >= 0, \sigma, z \in \mathbb{R}$$

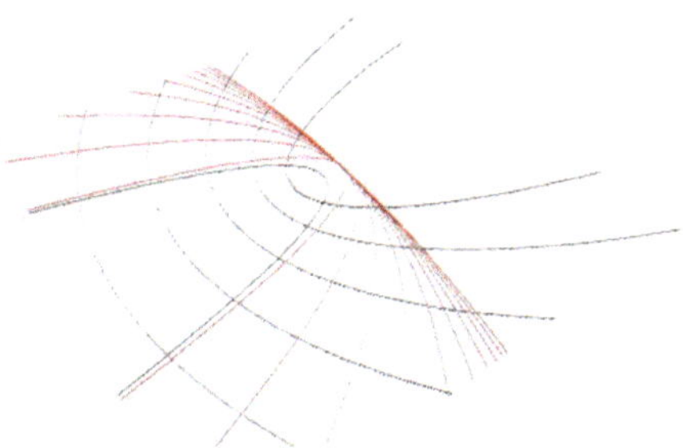

Orthogonal   yes

Jacobian and metric:

$$J(\sigma,\tau,z) \quad = \quad \begin{bmatrix} \tau & \sigma \\ -\sigma & \tau \end{bmatrix}$$

$$J^{-1}(\sigma,\tau,z) \quad = \quad \begin{bmatrix} \dfrac{\tau}{\sigma^2+\tau^2} & -\dfrac{\sigma}{\sigma^2+\tau^2} \\ \dfrac{\sigma}{\sigma^2+\tau^2} & \dfrac{\tau}{\sigma^2+\tau^2} \end{bmatrix}$$

$$\det\!\left(J(\sigma,\tau,z)\right) = \sigma^2 + \tau^2$$

Metric tensor (covariant) and derivatives:

$$[g_{ij}] \quad = \quad \begin{bmatrix} \sigma^2+\tau^2 & 0 \\ 0 & \sigma^2+\tau^2 \end{bmatrix}$$

$$\frac{\partial}{\partial\sigma}[g_{ij}] \quad = \quad \begin{bmatrix} 2\sigma & 0 \\ 0 & 2\,\sigma \end{bmatrix}$$

$$\frac{\partial}{\partial\tau}[g_{ij}] \quad = \quad \begin{bmatrix} 2\,\tau & 0 \\ 0 & 2\,\tau \end{bmatrix}$$

Inverse metric tensor (contravariant) and derivatives:

$$[g^{ij}] \quad = \quad \begin{bmatrix} \dfrac{1}{\sigma^2+\tau^2} & 0 \\ 0 & \dfrac{1}{\sigma^2+\tau^2} \end{bmatrix}$$

$$\frac{\partial}{\partial\sigma}[g^{ij}] \;=\; \begin{bmatrix} -\dfrac{2\sigma}{(\sigma^2+\tau^2)^2} & 0 \\[3mm] 0 & -\dfrac{2\sigma}{(\sigma^2+\tau^2)^2} \end{bmatrix}$$

$$\frac{\partial}{\partial\tau}[g^{ij}] \;=\; \begin{bmatrix} -\dfrac{2\tau}{(\sigma^2+\tau^2)^2} & 0 \\[3mm] 0 & -\dfrac{2\tau}{(\sigma^2+\tau^2)^2} \end{bmatrix}$$

Scale factors and Christoffel symbols:

$$Scale\ factors:\ \left\{ h_\sigma = h_\tau = \sqrt{\sigma^2+\tau^2} \right.$$

Non-zero Christoffel symbols of the 2nd kind and derivatives:

$$\Gamma^\sigma_{\sigma\sigma} = \frac{\sigma}{\sigma^2+\tau^2}$$

$$\Gamma^\sigma_{\sigma\tau} = \frac{\tau}{\sigma^2+\tau^2}$$

$$\Gamma^\sigma_{\tau\tau} = -\frac{\sigma}{\sigma^2+\tau^2}$$

$$\Gamma^\tau_{\sigma\sigma} = -\frac{\tau}{\sigma^2+\tau^2}$$

$$\Gamma^\tau_{\sigma\tau} = \frac{\sigma}{\sigma^2+\tau^2}$$

$$\Gamma^\tau_{\tau\tau} = \frac{\tau}{\sigma^2+\tau^2}$$

Unit Basis Vectors:

$$\hat{e}_\sigma = \begin{pmatrix} \dfrac{\tau}{\sqrt{\sigma^2 + \tau^2}} \\ \dfrac{-\sigma}{\sqrt{\sigma^2 + \tau^2}} \end{pmatrix} \qquad \hat{e}_\tau = \begin{pmatrix} \dfrac{\sigma}{\sqrt{\sigma^2 + \tau^2}} \\ \dfrac{\tau}{\sqrt{\sigma^2 + \tau^2}} \end{pmatrix}$$

Differential operators:

$$div\, V(V_\sigma, V_\tau) = \frac{1}{\sigma^2 + \tau^2}\left(\frac{\partial}{\partial\sigma}\left(\sqrt{\sigma^2 + \tau^2}\, V_\sigma\right) + \frac{\partial}{\partial\tau}\left(\sqrt{\sigma^2 + \tau^2}\, V_\tau\right) \right)$$

$$\Delta\, \Theta(\sigma, \tau) = \frac{1}{\sigma^2 + \tau^2}\left(\frac{\partial^2\Theta}{\partial\sigma^2} + \frac{\partial^2\Theta}{\partial\tau^2} \right)$$

I.A.2.12 Parabolic 3D Coordinates

References:

[Moon/Spencer]: p. 34

Coordinate definition:

$$Cartesian\ functions\ \begin{cases} x^1(\sigma, \tau, \phi) = \sigma\tau\, cos(\phi) \\ x^2(\sigma, \tau, \phi) = \sigma\tau\, sin(\phi) \\ x^3(\sigma, \tau, \phi) = \frac{1}{2}(\tau^2 - \sigma^2) \end{cases}$$

$$\sigma, \tau \in [0, \infty), \phi \bmod 2\pi$$

Orthogonal   yes

Jacobian and metric:

$$J(\sigma,\tau,\phi) \;=\; \begin{bmatrix} \tau\,cos(\phi) & \sigma\,cos(\phi) & -\sigma\tau\,sin(\phi) \\ \tau\,sin(\phi) & \sigma\,sin(\phi) & \sigma\tau\,cos(\phi) \\ -\sigma & \tau & 0 \end{bmatrix}$$

$$J^{-1}(\sigma,\tau,\phi) \;=\; \begin{bmatrix} \dfrac{\tau\,cos(\phi)}{\sigma^2+\tau^2} & \dfrac{\tau\,sin(\phi)}{\sigma^2+\tau^2} & -\dfrac{\sigma}{\sigma^2+\tau^2} \\[2mm] \dfrac{\sigma\,cos(\phi)}{\sigma^2+\tau^2} & \dfrac{\sigma\,sin(\phi)}{\sigma^2+\tau^2} & \dfrac{\tau}{\sigma^2+\tau^2} \\[2mm] -\dfrac{sin(\phi)}{\sigma\tau} & \dfrac{cos(\phi)}{\sigma\tau} & 0 \end{bmatrix}$$

$$det\big(J(\sigma,\tau,\phi)\big) = -(\sigma^2+\tau^2)\,\sigma\tau$$

Metric tensor (covariant) and derivatives:

$$[g_{ij}] \;=\; \begin{bmatrix} \sigma^2+\tau^2 & 0 & 0 \\ 0 & \sigma^2+\tau^2 & 0 \\ 0 & 0 & \sigma^2\tau^2 \end{bmatrix}$$

$$\frac{\partial}{\partial\sigma}[g_{ij}] \;=\; \begin{bmatrix} 2\sigma & 0 & 0 \\ 0 & 2\,\sigma & 0 \\ 0 & 0 & 2\,\sigma\tau^2 \end{bmatrix}$$

$$\frac{\partial}{\partial\tau}[g_{ij}] \;=\; \begin{bmatrix} 2\,\tau & 0 & 0 \\ 0 & 2\,\tau & 0 \\ 0 & 0 & 2\sigma^2\tau \end{bmatrix}$$

$$\frac{\partial}{\partial\phi}[g_{ij}] \;=\; [0]$$

Inverse metric tensor (contravariant) and derivatives:

$$[g^{ij}] = \begin{bmatrix} \dfrac{1}{\sigma^2 + \tau^2} & 0 & 0 \\[2ex] 0 & \dfrac{1}{\sigma^2 + \tau^2} & 0 \\[2ex] 0 & 0 & \dfrac{1}{\sigma^2\tau^2} \end{bmatrix}$$

$$\frac{\partial}{\partial\sigma}[g^{ij}] = \begin{bmatrix} -\dfrac{2\sigma}{(\sigma^2 + \tau^2)^2} & 0 & 0 \\[2ex] 0 & -\dfrac{2\sigma}{(\sigma^2 + \tau^2)^2} & 0 \\[2ex] 0 & 0 & -\dfrac{2}{\sigma^3\,\tau^2} \end{bmatrix}$$

$$\frac{\partial}{\partial\tau}[g^{ij}] = \begin{bmatrix} -\dfrac{2\tau}{(\sigma^2 + \tau^2)^2} & 0 & 0 \\[2ex] 0 & -\dfrac{2\tau}{(\sigma^2 + \tau^2)^2} & 0 \\[2ex] 0 & 0 & -\dfrac{2}{\sigma^2\,\tau^3} \end{bmatrix}$$

$$\frac{\partial}{\partial\phi}[g^{ij}] = [0]$$

Scale factors and Christoffel symbols:

$$\textit{Scale factors:} \quad \begin{cases} h_\sigma = h_\tau = \sqrt{\sigma^2 + \tau^2} \\ h_\phi = \sigma\tau \end{cases}$$

Non-zero Christoffel symbols of the 2nd kind and derivatives:

$$\Gamma^{\sigma}_{\sigma\sigma} = \frac{\sigma}{\sigma^2 + \tau^2}$$

$$\Gamma^{\sigma}_{\sigma\tau} = \frac{\tau}{\sigma^2 + \tau^2}$$

$$\Gamma^{\sigma}_{\tau\tau} = -\frac{\sigma}{\sigma^2 + \tau^2}$$

$$\Gamma^{\sigma}_{\phi\phi} = -\frac{\sigma\tau^2}{\sigma^2 + \tau^2}$$

$$\Gamma^{\tau}_{\sigma\sigma} = -\frac{\tau}{\sigma^2 + \tau^2}$$

$$\Gamma^{\tau}_{\sigma\tau} = \frac{\sigma}{\sigma^2 + \tau^2}$$

$$\Gamma^{\tau}_{\tau\tau} = \frac{\tau}{\sigma^2 + \tau^2}$$

$$\Gamma^{\tau}_{\phi\phi} = -\frac{\sigma^2\tau}{\sigma^2 + \tau^2}$$

$$\Gamma^{\phi}_{\sigma\phi} = \frac{1}{\sigma}$$

$$\Gamma^{\phi}_{\tau\phi} = \frac{1}{\tau}$$

Unit Basis Vectors:

$$\hat{e}_{\sigma} = \begin{pmatrix} \dfrac{\tau\,cos(\phi)}{\sqrt{\sigma^2 + \tau^2}} \\[2mm] \dfrac{\tau\,sin(\phi)}{\sqrt{\sigma^2 + \tau^2}} \\[2mm] -\dfrac{\sigma}{\sqrt{\sigma^2 + \tau^2}} \end{pmatrix} \qquad \hat{e}_{\tau} = \begin{pmatrix} \dfrac{\sigma\,cos(\phi)}{\sqrt{\sigma^2 + \tau^2}} \\[2mm] \dfrac{\sigma\,sin(\phi)}{\sqrt{\sigma^2 + \tau^2}} \\[2mm] \dfrac{\tau}{\sqrt{\sigma^2 + \tau^2}} \end{pmatrix} \qquad \hat{e}_{\phi} = \begin{pmatrix} -sin(\phi) \\ cos(\phi) \\ 0 \end{pmatrix}$$

Differential operators:

$$div\, V\left(V_\sigma, V_\tau, V_\phi\right) = \frac{1}{\sigma\tau(\sigma^2 + \tau^2)}\left(\frac{\partial}{\partial\sigma}\left(\sigma\tau\sqrt{\sigma^2 + \tau^2}\,V_\sigma\right) + \frac{\partial}{\partial\tau}\left(\sigma\tau\sqrt{\sigma^2 + \tau^2}\,V_\tau\right)\right.$$

$$\left. + \frac{\partial}{\partial\phi}\left((\sigma^2 + \tau^2)\,V_\phi\right)\right)$$

$$= \frac{1}{\sigma\tau(\sigma^2 + \tau^2)}\left(\tau\frac{\partial}{\partial\sigma}\left(\sigma\sqrt{\sigma^2 + \tau^2}\,V_\sigma\right)\right.$$

$$\left. + \sigma\frac{\partial}{\partial\tau}\left(\tau\sqrt{\sigma^2 + \tau^2}\,V_\tau\right)\right) + \frac{1}{\sigma\tau}\frac{\partial V_\phi}{\partial\phi}$$

$$curl\, V\left(V_\sigma, V_\tau, V_\phi\right) = \frac{1}{\sigma\tau\sqrt{\sigma^2 + \tau^2}}\left(\frac{\partial}{\partial\tau}\left(\sigma\tau\,V_\phi\right) - \frac{\partial}{\partial\phi}\left(\sqrt{\sigma^2 + \tau^2}\,V_\tau\right)\right)\hat{e}_\sigma$$

$$+ \frac{1}{\sigma\tau\sqrt{\sigma^2 + \tau^2}}\left(\frac{\partial}{\partial\phi}\left(\sqrt{\sigma^2 + \tau^2}\,V_\sigma\right) - \frac{\partial}{\partial\sigma}\left(\sigma\tau\,V_\phi\right)\right)\hat{e}_\tau$$

$$+ \frac{1}{\sigma^2 + \tau^2}\left(\frac{\partial}{\partial\sigma}\left(\sqrt{\sigma^2 + \tau^2}\,V_\tau\right) - \frac{\partial}{\partial\tau}\left(\sqrt{\sigma^2 + \tau^2}\,V_\sigma\right)\right)\hat{e}_\phi$$

$$\Delta\,\Theta(\sigma, \tau, \phi) = \frac{1}{\sigma\tau(\sigma^2 + \tau^2)}\left(\left(\frac{\partial}{\partial\sigma}\left(\sigma\tau\,\frac{\partial\Theta}{\partial\sigma}\right)\right) + \left(\frac{\partial}{\partial\tau}\left(\sigma\tau\,\frac{\partial\Theta}{\partial\tau}\right)\right)\right)$$

$$+ \frac{1}{\sigma^2\tau^2}\left(\frac{\partial^2\Theta}{\partial\phi^2}\right)$$

I.A.2.13 Parabolic Cylindrical Coordinates

References:

[Neutsch]: p. 1186

Coordinate definition:

$$\textit{Cartesian functions} \quad \begin{cases} x^1(\sigma,\tau,z) = \sigma\tau \\ x^2(\sigma,\tau,z) = \dfrac{1}{2}(\tau^2 - \sigma^2) \\ x^3(\sigma,\tau,z) = z \end{cases}$$

$$\tau \in [\, 0, \infty),\, \sigma, z \in \mathbb{R}$$

Orthogonal   yes

Jacobian and metric:

$$\mathcal{J}_{(\sigma,\tau,z)} \quad = \quad \begin{bmatrix} \tau & \sigma & 0 \\ -\sigma & \tau & 0 \\ 0 & 0 & 1 \end{bmatrix}$$

$$\mathcal{J}^{-1}{}_{(\sigma,\tau,z)} \quad = \quad \begin{bmatrix} \dfrac{\tau}{\sigma^2 + \tau^2} & -\dfrac{\sigma}{\sigma^2 + \tau^2} & 0 \\ \dfrac{\sigma}{\sigma^2 + \tau^2} & \dfrac{\tau}{\sigma^2 + \tau^2} & 0 \\ 0 & 0 & 1 \end{bmatrix}$$

$$det\left(\mathcal{J}_{(\sigma,\tau,z)}\right) = \sigma^2 + \tau^2$$

Metric tensor (covariant) and derivatives:

$$\left[g_{ij}\right] \quad = \quad \begin{bmatrix} \sigma^2 + \tau^2 & 0 & 0 \\ 0 & \sigma^2 + \tau^2 & 0 \\ 0 & 0 & 1 \end{bmatrix}$$

$$\frac{\partial}{\partial\sigma}\left[g_{ij}\right] \quad = \quad \begin{bmatrix} 2\,\sigma & 0 & 0 \\ 0 & 2\,\sigma & 0 \\ 0 & 0 & 0 \end{bmatrix}$$

$$\frac{\partial}{\partial\tau}\left[g_{ij}\right] \quad = \quad \begin{bmatrix} 2\,\tau & 0 & 0 \\ 0 & 2\,\tau & 0 \\ 0 & 0 & 0 \end{bmatrix}$$

$$\frac{\partial}{\partial z}\left[g_{ij}\right] \quad = \quad [0]$$

Inverse metric tensor (contravariant) and derivatives:

$$[g^{ij}] = \begin{bmatrix} \dfrac{1}{\sigma^2 + \tau^2} & 0 & 0 \\[2ex] 0 & \dfrac{1}{\sigma^2 + \tau^2} & 0 \\[2ex] 0 & 0 & 1 \end{bmatrix}$$

$$\frac{\partial}{\partial \sigma}[g^{ij}] = \begin{bmatrix} -\dfrac{2\sigma}{(\sigma^2 + \tau^2)^2} & 0 & 0 \\[2ex] 0 & -\dfrac{2\sigma}{(\sigma^2 + \tau^2)^2} & 0 \\[2ex] 0 & 0 & 0 \end{bmatrix}$$

$$\frac{\partial}{\partial \tau}[g^{ij}] = \begin{bmatrix} -\dfrac{2\tau}{(\sigma^2 + \tau^2)^2} & 0 & 0 \\[2ex] 0 & -\dfrac{2\tau}{(\sigma^2 + \tau^2)^2} & 0 \\[2ex] 0 & 0 & 0 \end{bmatrix}$$

$$\frac{\partial}{\partial z}[g^{ij}] = [0]$$

Scale factors and Christoffel symbols:

$$Scale\ factors: \begin{cases} h_\sigma = h_\tau = \sqrt{\sigma^2 + \tau^2} \\ h_z = 1 \end{cases}$$

Non-zero Christoffel symbols of the 2nd kind and derivatives:

$$\Gamma^{\sigma}_{\sigma\sigma} = \frac{\sigma}{\sigma^2 + \tau^2}$$

$$\Gamma^{\sigma}_{\sigma\tau} = \frac{\tau}{\sigma^2 + \tau^2}$$

$$\Gamma^{\sigma}_{\tau\tau} = -\frac{\sigma}{\sigma^2 + \tau^2}$$

$$\Gamma^{\tau}_{\sigma\sigma} = -\frac{\sigma}{\sigma^2 + \tau^2}$$

$$\Gamma^{\tau}_{\sigma\tau} = \frac{\sigma}{\sigma^2 + \tau^2}$$

$$\Gamma^{\tau}_{\tau\tau} = \frac{\tau}{\sigma^2 + \tau^2}$$

Unit Basis Vectors:

$$\hat{e}_{\sigma} = \begin{pmatrix} \dfrac{\tau}{\sqrt{\sigma^2 + \tau^2}} \\ -\dfrac{\sigma}{\sqrt{\sigma^2 + \tau^2}} \\ 0 \end{pmatrix} \quad \hat{e}_{\tau} = \begin{pmatrix} \dfrac{\sigma}{\sqrt{\sigma^2 + \tau^2}} \\ \dfrac{\tau}{\sqrt{\sigma^2 + \tau^2}} \\ 0 \end{pmatrix} \quad \hat{e}_{z} = \begin{pmatrix} 0 \\ 0 \\ 1 \end{pmatrix}$$

Differential operators:

$$div\, V(V_{\sigma}, V_{\tau}, V_z) = \frac{1}{\sigma^2 + \tau^2}\left(\frac{\partial}{\partial\sigma}\left(\sqrt{\sigma^2 + \tau^2}\, V_{\sigma} \right) + \frac{\partial}{\partial\tau}\left(\sqrt{\sigma^2 + \tau^2}\, V_{\tau} \right) \right) + \frac{\partial V_z}{\partial z}$$

$$= \frac{1}{\sqrt{\sigma^2 + \tau^2}}\left(\frac{\partial V_{\sigma}}{\partial\sigma} + \frac{\partial V_{\tau}}{\partial\tau} \right) + \frac{1}{(\sigma^2 + \tau^2)^{\frac{3}{2}}}\left(\sigma\, V_{\sigma} + \tau\, V_{\tau} \right) + \frac{\partial V_z}{\partial z}$$

$$curl\, V(V_{\sigma}, V_{\tau}, V_z) = -\frac{1}{\sqrt{\sigma^2 + \tau^2}}\left(\frac{\partial}{\partial z}\left(\sqrt{\sigma^2 + \tau^2}\, V_{\tau} \right) - \frac{\partial}{\partial\tau}\left(V_z \right) \right) \hat{e}_{\sigma}$$

$$+ \frac{1}{\sqrt{\sigma^2 + \tau^2}}\left(\frac{\partial}{\partial z}\left(\sqrt{\sigma^2 + \tau^2}\, V_{\sigma} \right) - \frac{\partial}{\partial\tau}\left(V_z \right) \right) \hat{e}_{\tau}$$

$$+ \frac{1}{\sqrt{\sigma^2 + \tau^2}}\left(\frac{\partial}{\partial\sigma}\left(\sqrt{\sigma^2 + \tau^2}\, V_{\tau} \right) - \frac{\partial}{\partial\tau}\left(\sqrt{\sigma^2 + \tau^2}\, V_{\sigma} \right) \right) \hat{e}_{z}$$

$$\Delta\,\Theta(\sigma,\tau,z) = \frac{1}{\sigma^2 + \tau^2}\left(\frac{\partial^2\Theta}{\partial\sigma^2} + \frac{\partial^2\Theta}{\partial\tau^2}\right) + \frac{\partial^2\Theta}{\partial z^2}$$

I.A.2.14 Prolate Spheroidal Coordinates

References:

[Neutsch]: p. 1207

Coordinate definition:

$$Cartesian\ functions \begin{cases} x^1(\xi,\eta,\phi) = a\ sinh(\xi)\ sin(\eta)\ cos(\phi) \\ x^2(\xi,\eta,\phi) = a\ sinh(\xi)\ sin(\eta)\ sin(\phi) \\ x^3(\xi,\eta,\phi) = a\ cosh(\xi)\ cos(\eta) \end{cases}$$

$$\xi \in [\,0,\infty),\ \ \eta \in [0,\pi],\ \ \ \phi \in [0,2\pi), a > 0, fixed$$

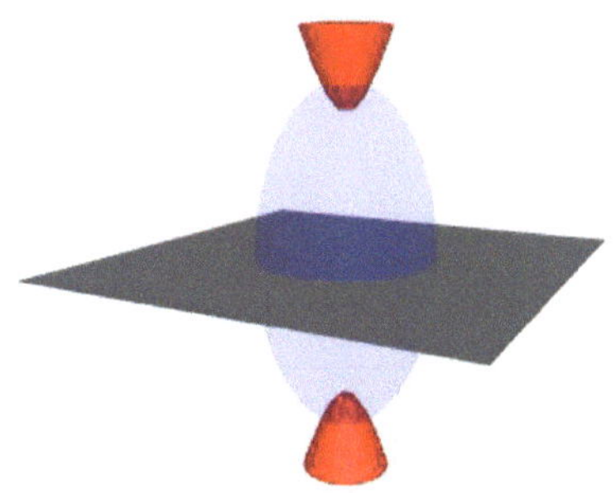

Orthogonal   yes

Jacobian and metric:

$$
\mathcal{J}_{(\xi,\eta,\phi)} =
\begin{cases}
[1,1]: a\cos(\phi)\cosh(\xi)\sin(\eta) \\
[1,2]: a\cos(\eta)\cos(\phi)\sinh(\xi) \\
[1,3]: -a\sin(\eta)\sin(\phi)\sinh(\xi) \\
[2,1]: a\cosh(\xi)\sin(\eta)\sin(\phi) \\
[2,2]: a\cos(\eta)\sin(\phi)\sinh(\xi) \\
[2,3]: a\cos(\phi)\sin(\eta)\sin(\xi) \\
[3,1]: a\cos(\eta)\sinh(\xi) \\
[3,2]: -a\cosh(\xi)\sin(\eta) \\
[3,3]: 0
\end{cases}
$$

$$
\mathcal{J}^{-1}{}_{(\xi,\eta,\phi)} =
\begin{cases}
[1,1]: \dfrac{\cos(\phi)\cosh(\xi)\sin(\eta)}{a\left(\sin^2(\eta)+\sinh^2(\xi)\right)} \\[3mm]
[1,2]: \dfrac{\cosh(\xi)\sin(\eta)\sin(\phi)}{a\left(\sin^2(\eta)+\sinh^2(\xi)\right)} \\[3mm]
[1,3]: \dfrac{\cos(\eta)\,\sinh(\xi)}{a\left(\sin^2(\eta)+\sinh^2(\xi)\right)} \\[3mm]
[2,1]: \dfrac{\cos(\eta)\,\cos(\phi)\,\sinh(\xi)}{a\left(\sin^2(\eta)+\sinh^2(\xi)\right)} \\[3mm]
[2,2]: \dfrac{\cos(\eta)\,\sin(\phi)\,\sinh(\xi)}{a\left(\sin^2(\eta)+\sinh^2(\xi)\right)} \\[3mm]
[2,3]: \dfrac{-\cosh(\xi)\,\sin(\eta)}{a\left(\sin^2(\eta)+\sinh^2(\xi)\right)} \\[3mm]
[3,1]: \dfrac{-\sin(\phi)}{a\,\sin(\eta)\,\sinh(\xi)} \\[3mm]
[3,2]: \dfrac{\cos(\phi)}{a\,\sin(\eta)\,\sinh(\xi)} \\[3mm]
[3,3]: 0
\end{cases}
$$

$$
\det\!\left(\mathcal{J}_{(\xi,\eta,\phi)}\right) = a^3\left(\sin^2(\eta)+\sinh^2(\xi)\right)\sin(\eta)\sinh(\xi)
$$

Metric tensor (covariant) and derivatives:

$$[g_{ij}] \quad = \quad diag \begin{Bmatrix} a^2(\sin^2(\eta) + \sinh^2(\xi)) \\ a^2(\sin^2(\eta) + \sinh^2(\xi)) \\ a^2 \sin^2(\eta) \sinh^2(\xi) \end{Bmatrix}$$

$$\frac{\partial}{\partial \xi}[g_{ij}] \quad = \quad \begin{bmatrix} a^2 \sinh(2\xi) & 0 & 0 \\ 0 & a^2 \sinh(2\xi) & 0 \\ 0 & 0 & a^2 \sin^2(\eta) \sinh(2\xi) \end{bmatrix}$$

$$\frac{\partial}{\partial \eta}[g_{ij}] \quad = \quad \begin{bmatrix} a^2 \sin(2\eta) & 0 & 0 \\ 0 & a^2 \sin(2\eta) & 0 \\ 0 & 0 & a^2 \sin(2\eta) \sinh^2(\xi) \end{bmatrix}$$

$$\frac{\partial}{\partial \phi}[g_{ij}] \quad = \quad [0]$$

Inverse metric tensor (contravariant) and derivatives:

$$[g^{ij}] \quad = \quad diag \begin{Bmatrix} \dfrac{1}{a^2(\sin^2(\eta) + \sinh^2(\xi))} \\[2mm] \dfrac{1}{a^2(\sin^2(\eta) + \sinh^2(\xi))} \\[2mm] \dfrac{1}{a^2 \sin^2(\eta) \sinh^2(\xi)} \end{Bmatrix}$$

$$\frac{\partial}{\partial \xi}[g^{ij}] \quad = \quad diag \begin{Bmatrix} -\dfrac{2\cosh(\xi)\sinh(\xi)}{a^2\left(\sin^2(\eta) + \sinh^2(\xi)\right)^2} \\[3mm] -\dfrac{2\cosh(\xi)\sinh(\xi)}{a^2\left(\sin^2(\eta) + \sinh^2(\xi)\right)^2} \\[3mm] -\dfrac{2\cosh(\xi)}{a^2 \sin^2(\eta) \sinh^3(\xi)} \end{Bmatrix}$$

$$\frac{\partial}{\partial \eta}[g^{ij}] \quad = \quad diag \begin{cases} -\dfrac{4\,sin(2\eta)}{a^2\big(cos(2\eta) - cosh(2\xi)\big)^2} \\[2ex] -\dfrac{4\,sin(2\eta)}{a^2\big(cos(2\eta) - cosh(2\xi)\big)^2} \\[2ex] -\dfrac{2\,cos(\eta)}{a^2\,sin^3(\eta)\,sinh^2(\xi)} \end{cases}$$

$$\frac{\partial}{\partial \phi}[g_{ij}] \quad = \quad [0]$$

Scale factors and Christoffel symbols:

$$Scale\ factors: \begin{cases} h_\xi = h_\eta = a\,\sqrt{sin^2(\eta) + sinh^2(\xi)} \\ h_\phi = a\,sin(\eta)\,sinh(\xi) \end{cases}$$

Non-zero Christoffel symbols of the 2nd kind and derivatives:

$$\Gamma^\xi_{\xi\xi} \quad = \quad \frac{cosh(\xi)\,sinh(\xi)}{sin^2(\eta) + sinh^2(\xi)}$$

$$\Gamma^\xi_{\xi\eta} \quad = \quad \frac{cos(\eta)\,sin(\eta)}{sin^2(\eta) + sinh^2(\xi)}$$

$$\Gamma^\xi_{\eta\eta} \quad = \quad -\frac{cosh(\xi)\,sinh(\xi)}{sin^2(\eta) + sinh^2(\xi)}$$

$$\Gamma^\xi_{\phi\phi} \quad = \quad -\frac{cosh(\xi)\,sinh(\xi)\,sin^2(\eta)}{sin^2(\eta) + sinh^2(\xi)}$$

$$\Gamma^\eta_{\xi\xi} \quad = \quad -\frac{cos(\eta)\,sin(\eta)}{sin^2(\eta) + sinh^2(\xi)}$$

$$\Gamma^{\eta}_{\xi\eta} = \frac{\cosh(\xi)\,\sinh(\xi)}{\sin^2(\eta) + \sinh^2(\xi)}$$

$$\Gamma^{\eta}_{\eta\eta} = \frac{\cos(\eta)\,\sin(\eta)}{\sin^2(\eta) + \sinh^2(\xi)}$$

$$\Gamma^{\eta}_{\phi\phi} = -\frac{\cos(\eta)\,\sin(\eta)\,\sinh^2(\xi)}{\sin^2(\eta) + \sinh^2(\xi)}$$

$$\Gamma^{\phi}_{\xi\phi} = \frac{\cosh(\xi)}{\sinh(\xi)}$$

$$\Gamma^{\phi}_{\eta\phi} = \frac{\cos(\eta)}{\sin(\eta)}$$

Unit Basis Vectors:

$$\hat{e}_{\xi} = \begin{pmatrix} \dfrac{\cos(\phi)\,\cosh(\xi)\,\sin(\eta)}{\sqrt{\sin^2(\eta)+\sinh^2(\xi)}} \\[2mm] \dfrac{\cosh(\xi)\,\sin(\eta)\,\sin(\phi)}{\sqrt{\sin^2(\eta)+\sinh^2(\xi)}} \\[2mm] \dfrac{\cos(\eta)\,\sinh(\xi)}{\sqrt{\sin^2(\eta)+\sinh^2(\xi)}} \end{pmatrix} \qquad \hat{e}_{\eta} = \begin{pmatrix} \dfrac{\cos(\eta)\,\cos(\phi)\sinh(\xi)}{\sqrt{\sin^2(\eta)+\sinh^2(\xi)}} \\[2mm] \dfrac{\cos(\eta)\,\sin(\phi)\sinh(\xi)}{\sqrt{\sin^2(\eta)+\sinh^2(\xi)}} \\[2mm] -\dfrac{\cosh(\xi)\,\sin(\eta)}{\sqrt{\sin^2(\eta)+\sinh^2(\xi)}} \end{pmatrix}$$

$$\hat{e}_{\phi} = \begin{pmatrix} -\sin(\phi) \\ \cos(\phi) \\ 0 \end{pmatrix}$$

Differential operators:

$$\operatorname{div} V\left(V_{\xi}, V_{\eta}, V_{\phi}\right)$$

$$= \frac{1}{a\left(\sin^2(\eta) + \sinh^2(\xi)\right)\sin(\eta)\sinh(\xi)}\left(\frac{\partial}{\partial\eta}\left(\sqrt{\sin^2(\eta) + \sinh^2(\xi)}\;\sin(\eta)\sinh(\xi)\;V_{\eta}\right)\right.$$

$$+ \frac{\partial}{\partial\xi}\left(\sqrt{\sin^2(\eta) + \sinh^2(\xi)}\;\sin(\eta)\sinh(\xi)\,V_{\xi}\right)$$

$$\left. + \frac{\partial}{\partial\phi}\left(\left(\sin^2(\eta) + \sinh^2(\xi)\right)V_{\phi}\right)\right)$$

$$= \frac{1}{a\left(\sin^2(\eta) + \sinh^2(\xi)\right)\sin(\eta)\sinh(\xi)}\left(\frac{\partial}{\partial\eta}\left(\sqrt{\sin^2(\eta) + \sinh^2(\xi)}\;\sin(\eta)\sinh(\xi)\;V_{\eta}\right)\right.$$

$$\left.+ \frac{\partial}{\partial\xi}\left(\sqrt{\sin^2(\eta) + \sinh^2(\xi)}\;\sin(\eta)\sinh(\xi)\,V_{\xi}\right)\right) + \frac{1}{\sin(\eta)\sinh(\xi)}\frac{\partial}{\partial\phi}\left(V_{\phi}\right)$$

$$\operatorname{curl} V\left(V_{\xi}, V_{\eta}, V_{\phi}\right)$$

$$= \left(\frac{1}{a\sqrt{\sin^2(\eta) + \sinh^2(\xi)}\,\sin(\eta)}\frac{\partial}{\partial\eta}\left(\sin(\eta)\;V_{\phi}\right)\right.$$

$$\left.- \frac{1}{a\sin(\eta)\sinh(\xi)}\frac{\partial V_{\eta}}{\partial\phi}\right)\hat{e}_{\xi}$$

$$- \left(\frac{1}{a\sqrt{\sin^2(\eta) + \sinh^2(\xi)}\,\sinh(\xi)}\frac{\partial}{\partial\xi}\left(\sinh(\xi)\,V_{\phi}\right)\right.$$

$$\left.+ \frac{1}{a\sin(\eta)\sinh(\xi)}\frac{\partial V_{\xi}}{\partial\phi}\right)\hat{e}_{\eta}$$

$$- \left(\frac{1}{a\left(\sin^2(\eta) + \sinh^2(\xi)\right)}\frac{\partial}{\partial\eta}\left(\sqrt{\sin^2(\eta) + \sinh^2(\xi)}\;V_{\xi}\right)\right.$$

$$\left.+ \frac{1}{a\left(\sin^2(\eta) + \sinh^2(\xi)\right)}\frac{\partial}{\partial\xi}\left(\sqrt{\sin^2(\eta) + \sinh^2(\xi)}\,V_{\eta}\right)\right)\hat{e}_{\phi}$$

$$\Delta\,\Theta(\xi,\eta,\phi) = \frac{1}{a^2(sin^2(\eta) + sinh^2(\xi))\,sinh(\xi)}\frac{\partial}{\partial\xi}\left(\frac{\partial}{\partial\xi}(sinh(\xi)\,\Theta)\right)$$

$$+\frac{1}{a^2\,(sin^2(\eta) + sinh^2(\xi))\,sin(\eta)}\frac{\partial}{\partial\eta}\left(\frac{\partial}{\partial\eta}(F(\xi,\eta,\phi))\,sin(\eta)\right)$$

$$+\frac{1}{a^2\,sin^2(\eta)\,sinh^2(\xi)}\frac{\partial^2\Theta}{\partial\phi^2}$$

I.A.2.15 Toroidal Coordinates

References:

[Moon/Spencer]: p. 112

Coordinate definition:

$$\textit{Cartesian functions} \quad \begin{cases} x^1(u,v,\varphi) = a\,\dfrac{sinh(v)\ cos(\varphi)}{cosh(v) - cos(u)} \\[2mm] x^2(u,v,\varphi) = a\,\dfrac{sinh(v)\ sin(\varphi)}{cosh(v) - cos(u)} \\[2mm] x^3(u,v,\varphi) = a\,\dfrac{sin(u)}{cosh(v) - cos(u)} \end{cases}$$

$$v >= 0, u \in [-\pi, \pi], \varphi \in [0, 2\pi)$$

Orthogonal   yes

Jacobian and metric:

$$J_{(u,v,\varphi)} = \begin{cases} [1,1] & -\dfrac{a\cos(\varphi)\sin(u)\sinh(v)}{(\cos(u)-\cosh(v))^2} \\[2ex] [1,2] & -\dfrac{a\,(\cos(u)\cosh(v)-1)\cos(\varphi)}{(\cos(u)-\cosh(v))^2} \\[2ex] [1,3] & \dfrac{a\sin(\varphi)\sinh(v)}{\cos(u)-\cosh(v)} \\[2ex] [2,1] & -\dfrac{a\sin(u)\sin(\varphi)\sinh(v)}{(\cos(u)-\cosh(v))^2} \\[2ex] [2,2] & -\dfrac{a\,(\cos(u)\cosh(v)-1)\sin(\varphi)}{(\cos(u)-\cosh(v))^2} \\[2ex] [2,3] & -\dfrac{a\cos(\varphi)\sinh(v)}{\cos(u)-\cosh(v)} \\[2ex] [3,1] & \dfrac{a\,(\cos(u)\cosh(v)-1)}{(\cos(u)-\cosh(v))^2} \\[2ex] [3,2] & -\dfrac{a\sin(u)\sinh(v)}{(\cos(u)-\cosh(v))^2} \\[2ex] [3,3] & 0 \end{cases}$$

$$J^{-1}{}_{(u,v,\varphi)} = \begin{cases} [1,1] & -\dfrac{\sin(u)\ \cos(\varphi)\ \sinh(v)}{a} \\[2ex] [1,2] & -\dfrac{\sin(u)\ \sin(\varphi)\ \sinh(v)}{a} \\[2ex] [1,3] & \dfrac{\cos(u)\ \cosh(v)-1}{a} \\[2ex] [2,1] & -\dfrac{(\cos(u)\ \cosh(v)-1)\ \cos(\varphi)}{a} \\[2ex] [2,2] & -\dfrac{(\cos(u)\ \cosh(v)-1)\ \sin(\varphi)}{a} \\[2ex] [2,3] & -\dfrac{\sin(u)\ \sinh(v)}{a} \\[2ex] [3,1] & \dfrac{(\cos(u)-\cosh(v))\ \sin(\varphi)}{a\,\sinh(v)} \\[2ex] [3,2] & -\dfrac{(\cos(u)-\cosh(v))\ \cos(\varphi)}{a\,\sinh(v)} \\[2ex] [3,3] & 0 \end{cases}$$

$$det\big(J_{(u,v,\varphi)}\big) = \frac{a^3\,\sinh(v)}{(\cos(u)-\cosh(v))^3}$$

Metric tensor (covariant) and derivatives:

$$[g_{ij}] \quad = \quad diag \left\{ \begin{array}{c} \dfrac{a^2}{(cosh(v) - cos(u)\,)^2} \\[2ex] \dfrac{a^2}{(cosh(v) - cos(u)\,)^2} \\[2ex] \dfrac{a^2\,sinh^2(v)}{(cosh(v) - cos(u)\,)^2} \end{array} \right.$$

$$\frac{\partial}{\partial u}[g_{ij}] \quad = \quad diag \left\{ \begin{array}{c} \dfrac{2\,a^2\,sin(u)}{(cos(u) - cosh(v))^3} \\[2ex] \dfrac{2\,a^2\,sin(u)}{(cos(u) - cosh(v))^3} \\[2ex] \dfrac{2\,a^2\,sin(u)\,sinh^2(v)}{(cos(u) - cosh(v))^3} \end{array} \right.$$

$$\frac{\partial}{\partial v}[g_{ij}] \quad = \quad diag \left\{ \begin{array}{c} \dfrac{2\,a^2\,sinh(v)}{(cos(u) - cosh(v))^3} \\[2ex] \dfrac{2\,a^2\,sinh(v)}{(cos(u) - cosh(v))^3} \\[2ex] \dfrac{2\,a^2\,sinh(v)\,\big(cos(u)\,cosh(v - 1)\big)}{(cos(u) - cosh(v))^3} \end{array} \right.$$

$$\frac{\partial}{\partial \phi}[g_{ij}] \quad = \quad [0]$$

Inverse metric tensor (contravariant) and derivatives:

$$[g^{ij}] \quad = \quad \begin{bmatrix} \dfrac{(cosh(v) - cos(u)\,)^2}{a^2} & 0 & 0 \\[3ex] 0 & \dfrac{(cosh(v) - cos(u)\,)^2}{a^2} & 0 \\[3ex] 0 & 0 & \dfrac{(cosh(v) - cos(u)\,)^2}{a^2\,sinh(v)} \end{bmatrix}$$

$$\frac{\partial}{\partial u}[g^{ij}] = \quad diag \left\{ \begin{array}{l} -\dfrac{2\ (cos(u) - cosh(v))\ sin(u)}{a^2} \\[2ex] -\dfrac{2\ (cos(u) - cosh(v))\ sin(u)}{a^2} \\[2ex] -\dfrac{2\ (cos(u) - cosh(v))\ sin(u)}{a^2\ sinh^2(v)} \end{array} \right.$$

$$\frac{\partial}{\partial v}[g^{ij}] = \quad diag \left\{ \begin{array}{l} -\dfrac{2\ (cos(u) - cosh(v))\ sinh(v)}{a^2} \\[2ex] -\dfrac{2\ (cos(u) - cosh(v))\ sinh(v)}{a^2} \\[2ex] -\dfrac{2\ \left((c(u) - ch(v))ch(v) + sh^2(v)\right)(c(u) - ch(v))}{a^2\ sinh^3(v)} \end{array} \right.$$

$$\frac{\partial}{\partial \phi}[g^{ij}] = \quad [0]$$

Scale factors and Christoffel symbols:

$$Scale\ factors: \left\{ \begin{array}{l} h_u = h_v = \dfrac{a}{cosh(v) - cos(u)} \\[2ex] h_\varphi = \dfrac{a\ sinh(v)}{cosh(v) - cos(u)} \end{array} \right.$$

Non-zero Christoffel symbols of the 2nd kind:

$$\Gamma^u_{uu} = \frac{sin(u)}{cos(u) - cosh(v)}$$

$$\Gamma^u_{uv} = \frac{sinh(v)}{cos(u) - cosh(v)}$$

$$\Gamma^u_{vv} = -\frac{sin(u)}{cos(u) - cosh(v)}$$

$$\Gamma^u_{\phi\phi} = -\frac{sin(u)\ sinh^2(v)}{cos(u) - cosh(v)}$$

$$\Gamma^{v}_{uu} = -\frac{sinh(v)}{(cos(u) - cosh(v))}$$

$$\Gamma^{v}_{uv} = \frac{sin(u)}{cos(u) - cosh(v)}$$

$$\Gamma^{v}_{vv} = \frac{sinh(v)}{cos(u) - cosh(v)}$$

$$\Gamma^{v}_{\phi\phi} = -\frac{(cos(u)\,cosh(v) - 1)\,sinh(v)}{cos(u) - cosh(v)}$$

$$\Gamma^{\phi}_{u\phi} = \frac{sin(u)}{cos(u) - cosh(v)}$$

$$\Gamma^{\phi}_{v\phi} = \frac{(cos(u)\,cosh(v) - 1)}{((cos(u) - cosh(v))\,sinh(v))}$$

Unit Basis Vectors:

$$\hat{e}_u = \begin{pmatrix} \dfrac{cos(\varphi)sin(u)sinh(v)}{cos(u) - cosh(v)} \\[2ex] \dfrac{sin(u)sin(\varphi)sinh(v)}{cos(u) - cosh(v)} \\[2ex] \dfrac{1 - cos(u)cosh(v)}{cos(u) - cosh(v)} \end{pmatrix}$$

$$\hat{e}_v = \begin{pmatrix} \dfrac{(cos(u)\,cosh(v) - 1)\,cos(\varphi)}{cos(u) - cosh(v)} \\[2ex] \dfrac{(cos(u)cosh(v) - 1)sin(\varphi)}{cos(u) - cosh(v)} \\[2ex] \dfrac{sin(u)sinh(v)}{cos(u) - cosh(v)} \end{pmatrix}$$

$$\hat{e}_\varphi = \begin{pmatrix} -sin(\varphi) \\ cos(\varphi) \\ 0 \end{pmatrix}$$

Differential operators:

$$div\ V\left(V_u, V_v, V_\varphi\right) = \frac{(cosh(v) - cos(u))^3}{a\ sinh(v)}\left(\frac{\partial}{\partial u}\left(\frac{sinh(v)}{(cos(u) - cosh(v))^2}\ V_u\right)\right.$$

$$+\frac{\partial}{\partial v}\left(\frac{sinh(v)}{(cosh(v) - cos(u))^2}\ V_v\right)$$

$$\left.+\frac{\partial}{\partial \phi}\left(\frac{1}{(cosh(v) - cos(u))^2}\ V_\varphi\right)\right)$$

$$curl\ V\left(V_u, V_\sigma, V_\varphi\right) = \frac{(cosh(v) - cos(u))^2}{a\ sinh(v)}\left(\frac{\partial}{\partial v}\left(\frac{sinh(v)}{cosh(v) - cos(u)}\ V_\varphi\right)\right.$$

$$\left.-\frac{\partial}{\partial \phi}\left(\frac{1}{cosh(v) - cos(u)}\ V_v\right)\right)\hat{e}_u$$

$$-\frac{(cosh(v) - cos(u))^2}{a\ sinh(v)}\left(\frac{\partial}{\partial u}\left(\frac{sinh(v)}{cosh(v) - cos(u)}\ V_\varphi\right)\right.$$

$$\left.-\frac{\partial}{\partial \phi}\left(\frac{1}{cosh(v) - cos(u)}\ V_u\right)\right)\hat{e}_v$$

$$+\frac{(cosh(v) - cos(u))^2}{a}\left(\frac{\partial}{\partial u}\left(\frac{1}{cosh(v) - cos(u)}\ V_v\right)\right.$$

$$\left.-\frac{\partial}{\partial v}\left(\frac{1}{cosh(v) - cos(u)}\ V_u\right)\right)\hat{e}_\varphi$$

$$\nabla^2\Theta(\mu, v, \varphi) = \frac{(cosh(v) - cos(u))^3}{a^3\ sinh(v)}\frac{\partial}{\partial u}\left(\frac{sinh(v)}{cosh(v) - cos(u)}\frac{\partial\Theta}{\partial u}\right.$$

$$\left.+ sinh(v)\frac{\partial}{\partial v}\left(\frac{1}{cosh(v) - cos(u)}\frac{\partial\Theta}{\partial v}\right)\right)$$

$$+\frac{(cosh(v) - cos(u))^2}{a^2\ sinh^2(v)}\frac{\partial^2\Theta}{\partial\varphi^2}$$

I.A.2.16 6-Sphere Coordinates

References:

https://mathworld.wolfram.com/6-SphereCoordinates.html

Coordinate definition:

$$Cartesian\ functions \begin{cases} x^1(u,v,w) = \dfrac{u}{u^2 + v^2 + w^2} \\[2mm] x^2(u,v,w) = \dfrac{v}{u^2 + v^2 + w^2} \\[2mm] x^3(u,v,w) = \dfrac{w}{u^2 + v^2 + w^2} \end{cases}$$

$$u, v, w \ \in \mathbb{R}$$

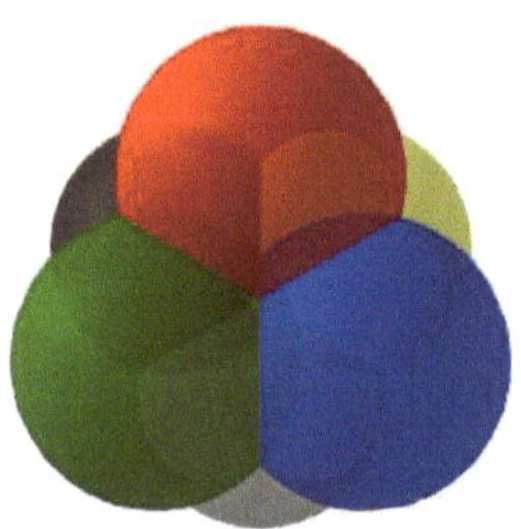

Orthogonal   yes

Jacobian and metric:

$$J_{(u,v,\varphi)} = \begin{bmatrix} -\dfrac{u^2 - v^2 - w^2}{(u^2 + v^2 + w^2)^2} & -\dfrac{2uv}{(u^2 + v^2 + w^2)^2} & -\dfrac{2uw}{(u^2 + v^2 + w^2)^2} \\[3mm] -\dfrac{2uv}{(u^2 + v^2 + w^2)^2} & \dfrac{u^2 - v^2 + w^2}{(u^2 + v^2 + w^2)^2} & -\dfrac{2vw}{(u^2 + v^2 + w^2)^2)} \\[3mm] -\dfrac{2uw}{(u^2 + v^2 + w^2)^2} & -\dfrac{2vw}{(u^2 + v^2 + w^2)^2} & \dfrac{u^2 + v^2 - w^2}{(u^2 + v^2 + w^2)^2} \end{bmatrix}$$

$$J^{-1}_{(u,v,\varphi)} = \begin{bmatrix} -u^2 + v^2 + w^2 & -2uv & -2uw \\ -2uv & u^2 - v^2 + w^2 & -2vw \\ -2uw & -2vw & u^2 + v^2 - w^2 \end{bmatrix}$$

$$det\big(J_{(u,v,\varphi)} = -\dfrac{1}{(u^2 + v^2 + w^2)^3}$$

Metric tensor (covariant) and derivatives:

$$[g_{ij}] = \begin{bmatrix} \dfrac{1}{(u^2 + v^2 + w^2)^2} & 0 & 0 \\[3mm] 0 & \dfrac{1}{(u^2 + v^2 + w^2)^2} & 0 \\[3mm] 0 & 0 & \dfrac{1}{(u^2 + v^2 + w^2)^2} \end{bmatrix}$$

$$\dfrac{\partial}{\partial u}[g_{ij}] = \begin{bmatrix} -\dfrac{4u}{(u^2 + v^2 + w^2)^3} & 0 & 0 \\[3mm] 0 & -\dfrac{4u}{(u^2 + v^2 + w^2)^3} & 0 \\[3mm] 0 & 0 & -\dfrac{4u}{(u^2 + v^2 + w^2)^3} \end{bmatrix}$$

$$\dfrac{\partial}{\partial v}[g_{ij}] = \begin{bmatrix} -\dfrac{4v}{(u^2 + v^2 + w^2)^3} & 0 & 0 \\[3mm] 0 & -\dfrac{4v}{(u^2 + v^2 + w^2)^3} & 0 \\[3mm] 0 & 0 & -\dfrac{4v}{(u^2 + v^2 + w^2)^3} \end{bmatrix}$$

$$\frac{\partial}{\partial w}\left[g_{ij}\right] \; = \; \begin{bmatrix} -\dfrac{4w}{(u^2 + v^2 + w^2)^3} & 0 & 0 \\[3mm] 0 & -\dfrac{4w}{(u^2 + v^2 + w^2)^3} & 0 \\[3mm] 0 & 0 & -\dfrac{4w}{(u^2 + v^2 + w^2)^3} \end{bmatrix}$$

Inverse metric tensor (contravariant) and derivatives:

$$\left[g^{ij}\right] \; = \; \begin{bmatrix} (u^2 + v^2 + w^2)^2 & 0 & 0 \\ 0 & (u^2 + v^2 + w^2)^2 & 0 \\ 0 & 0 & (u^2 + v^2 + w^2)^2 \end{bmatrix}$$

$$\frac{\partial}{\partial u}\left[g^{ij}\right] \; = \; \begin{bmatrix} 4u\,(u^2 + v^2 + w^2) & 0 & 0 \\ 0 & 4u\,(u^2 + v^2 + w^2) & 0 \\ 0 & 0 & 4u\,(u^2 + v^2 + w^2) \end{bmatrix}$$

$$\frac{\partial}{\partial v}\left[g^{ij}\right] \; = \; \begin{bmatrix} 4v\,(u^2 + v^2 + w^2) & 0 & 0 \\ 0 & 4v\,(u^2 + v^2 + w^2) & 0 \\ 0 & 0 & 4v\,(u^2 + v^2 + w^2) \end{bmatrix}$$

$$\frac{\partial}{\partial w}\left[g^{ij}\right] \; = \; \begin{bmatrix} 4w\,(u^2 + v^2 + w^2) & 0 & 0 \\ 0 & 4w\,(u^2 + v^2 + w^2) & 0 \\ 0 & 0 & 4w\,(u^2 + v^2 + w^2) \end{bmatrix}$$

Scale factors and Christoffel symbols:

$$Scale\ factors:\ \left\{ h_u = h_v = h_w = \frac{1}{u^2 + v^2 + w^2} \right.$$

Non-zero Christoffel symbols of the 2nd kind:

$$\Gamma^u_{uu} = -\frac{2u}{u^2 + v^2 + w^2}$$

$$\Gamma^u_{uv} = -\frac{2v}{u^2 + v^2 + w^2}$$

$$\Gamma^u_{uw} = -\frac{2w}{u^2 + v^2 + w^2}$$

$$\Gamma^u_{vv} = \frac{2u}{u^2 + v^2 + w^2}$$

$$\Gamma^u_{ww} = \frac{2u}{u^2 + v^2 + w^2}$$

$$\Gamma^v_{uu} = \frac{2v}{u^2 + v^2 + w^2}$$

$$\Gamma^v_{uv} = -\frac{2u}{u^2 + v^2 + w^2}$$

$$\Gamma^v_{vv} = -\frac{2v}{u^2 + v^2 + w^2}$$

$$\Gamma^v_{vw} = -\frac{2w}{u^2 + v^2 + w^2}$$

$$\Gamma^v_{ww} = \frac{2v}{u^2 + v^2 + w^2}$$

$$\Gamma^w_{uu} = \frac{2w}{u^2 + v^2 + w^2}$$

$$\Gamma^w_{uw} = -\frac{2u}{u^2 + v^2 + w^2}$$

$$\Gamma^w_{vv} = \frac{2w}{u^2 + v^2 + w^2}$$

$$\Gamma^w_{vw} = -\frac{2v}{u^2 + v^2 + w^2}$$

$$\Gamma^w_{ww} = -\frac{2w}{u^2 + v^2 + w^2}$$

Unit Basis Vectors:

$$\hat{e}_u = \begin{pmatrix} -\dfrac{u^2 - v^2 - w^2}{u^2 + v^2 + w^2} \\[2ex] -\dfrac{2uv}{u^2 + v^2 + w^2} \\[2ex] -\dfrac{2uw}{u^2 + v^2 + w^2} \end{pmatrix} \qquad \hat{e}_v = \begin{pmatrix} -\dfrac{2uv}{u^2 + v^2 + w^2} \\[2ex] \dfrac{u^2 - v^2 + w^2}{u^2 + v^2 + w^2} \\[2ex] -\dfrac{2vw}{u^2 + v^2 + w^2} \end{pmatrix}$$

$$\hat{e}_w = \begin{pmatrix} -\dfrac{2uw}{u^2 + v^2 + w^2} \\[2ex] -\dfrac{2vw}{u^2 + v^2 + w^2} \\[2ex] \dfrac{u^2 + v^2 - w^2}{u^2 + v^2 + w^2} \end{pmatrix}$$

Differential operators:

$$div\, V(V_u, V_v, V_w) = (u^2 + v^2 + w^2)^3 \left(\frac{\partial}{\partial u}\left(\frac{1}{(u^2 + v^2 + w^2)^2}\, V_u \right) \right.$$

$$\left. + \frac{\partial}{\partial v}\left(\frac{1}{(u^2 + v^2 + w^2)^2}\, V_v \right) + \frac{\partial}{\partial w}\left(\frac{1}{(u^2 + v^2 + w^2)^2}\, V_w \right) \right)$$

$$curl\, V(V_u, V_v, V_w) = (u^2 + v^2 + w^2)^2 \left(\frac{\partial}{\partial v}\left(\frac{1}{u^2 + v^2 + w^2}\, V_w \right) \right.$$

$$\left. - \frac{\partial}{\partial w}\left(-\frac{1}{u^2 + v^2 + w^2}\, V_v \right) \right) \hat{e}_u$$

$$- (u^2 + v^2 + w^2)^2 \left(\frac{\partial}{\partial u}\left(\frac{1}{u^2 + v^2 + w^2}\, V_w \right) \right.$$

$$\left. - \frac{\partial}{\partial w}\left(-\frac{1}{u^2 + v^2 + w^2}\, V_u \right) \right) \hat{e}_v$$

$$+ (u^2 + v^2 + w^2)^2 \left(\frac{\partial}{\partial u}\left(\frac{1}{u^2 + v^2 + w^2}\, V_v \right) \right.$$

$$\left. - \frac{\partial}{\partial v}\left(\frac{1}{u^2 + v^2 + w^2}\, V_u \right) \right) \hat{e}_w$$

$$\nabla^2 \Theta(u, v, w) = (u^2 + v^2 + w^2)^3 \left(\frac{\partial}{\partial u}\left(\frac{1}{u^2 + v^2 + w^2}\, \frac{\partial \Theta}{\partial u} \right) \right.$$

$$\left. + \frac{\partial}{\partial v}\left(\frac{1}{u^2 + v^2 + w^2}\, \frac{\partial \Theta}{\partial v} \right) + \frac{\partial}{\partial w}\left(\frac{1}{u^2 + v^2 + w^2}\, \frac{\partial \Theta}{\partial w} \right) \right)$$

I.B Bibliography

[ANTONI]: Markus Antoni: Calculus with curvilinear coordinates (2019)

ISBN 978-3-030-00415-6

ISBN 978-3-030-00416-3

This book is a good introduction to calculus in curvilinear coordinates. There are many examples and exercises with solutions.

[Arfken]: Arfken, Weber, Harris: Mathematical methods for physicists, Seventh edition (2013)

ISBN: 978-0-12-384654-9

Especially, chapter 4.3 (Tensors in general coordinates) is of most interest.

[FEUERBACHER]: Björn Feuerbacher: Tutorium Mathematische Methoden der Elektrodynamik

(In German!) Although ESN is explicitly not used in this book, numerous divergence, curl, and gradient equations are presented in "almost" ESN notation.

[GRINFELD]: Pavel Grinfeld: Introduction to tensor analysis and the calculus of moving surfaces (2013)

ISBN 978-1-4614-7866-9

Recommended resource for tensor calculus in classical mechanics. There are also video lessons on youtube, based on his book. Especially chapter 8.9 is interesting, as velocity and acceleration are derived using tensor calculus.

[ISLAM]: Nasrul Islam: Tensors and their application (2006)

ISBN (13): 978-81-224-2700-4

Especially chapter 8 (Analytical mechanics) contains a demonstration on how tensor calculus is used in mechanics.

[KAY]: Schaum's Outlines: David C.Kay: Tensor Calculus (1988)

ISBN 0-07-033484-6

Provides a good introduction to ESN with numerous examples. Anyone who wants to deal with tensors is well advised here.

[MISNER]: Ch.W. Misner, Kip.S. Thorn, J.A. Wheeler: Gravitation (2017)

ISBN: 978-0-691-17779-3

[MOON/SPENCER]: P.Moon / D.E.Spener: Field Theory Handbook, 2nd $edition, Springer\, Verlag$

ISBN-13: 978-3-540-18430-0 001: 10.1007/978-3-642-83243-7

e-ISBN-13: 978-3-642-83243-7

[NEUTSCH]: Wolfram Neutsch: Coordinates (1996),

ISBN-10: 3110148528
ISBN-13: 978-3110148527

Although this book is very expensive, I recommend the lecture. It gives a deep theoretical insight of coordinate systems along with recommendations where you could use which coordinate system.

[NOLTING]: Wolfgang Nolting: Grundkurs Theoretische Physik 1
11. Auflage, 2018
ISBN:978-3-662-57583-3

[PFEIFFER]: Hans-Friedrich Pfeiffer: Introduction to Einstein Summation Notation

ISBN 978-3-7412-9257-6

Most parts of chapter $I.2.1$ about ESN are included in this book. Additionally, it contains the proofs omitted in chapter $I.2.1$.

[RASBAND]: S. Neil Rasband: Dynamics (1983)

ISBN 0-471-87398-5

In the first chapter, Rasband uses tensor calculus for velocity and acceleration. It is a good resource to see how tensor calculus can be done in classical mechanics.

[WERNER]: Wolfgang Werner: Vektoren und Tensoren als universelle Sprache in Physik und Technik, Band 1

ISBN-10 3658252715

ISBN-13 978-3658252717

This book gives a good introduction to tensor calculation. The only disadvantage is that the author explicitly does not use ESN – he uses an own invented system for indicating superscripts and subscripts. In my opinion, this is sometimes difficult to understand. The reader should translate these equations into ESN.

The second book (same title) covers tensors in classical mechanics and relativistic physics. Recommendable!

Internet sites

The following list contains internet-sites to different themes covered in this book. The list can be downloaded using:

http://www.hans-friedrich-pfeiffer.de/PGTCT/ListOfInternetSites.txt

Choleski decomposition

https://en.wikipedia.org/wiki/Cholesky_decomposition

Christoffel symbols of the 2nd kind

https://drive.google.com/file/d/1peg6n2aD3yowiTmGIrfU36MFTfhgupX0/view

Directional derivative

https://www.whitman.edu/mathematics/calculus_online/section14.05.html

ESN

There is a wealth of information about the ESN on the internet. From the many sources, the following are listed:

Alan H. Barr: The Einstein Summation Notation

http://vucoe.drbriansullivan.com/wp-content/uploads/Einstein-Summation-Notation.pdf

The script describes some interesting suggestions for extensions to the ESN. These extensions also allow for index notations that could not previously be represented as ESN. As an example, the extensions are considered using quaternions.

Albert Einstein: Die Grundlage der allgemeinen Relativitätstheorie

http://www.alberteinstein.info/gallery/pdf/CP6Doc30_pp284-339.pdf

(In German) In this original edition you can see how A. Einstein uses the summation named after him.

General coordinate transformations

https://javierrubioblog.com/wp-content/uploads/2015/12/chapter4.pdf

This site introduces general coordinate systems.

Helical coordinate system

https://www.researchgate.net/publication/280656589_The_use_of_helical_coordinate_systems

This article describes an application of the helical coordinate system in physics.

Jacobi and Dirac-Delta-distribution

The following internet-sites describe coordinate transformations done by using the Dirac-Delta distribution. The authors describe an incredible connection between Jacobi-matrix and the δ-distribution.

https://iopscience.iop.org/article/10.1088/1361-6404/aba455
https://iopscience.iop.org/article/10.1088/1361-6404/abdca9/pdf

The geometry of vector calculus

https://bridge.math.oregonstate.edu/Book/GVC.html
This is a very recommendable site, explaining all kins of geometric elements.

Overview of coordinate systems

https://dl.acm.org/doi/pdf/10.1145/363162.363211

This article gives an overview over some coordinate systems. You will find these coordinate systems in the APPENDIX $I.A$

https://javierrubioblog.com/wp-content/uploads/2015/12/chapter4.pdf

Diagonal metric tensor

https://drive.google.com/file/d/1peg6n2aD3yowiTmGIrfU36MF TfhgupX0/view

Integrals

https://nucinkis-lab.cc.ic.ac.uk/HELM/workbooks/workbook_27/27_4_changing_coords.pdf

This paper points out the role of the Jacobi determinant in integrals when moving to curvilinear coordinates.

https://www.math.kit.edu/iana2/lehre/hm2info2012s/media/grueb5.pdf

This paper offers exercises in integral calculus in curvilinear coordinates.

Tensors

https://www.ita.uni-heidelberg.de/~dullemond/lectures/tensor/tensor.pdf

Introduction to tensor calculus.

https://cseweb.ucsd.edu/~gill/CILASite/Resources/15Chap11.pdf

Derives the equivalence between the definition of a tensor as a multilinear mapping and its transformation behavior.

https://itp.uni-frankfurt.de/~hees/cosmo-SS12/ART-Vorlesung.pdf

Written in German, this paper focuses on the covariant derivative of tensors.

https://grinfeld.org/books/An-Introduction-To-Tensor-Calculus/Chapter11.html

This paper has a focus on how to raise or lower indices with tensors.

http://www.damtp.cam.ac.uk/user/tong/vc/vc7.pdf

The paper focuses on tensors in $\mathbb{R}^3$ and defines tensors as a transformation law with orthogonal matrices.

Index

D

P

Q

R

S

T

U

V

Δ